An Introduction

The team may be gone yet the memories live on forever. That's the reason for a 25th anniversary book saluting hockey's Whalers.

From 1972 to 1997, the Whalers played 1,975 regular season games and 123 playoff hockey games. They began in the World Hockey Association in Boston and closed out their National Hockey League history in Hartford.

The first game was played in Boston Garden. The date was October 11, 1972. The result was a 4-3 win over the Philadelphia Blazers, the first of 555 games in the WHA. The finale was played April 13, 1997, a 2-1 decision over Tampa Bay at the Hartford Civic Center Coliseum. It was the last of 1,420 games played over 18 years in the NHL.

The high-water marks range from the first WHA championship in 1973 to winning the Adams Division title in 1987. The disappointments, and there were many, reached a low point on March 26, 1997 when an owner's ego and economic miscalculations resulted in locking the door on major league hockey in southern New England.

Twenty-five years of hockey.

Twenty-five years of memories.

Twenty-five years of Whalers.

Few understand the love affair diehard loyalists had with this underachieving cast docked a short slapshot from the Connecticut River where I-84 and I-91 intersect.

For better or for worse, there was always hope.

Moments after another season ended in failure, the frustrations would pass during the summer. When the leaves on the oaks, maples and elms changed their hues in southern New England, training camp beckoned. The icemen in Hartford were back for another season.

Would this year's top pick make a favorable impression?

Will the newcomers obtained in the off-season swaps supply the necessary scoring punch?

Do we budget for playoffs tickets this season?

These days, what's left for hockey fans of the Whalers are memories. This book is a complete look at the franchise, a biographical, statistical and pictorial measure of the heroes and villains who played, made the line changes or paid the bills for the WHA-NHL Whalers.

Several players chronicled in this book will be scoring goals or trying to prevent others from doing the same in 1997-98 and beyond.

Some will be Carolina Hurricanes.

Others will be playing for teams around the league.

No matter the sweater, they will be **Forever Whalers**.

For Patrick and Nolan, two youngsters like many others in
New England who remember mom and dad taking them to
see the Whalers play.

Published by Glacier Publishing
40 Oak Street
Southington, CT 06489
(860) 621-7644

Manufactured in the United States of America
ISBN 0-9650315-3-5
      1. Hockey players - biography
      2. Hockey team - Whalers
      3. Hockey history - NHL-WHA

Other Books Available Through Glacier Publishing
Same Game, Different Name (0-9650315-1-9)
Whalers Trivia Compendium (0-9650315-0-0)
All-Star Dads (0-9650315-4-3)

Photo Credits
   Many pictures used in this book were supplied by the players
from their personal collections. Others were issued annually by
the New England and/or Hartford Whalers.

*Jack Lautier chronicled the Whalers for a number of
publications for close to two decades. He has authored and
published books on baseball, professional hockey and
parenting, including* Fenway Voices, *an oral history of the
Boston Red Sox,* 15 Years of Whalers Hockey, Whalers Trivia
Compendium, Same Game Different Name *(the history of the
World Hockey Association) and* All-Star Dads.

Whalers 4, Blazers 3

October 12, 1972

at Boston Garden

PHILADELPHIA	2 0 1 -	3
NEW ENGLAND	1 2 1 -	4

FIRST PERIOD

1, Philadelphia, Sanderson 1 (O'Donoghue, Cardiff), 6:52.

2, Philadelphia, Plumb 1 (Sanderson, Lacroix), 12:24 (pp).

3, New England, Williams 1 (Earl, Cunniff), 13:42.

Penalties: Hutchison, Philadelphia (interference), 2:22; Herriman, Philadelphia, major-minor (fighting, roughing), 8:48; Hyndman, New England, major-minor (fighting, roughing), 8:48; Hurley, New England (tripping), 11:50; Selwood, New England (cross-checking), 17:52.

SECOND PERIOD

4, New England, Sarrazin 1 (Webster), 1:29 (pp).

5, New England, Cunniff 1 (Green) 8:45.

Penalties: Bennett, Philadelphia (tripping), :34; New England, bench minor (too many men), 5:08; Hurley, New England (holding), 12:54; Smith, New England (slashing), 19:41; Lacroix, Philadelphia (slashing), 19:41.

THIRD PERIOD

6, Philadelphia, Migneault 1 (Rouleau), 11:04.

7, New England, Pleau 1 (Webster, Sarrazin), 17:49.

Penalties: Hutchison, Philadelphia (elbowing), 2:43; Sheehy, New England (slashing), 2:43.

SHOTS ON GOAL

Philadelphia   12 7 10 - 29

New England   6 8   9 - 23

POWER-PLAY CHANCES

Philadelphia 1-5, New England 1-1

GOALIES

Philadelphia, Parent (19 saves)

New England, Smith (26 saves)

ATTENDANCE

14,552

REFEREE

Bill Friday

A

Christer Abrahamsson played 102 games for the WHA Whalers and was among the first Europeans to cross the Atlantic to play in North America. He played in a career-high 45 games in 1976-77, his last year in New England.

Christer Abrahamsson

One of the first Europeans to play for the Whalers, this goaltender crossed the Atlantic with his twin brother Thommy, a defenseman, to sign with New England in 1974 and went on to play three seasons in the WHA. His best year was 1975-76 when he posted a 2.62 goals-against average in his first 23 appearances before a series of injuries limited his effectiveness. Posted a 3.58 GAA in 102 games with the Whalers with his last year being 1976-77. Played for Sweden in the 1972 Olympic Games.

Thommy Abrahamsson

A speedy skater, this defenseman left Sweden with his twin brother Christer, a goalie, to sign with the Whalers in 1974. He played 202 games in the WHA, going 28-67-95 in three years. Captained the Swedish National Team and also Leksand Club of the Swedish Elite League. In 1980-81, Abrahamsson returned to the Whalers, signing as a free agent on May 23, 1980. He played 32 games in the NHL, but at 34, his skills had diminished.

Greg Adams

A forward who logged 545 NHL games with stops in Philadelphia, Hartford, Washington, Edmonton, Quebec, Vancouver and Detroit, Adams came to the Whalers from the Flyers in a three-team swap involving Mark Howe on August 19, 1982. In 79 games in 1982-83, Adams had 10-13-23 along with 216 penalty minutes. Hartford shipped Adams to Washington on October 3, 1983 for rugged winger Torrie Robertson. Both were teammates in Canadian juniors at Victoria of the Western League.

Jim Agnew

A hard-hitting defenseman who started with Vancouver, Agnew signed on July 24, 1992 with the Whalers as a free agent. After being unable to crack the logjam of veterans on defense with the Canucks, Agnew, tabbed by Vancouver 157th overall out of Brandon in the 1984 Entry Draft, had a strong opportunity to be in Hartford's six-member unit in 1992-93. However, he played only 16 games (with 68 penalty minutes) before his career prematurely ended with a severe knee injury. After sitting out a year, Agnew attempted a comeback in 1994-95. Though the heart was willing, Agnew's left knee could not withstand the daily pounding. Overall, Agnew played in 65 NHL games with one assist and 189 penalty minutes.

Kevin Ahearn

A member of the 1973 AVCO Cup championship, Ahearn was one of several Boston-based players to sign with the New England Whalers in 1972-73. In 78 games, Ahearn had 20-22-42 in his only campaign with the Whalers. He played three seasons at Boston College and was a member of the 1972 U.S. Olympic Team.

Defenseman Thommy Abrahamsson played for the Whalers in both leagues. His last hitch in Hartford was 32 games in 1980-81, going 6-11-17 with highlight being the game-winning marker in a 9-7 shootout win over Pittsburgh on December 26, 1980.

Russ Anderson captained the Whalers in 1982-83. Injuries limited Anderson to 79 games over two seasons. His only goal for Hartford was the game-winner in a 4-3 comeback decision over Boston on February 24, 1982.

Steve Alley

Originally drafted by Chicago (141st overall in 1973), Alley played at the University of Wisconsin and also served a hitch with the U.S. Olympic Team in 1976 before turning pro. He signed with the WHA and played with the Birmingham Bulls. Claimed by Hartford in the 1979 Dispersal Draft of WHA players, Alley played with the Whalers in 1979-80 and 1980-81, a total of 15 games.

Ray Allison

A high-scoring winger in Canadian juniors, Allison was Hartford's first-ever NHL draft pick, being tabbed 18th overall in 1979. Allison was one of four first-round picks off the Brandon roster that year (Brian Propp, Laurie Boschman and Brad McCrimmon were the others) following an incredible junior season where the Wheat Kings went 58-5-9 in 1978-79 en route to the Memorial Cup championship. Allison played in 70 games with the Whalers, going 17-12-29. All but one of his points were tallied his rookie year. Moved on to Philadelphia in a multi-player swap involving Rick MacLeish on July 3, 1981. Allison wound up playing 238 NHL games, the most coming in 1982-83 when he had 21-30-51 in 67 games with the Flyers.

John Anderson

A sharp acquisition by general manager Emile Francis, the Whalers obtained Anderson from Quebec on March 8, 1986 in exchange for defenseman Risto Siltanen. A proven goal scorer, Anderson was inserted on Hartford's top line with Ron Francis and Kevin Dineen and helped fuel a run to end a five-year playoff absence. Anderson had 5-8-13 in 10 playoff games that year, setting a team record for most helpers. Drafted by Toronto 11th overall in 1977, Anderson hit the 30-goal plateau six times during his career, winding up with 282-349-631 in 814 NHL games.

Russ Anderson

Obtained from Pittsburgh on December 29, 1981 when the Whalers shuttled Rick MacLeish to the Penguins, this hard-hitting defenseman became the fifth captain in team history, serving that post in 1982-83. Limited to just 82 games over two seasons because of injuries, Anderson, drafted 31st overall in 1975 by Pittsburgh, made his only goal in Hartford annals memorable, a third-period strike on February 24, 1982 to beat Boston 4-3. Anderson finished his career with the Los Angeles Kings in 1984-85, a 10-year career hitch of 519 NHL games. At Minnesota in 1973-74, Anderson teammed with future Whaler Warren Miller in helping the Golden Gophers to the NCAA championship.

Mikael Andersson

Tabbed in the 1989 Waiver Draft from Buffalo by general manager Eddie Johnston, the Swedish forward proved to be one of EJ's better acquisitions, emerging as a dependable checker for the Whalers over a three-year period, including 18-29-47 in 1991-92. The Sabres drafted Andersson 18th overall in 1984 out of Vastra Frolunda of the Swedish League. Hartford, during Brian Burke's one-year stewardship at general manager, opted not to resign Andersson who hooked on with the Tampa Bay Lightning, becoming that team's first free agent signing on July 8, 1992. Andersson completed his fourth year with the Lightning in 1995-96, reaching the 500-game milestone in his career.

Mike Antonovich

One of several Whalers to play for the franchise in both the WHA and NHL, Antonovich played the left side on a line with Dave Keon and John McKenzie, one of the most effective trios in club annals. Though a diminutive 5-foot-8, 165 pounds, "Antman" used his explosive speed to create chances and beat rivals to loose pucks in the corners. Antonovich also played in the NHL for Minnesota and New Jersey, a total of 87 games. Drafted by the Fightning Saints in 1972, Antonovich launched what would be a 486-game tenure in the WHA. Joined the Whalers on February 5, 1977 along with Bill Butters from Edmonton in a swap for Brett Callighen and Ron Busniuk. Led New England in playoff scoring in 1978, going 10-7-17 as the Whalers reached the AVCO Finals against Winnipeg.

John Anderson had the touch around the net to finish, scoring 72 goals for the Whalers in three-plus seasons. Anderson came to Hartford from Quebec on March 8, 1986 for defenseman Risto Siltanen.

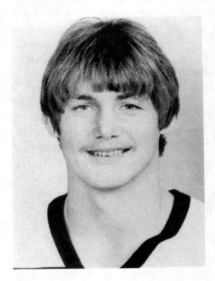

In 1979, **Ray Allison** was the first-ever NHL draft pick by the Whalers. He was a regular as a rookie in 1979-80, scoring 16 goals including the game-winner in Hartford's last game played in Georgia against the Flames, a 6-5 win over Atlanta on February 23, 1980.

Favorably compared to Larry Robinson in juniors, defenseman **Fred Arthur** was taken eighth overall by the Whalers in the 1980 Entry Draft. Within a year, Arthur was packaged, along with 1979 top pick **Ray Allison**, to the Philadelphia Flyers for one-time sniper **Rick MacLeish**. Today, Arthur is a doctor in Canada.

*Forward **Mike Antonovich**, who led the Whalers in playoff scoring in 1978, lets one fly against Quebec as J.C. Tremblay of the Nordiques defends.*

Danny Arndt

A speedy winger who opted to sign with New England of the WHA over Chicago of the NHL, Arndt played 115 games with the Whalers over two seasons with modest totals. His biggest role in team history came off the ice, that being a transaction on January 19, 1976 when he was traded along with cash and future considerations to Edmonton for Dave Keon, John McKenzie, Jack Carlson, Steve Carlson and Dave Dryden. Arndt played one game with the Oilers and eventually finished his career with Birmingham in 1977-78.

Fred Arthur

Touted as one of the best amateurs in the 1980 Entry Draft, the Whalers sensed they had landed a cornerstone for their defense when general manager Jack Kelley selected the 6-foot-5, 210-pound rear guard with the eighth overall pick. Though Arthur capped a great junior career with a strong postseason in helping the Cornwall Royals to the 1980 Memorial Cup championship, he never achieved the promise projected by NHL scouts in a harvest which produced prolific scorers at the blue line such as Paul Coffey (by Edmonton at No. 6), Larry Murphy (by Los Angeles at No. 4) and Dave Babych (by Winnipeg at No. 2). Arthur, meanwhile, played just three games in Hartford before being packaged in a deal with Ray Allison to Philadelphia on July 3, 1981 which brought one-time 50-goal scorer Rick MacLeish to the Whalers. Arthur went 1-8-9 in 77 games with Philadelphia over a two-year period, exiting the NHL after 1982-83 to pursue a career in medicine. Today, Arthur is an established physician in his native Ontario.

B

Dave Babych

One of the best skaters in the NHL, this defenseman played 349 games at the blue line once arriving from Winnipeg on November 21, 1985 in exchange for Ray Neufeld. Drafted second overall in 1980 by the Jets (Montreal used the No. 1 pick to take Doug Wickenheiser), Babych possessed offensive ability from the point. His biggest goal in team annals came on April 29, 1986 in the Stanley Cup playoffs, a blast from just inside the blue line with 2:43 left in regulation to push Game 7 against Montreal into overtime (the Whalers would lose this epic match 2-1 at 5:55 of overtime on a goal by Claude Lemieux). A wrist injury limited Babych's final year (1990-91) in Hartford to just eight games. Claimed by Minnesota in the Expansion Draft, Babych wound up moving on to Vancouver in a deal on June 22, 1991 for Tom Kurvers. With the Canucks, Babych played in his 1,000th NHL game and helped Vancouver to the Cup Finals in 1994. Also holds the distinction as one of several brother acts to play for the Whalers (older brother Wayne played 41 games with Hartford).

Wayne Babych

The Whalers obtained this one-time 50-goal scorer from Quebec for Greg Malone on January 17, 1986 but his stay in Hartford was limited to just 41 games because of a knee

__Dave Babych__, who has played in over 1,000 NHL games, was a rock on defense during a 349-game hitch over six seasons in Hartford. His biggest goal came in Game 7 of the 1986 Stanley Cup playoffs against Montreal, a blast from the left point with 2:43 left in regulation to force overtime.

injury. A gritty winger, Wayne was the third overall pick in the 1978 Entry Draft (following Bobby Smith by Minnesota and Ryan Walter by Washington) and wound up playing 519 NHL games, mostly with St. Louis. His best year was 1980-81 when he potted 54 goals with the Blues.

Jergus Baca

Hartford selected this Czech defenseman in 1990 with the 141st overall pick. A first-team selection with the Czech National Team in 1989 and 1990, Baca had two assists in 10 career games with the Whalers, his only appearances in the NHL. His mobility and offensive skills proved most effective in the American Hockey League where he helped Springfield to the 1991 Calder Cup championship.

Ralph Backstrom

The NHL's Calder Trophy winner in 1958-59 capped a 19-year career in pro hockey with the Whalers, playing 115 games over two seasons including his finale in 1976-77. The classy center played on six Stanley Cup winners in Montreal where his speed and anticipation enhanced his forechecking skills as the third-string center behind Jean Beliveau and Henri Richard. Also played for Los Angeles and Chicago in the NHL. Jumped to the WHA in 1973-74 and helped the surprising Chicago Cougars to the AVCO Finals in a season where he, Pat Stapleton and Dave Dryden wound up as part-owners of a beleaguered franchise which needed a bailout from the league treasury. In 1974, he played for Team Canada, a WHA All-Star delegation which played the Soviet Union in an eight-game series. By most accounts, Backstrom was the standout of the Series, going 4-4-8. After two seasons in Chicago, Backstrom was the mainstay of the ill-fated expansion Denver Spurs in 1975-76, a club that disbanded after 41 games which included a two-game trial as the Ottawa Civics. When the hockey team folded on January 16, 1976, New England secured Backstrom's contract along with linemate Don Borgeson. Backstrom's arrival added another experienced player in the pivot which already included Dave Keon, Larry Pleau and Mike Rogers. Played in over 1,000 NHL games and 305 WHA games.

Reid Bailey

A defenseman who played in 40 NHL games, the Whalers signed Bailey as a free agent on December 9, 1983 which led to a 12-game trial during the 1983-84 season. The Flyers signed him out of the International League on November 20, 1978 which led to a 31-game stint over two years in Philadelphia including 12 in the Stanley Cup playoffs. Also played one game with Toronto in 1982-83.

*The AVCO Trophy rests between Whalers general managing partner **Howard L. Baldwin** and operations director **Jack Kelley**. The pair organized the New England Whalers of the World Hockey Association, a team that won the first-ever WHA championship in 1972-73.*

*Veteran **Ralph Backstrom** proved to be a valuable addition to the Whalers, arriving when the Denver-Ottawa franchise ceased operations midway in the 1975-76 campaign. Besides being the last Whaler to wear No. 9 before **Gordie Howe** did in club annals, Backstrom went 31-50-81 in 115 games in the pivot over two campaigns.*

The founders of the New England Whalers included (left to right): **William Barnes, John Coburn, Howard L. Baldwin and Godfrey Wood.** *The foursome landed the franchise in November, 1971 and scrambled in the early going until Baldwin convinced Boston Celtics owner* **Robert Schmertz** *that his sports empire lacked a hockey team. Schmertz did something about it in a hurry. He bankrolled the newest hockey team in Boston and wound up putting together the best organization in the upstart league. The Whalers won a league-best 46 games in the regular season and then went 12-3 in the playoffs to win the championship.*

Howard L. Baldwin is considered the main founder of the Whalers and organized the financial resources to launch the franchise in 1972. Baldwin later served as president of the WHA and then played a pivotal role along with NHL president John Ziegler in the unification of professional hockey in 1979 when the NHL and WHA made peace. Baldwin oversaw the management structure of the Whalers in Hartford. He departed in 1988 when the franchise was sold to real estate developer Richard Gordon and insurance mogul Donald Conrad for $31 million. These days, Baldwin heads up the Pittsburgh Penguins.

Howard Baldwin

If Bob Schmertz supplied the financial capital to make the New England Whalers a reality, it was the vision and drive of Howard Baldwin who solidified the presence of the hockey club during its early years in the WHA, its move to Hartford in 1974-75 and eventual move into the NHL in 1979. Baldwin, joined by John Colburn, Godfrey Wood and William Barnes, founded the hockey organization when they pooled $25,000 to secure a charter membership in the WHA in 1972-73. One of Baldwin's best moves followed shortly by coaxing Schmertz, who owned the Boston Celtics at the time and collected sports franchises as if they were postage stamps, to come on board. "Our biggest break," Baldwin recalls, "was getting a guy like Bob Schmertz. I can honestly say that the hockey team would have never survived. There were so many things that he did above and beyond the call of duty. An entrepreneur in every sense of the word." Baldwin had no trouble landing venture capitalists in subsequent seasons. Another coup came when the team exited Boston for Connecticut and Baldwin tapped the rich corporate support in Greater Hartford. It was during this juncture that the Whalers survived the roof collapse of the Hartford Civic Center and Baldwin, also wearing the hat as WHA president, played a pivotal role in negotiations that brokered an expansion plan with the NHL in 1979 that would absorb four teams from his league. Baldwin headed up the Whalers until the franchise was sold. Hartford businessmen Donald Conrad and Richard Gordon paid, at the time, a record price of $31 million for a hockey team Baldwin had nurtured from birth and oversaw for 17 years. "It's the natural progression for the franchise, from corporate to individual ownership," Baldwin stated on September 8, 1988. "It's time to move on." In retrospect, Baldwin's exit proved to be the transaction that eventually led the hockey club to conclude its run in Hartford within the coming years. Baldwin, connected within the hockey community at large, had the cleverness and resources to keep the Whalers docked in Connecticut. Baldwin remained in hockey and soon allied with rental car mogul Morris Belzberg. The pair purchased the Minnesota North Stars in 1989 and virtually kept the franchise from moving to California. Once forced out of power when Norman Green took over controlling interest in the team and announced plans to relocate the club to Dallas, Baldwin, keen on operating a team, formed a group to purchase the Pittsburgh Penguins. The NHL approved the sale on November 19, 1991 when the DeBartolo family sold the hockey club to a triumvirate headed by Baldwin, Belzberg and Thomas Ruta. These days, Baldwin has retained a major interest in the Penguins while charting tougher waters in the entertainment business with the release of several motion pictures including *Sudden Death*, *From the Hip* and *Spellbinder*.

Norm Barnes

A defenseman who played with plenty of heart, Barnes appeared in 156 NHL games, a stretch that included 74 in Hartford over two campaigns with the last being 1981-82. The Whalers acquired Barnes along with Jack McIlhargey to shore up the blue line on November 2, 1980 in exchange for a second-round draft pick. Drafted 122nd overall by Philadelphia in 1973 out of Michigan State, Barnes spent several years in the Flyers system, sparking Maine to two Calder Cup titles (1978 and 1979) before finally getting his chance in the NHL. In 1979-80, Philadelphia went 35 consecutive games (25-0-10) without a loss en route to the league's best record. Among the standouts that season was Barnes who was named to the midseason All-Star Game, going 4-21-25 in 59 games before an abdominal injury shortened his year.

Dave Barr

Though only in Hartford for 30 games in 1986-87, this forward is one of a select few to score a goal in his first game (a 5-4 win over Buffalo on October 24, 1986) and his last (a 4-3 loss at Minnesota on January 10, 1987) as a Whaler. Barr signed as a free agent with Boston in 1981 and forged a career of over 600 NHL games, logging time with the St. Louis Blues, New York Rangers, Detroit Red Wings, New Jersey Devils and Dallas Stars. Hartford sent Tim Bothwell to St. Louis on October 21, 1986 to obtain Barr and then peddled the winger to Detroit for Randy Ladouceur on January 12, 1987.

*Coach **Don Blackburn** watches hockey legend **Bobby Hull** put on a Hartford jersey for the first time. The "Golden Jet" finished out his extensive career with the Whalers. Hull played a dozen overall games with Hartford, his last being a 4-3 overtime loss to Montreal in the Stanley Cup playoffs on April 11, 1980.*

Bill Bennett

One of three sons of former Boston netminder Harvey Bennett to play in the NHL, "Big" Bill was plucked from the Bruins by Hartford in the 1979 Expansion Draft. The 6-foot-5, 235-pound winger played in just 24 games with the Whalers in 1979-80 which was his last tour in the NHL.

Marc Bergevin

Drafted 59th overall by Chicago in 1983, this defenseman has appeared in over 700 games including a stretch of 79 over parts of two seasons in Hartford. Bergevin was a regular at the blue line in 1991-92 after reviving his career with two strong years in Springfield. The Whalers obtained Bergevin from the New York Islanders on October 30, 1990 for a fifth-round draft choice (Ryan Ruthie) in 1992. By then, the Montreal native had signed on with the expansion Tampa Bay Lightning. A member of Detroit's NHL record 62-win team in 1995-96 that reached the Stanley Cup Finals, Bergevin reached the 700-game plateau with St. Louis early in the 1996-97 campaign.

Bill Berglund

A goaltender who played briefly in the WHA with the Whalers, Berglund left the Montreal chain to sign with New England out of the American League. Played in five games from 1973 to 1975, virtually the third-string netminder behind Al Smith and Bruce Landon.

James Black

Hartford used its fifth-round pick in 1989 to select this center from the Portland Winterhawks. He appeared in 32 games over three seasons with the Whalers, going 4-6-10. Dealt to Minnesota on September 3, 1992 in a swap which brought center Mark Janssens to Hartford. Black has also played for Dallas and Buffalo in the NHL.

Don Blackburn

"Blackie" played 146 games over three seasons in the WHA with the Whalers. He jumped to New England after splitting 1972-73 with the New York Islanders and Minnesota North Stars. Blackburn also coached the team in the WHA as well as serving as Hartford's first NHL bench boss. His career highlight occurred in 1979-80 when

he directed Hartford's veteran cast which included Gordie Howe, Bobby Hull and Dave Keon to a berth in the Stanley Cup playoffs. The Whalers, picked by everyone to finish last among 21 teams, compiled a 27-34-19 record for 73 points for the 14th best mark in the league (and two points out of 11th). Blackie's tenure ended on February 20, 1981 when the Whalers, forced to go several games without ace defenseman Mark Howe, went on a 2-14-8 slide from January 2 to February 19 to fall 14 games under .500 to 15-29-16 and out of postseason contention.

Bob Bodak

The Whalers signed this forward as a free agent in 1988. In the only game he played for Hartford, the Whalers went into Philadelphia on December 14, 1989 and stunned the Flyers by winning 3-2. Bodak also played three games with Calgary in 1987-88.

Danny Bolduc

"Bulldog" logged three years with the Whalers in the WHA, a span of 88 games, but his hustle and skating skills are best remembered for the way he shadowed Marc Tardif of the Quebec Nordiques in the 1978 playoffs. His defensive play was critical to New England's surprise upset of the Nordiques in the semifinals. In five games, Bolduc limited Tardif, the WHA's scoring champ, to just two goals and nine shots. A contract impasse resulted in Bolduc jumping to the Detroit Red Wings in 1978-79. He also played with Calgary, a total of 102 NHL games before retiring after the 1983-84 season.

Don Borgeson

Joined the Whalers along with Ralph Backstrom when the Denver-Ottawa franchise ceased operations midway in the 1975-76 season. A right wing, Borgeson played in 31 games with the Whalers and wound up with the Calgary Cowboys after his tour of duty with New England.

Tim Bothwell

A defenseman who appeared in 502 NHL games, Bothwell played parts of two seasons in Hartford. The Whalers purchased his contract from St. Louis on October 4, 1985 and dealt him back to the Blues a year later in a trade for Dave Barr on October 21, 1986. Also played with the New York Rangers who he signed with as a free agent after two seasons at Brown University.

Dan Bourbonnais

Hartford drafted Bourbonnais with the 103rd pick in the 1981 Entry Draft from the Calgary Wranglers of the Western Hockey League. Like many forwards over the years with the Whalers, Bourbonnais had a couple of trials and was soon gone. He appeared in 59 games over two seasons, going 3-9-12 as a rookie in 24 games in 1981-82 and then 0-16-16 in 35 games in 1983-84, his last in the NHL.

Charlie Bourgeois

One of several former St. Louis players who were obtained by former Blues general manager Emile Francis, Bourgeois had a short tenure in Hartford. Obtained as insurance at the blue line on March 8, 1988 along with Mark Reeds in a swap of draft choices, Bourgeois played in his one and only game on March 15, 1988, an 8-6 loss to Calgary. It was the final NHL game of a 280-game tenure for Bourgeois who also had stops in Calgary and St. Louis. He signed as a free agent on April 19, 1981 with the Flames.

Pat Boutette

General Manager Jack Kelley may have made his best NHL trade on December 24, 1979 when he sent Bob Stephenson to Toronto to obtain durable Pat Boutette, a strike of good fortune considering Leafs general manager George "Punch" Imlach believed he was acquiring veteran goalie Wayne Stephenson (who was with Philadelphia at the time). A good mucker in the corners, Boutette added the "Bash" on a line with Mike "Dash" Rogers and Blaine "Stash" Stoughton which fueled the team's playoff run. Over the final 35 games in 1979-80, the "Bash-Dash-Stash" trio combined for 62 goals for and just 22 goals against. Following the 1980-81 season where Boutette averaged a point-per-game, he was lost as compensation along with promising center Kevin McClelland on June 29, 1981 to the Pittsburgh Penguins when Hartford, then under the direction of Larry Pleau, signed free agent goaltender Greg Millen. Boutette returned in 1984-85 for what would be his last of 10 NHL seasons in a deal for Ville Siren on November 16, 1984. It capped an impressive run for the 5-foot-8, 175-pound winger who played in 756 NHL games and was drafted 139th overall by Toronto in 1972.

Greg Britz

A forward with good speed who played in eight NHL games, Britz signed as a free agent with Hartford following a strong training camp on October 5, 1986. Shortly after his only game as a Whaler, a 7-2 loss at Boston on October 12, 1986, Britz was sent to the minors and never resurfaced in the NHL. Played four years at Harvard University where he caught the eyes of Toronto scouts who signed him as a free agent on November 2, 1983.

*Gritty **Pat Boutette** supplied solid two-way play during two stints with Hartford. "Bash" came to the Whalers from the Toronto Maple Leafs on December 24, 1979 for **Bob Stephenson** and was inserted on the club's top line with center **Mike Rogers** and winger **Blaine Stoughton**, a trio which fueled the Whalers run to the playoffs in 1980.*

Richard Brodeur

The "King" closed out his reign in pro hockey with the Whalers in 1987-88, a career which included 305 games in the WHA and 385 in the NHL. Hartford obtained the veteran netminder from Vancouver for Steve Weeks on March 8, 1988. Brodeur played six regular season games and four Stanley Cup games, wrapping up an extensive tenure that began in 1972-73 with the Quebec Nordiques. He anchored the Nordiques to the AVCO Trophy in 1977 and was the backbone for Vancouver's first trip to the Stanley Cup Finals in 1982. Originally drafted by the New York Islanders in 1972, Brodeur signed on with Quebec and became one of a handful of players to play each year of the rival league's existence. The Nordiques made Brodeur one of their priority picks upon joining the NHL in 1979 but he was then shuttled to the Islanders for goalie Goran Hogosta on August 15, 1979. Brodeur played in two games for the Islanders in 1979-80 and spent the bulk of the year in Indianapolis where he was voted the top goaltender in the Central League. Contemplating retirement if returned to the minors, Brodeur gained a rebirth on October 6, 1980 when he was dealt to Vancouver in a swap that also included fifth-round draft picks. With the Canucks, Brodeur had plenty of success with the focal point being the 1982 playoffs when he went 11-6 with a 2.70 goals-against average.

Jeff Brown

A dynamic performer on the power play, this defenseman helped revive Hartford's sagging special team unit during 1995-96 once coming over from Vancouver on December 19, 1995 in a multi-player deal. Brown, who was feuding with ex-Whaler Rick Ley who was coaching the Canucks at the time, soon became the centerpiece in a three-team swap. The Whalers sent forward Jocelyn Lemieux and a second-round pick in 1998 to New Jersey for center Jim Dowd and the Devils' second-round pick in 1997. Hartford then pitched Dowd, defenseman Frantisek Kucera and its own second-round pick in 1997 to the Canucks for Brown and Vancouver's third pick in 1998. Brown delivered with 7-31-38 in 48 games with the Whalers who eventually fell short in a bid to make the playoffs. In 1996-97, Brown was heavily counted on to be a mainstay at the blue line but suffered a back injury against Phoenix on Opening Night. The impairment required surgery and virtually scuttled his season. Originally taken 36th overall by Quebec in 1982, Brown began his NHL career with the Nordiques in 1985-86. Has tallied 20 or more goals three times during his career. Exited Quebec on December 13, 1989 for St. Louis in a swap for Tony Hrkac and Greg Millen. Shipped to the Canucks on March 21, 1994 along with Bret Hedican and Nathan Lafayette for Craig Janney and contributed to Vancouver's trip to the Stanley Cup Finals. Has played 687 NHL games with 150-406-556 over a dozen seasons.

Kevin Brown

The first English-born player in club history, Brown made his debut with the Whalers in 1996-97, playing 11 games after coming over in a deal with Anaheim on October 1, 1996 for Espen Knutsen, a center from Norway selected by Hartford as the 204th overall pick in the 1990 Entry Draft. Taken 87th overall in 1992 by the Los Angeles Kings, Brown is one of several pros to play in junior programs sponsored by Compuware. He played two years for the Detroit Junior Whalers capped by a 14-26-40 explosion in 17 playoff games in 1994. Has 3-7-10 in 41 career NHL games.

Rob Brown

The 1987 Canadian Junior Player of the Year played his best NHL hockey in Pittsburgh for the Penguins who selected the moody winger 67th overall in 1986. After rolling up an incredible 76 goals and 212 points in 1986-87 with Kamloops of the Western Hockey League, Brown quickly fit into Pittsburgh's high-octane attack. In his second year in the NHL, he wound up fifth in the scoring race (behind Mario Lemieux, Wayne Gretzky, Steve Yzerman and Bernie Nicholls) with 49 goals and 115 points in 1988-89. The Whalers, then under the stewardship of Eddie Johnston who was in charge of the Penguins when Brown was drafted, sent Scott Young, Hartford's top pick in 1986 who had yet to blossom, to his former team on December 21, 1990 to juice up the Whalers offense. Brown had 34 goals and 73 points in 86 games over parts of two seasons in Hartford before moving on to Chicago in a swap on January 24, 1992 which brought veteran defenseman Steve Konroyd to the Whalers. Brown, just 25, soon vanished from the NHL

*Defenseman **Jeff Brown** supplied a spark on the power play in 1995-96 once coming over from Vancouver in a trade and was expected to do the same throughout 1996-97. But when Brown, who cracked the 20-goal plateau three times in his career, suffered a back injury in the season opener against Phoenix and wound up missing the entire year, his absence was a major reason why the Whalers missed the playoffs in their final campaign. In 49 games for Hartford, Brown went 7-31-38.*

after trials with the Blackhawks, Dallas Stars and Los Angeles Kings. However, his offensive exploits have continued in the International League where he has won a couple of scoring titles.

Jack Brownschidle

The Whalers made a habit out of acquiring ex-St. Louis players during the Emile Francis era and claimed this lanky defenseman on waivers on March 2, 1984. Brownschidle made his first goal with Hartford a game-winner in a 4-1 decision over Buffalo on March 27, 1984. A two-time All-American at Notre Dame, Brownschidle was drafted by the Blues as the 99th overall pick in 1975. Overall, he played nine years and 494 NHL games, going 39-162-201. His last season was 1985-86 when he failed to score a point in nine games with the Whalers while spending most of the campaign in Binghamton of the American League. The older brother of Jeff Brownschidle who also played in Hartford, Jack logged 39 games with the Whalers.

Jeff Brownschidle

Hartford signed Jeff Brownschidle, the younger brother of Jack Brownschidle who also played for the Whalers, as a free agent on June 9, 1981. Jeff eventually made it to the Whalers for seven games over two seasons, getting his only NHL point on April 3, 1982 in a 3-3 tie against the New York Rangers.

*In 1992-93, the Whalers changed the look of their sweater and logo. Holding the jersey is general manager **Brian Burke** and goaltender **Sean Burke**. Rather than green being the key color in the design, the prominent shade became navy blue. A silver outline also was added to the crest to literally make it "jump off the uniform" as described in the team's yearly press guide.*

Jeff Brubaker

The Whalers, gearing up for their final WHA season, signed two Canadian juniors on June 9, 1978. One was Jordy Douglas and the other was Brubaker who had also been tabbed by Boston in the NHL draft as the 102nd overall pick. A rugged forward who had 20 goals and 307 penalty minutes with Peterborough of the Ontario League, Brubaker brought a physical presence to the rink throughout his career. He logged a dozen WHA games with the Whalers and also 46 NHL games. He was claimed by Montreal on October 5, 1981 in the Waiver Draft, a lottery where his name was often called. Brubaker exchanged teams four times in the preseason roster shuffle including twice in 1983. Overall, he played in 178 NHL games with Montreal, Calgary Flames, Toronto Maple Leafs, Edmonton Oilers, New York Rangers and Detroit Red Wings. His last season was 1988-89.

Brian Burke

As a former player, agent and assistant general manager in Vancouver, Brian Burke had the credentials and hope of the Civic Center faithful to lead Hartford out of the darkness when Richard Gordon hired him to take over hockey operations on May 26, 1992. Besides changing the team colors to a more prominent use of navy blue, Burke's most significant deal came on August 28, 1992 when he obtained goaltender Sean Burke (no relation) in a deal with New Jersey for Bobby Holik. At the Entry Draft table, Burke oversaw two proceedings. His top picks, Robert Petrovicky (1992) and Chris Pronger (1993), never

got the chance to blossom in Hartford. Like other high-profile choices, both were discarded when the managerial revolving door brought more change to the front office. Quite simply, Burke did not get the long-term chance to accomplish much in Hartford. He elected to leave the day-to-day operations for a significant position with the league office, maybe sensing that Hartford's NHL future was drawing to a close more than the fallout of a 52-loss season might with Gordon. Today, Burke is Commissioner Gary Bettman's right-hand man. He usually doles out penalties for on-ice aggressiveness when players cross over the guidelines in the NHL rule book. It's a bit of a twist for Burke whose main focus was to establish a physical, on-ice presence for his Whalers. "I wanted to build a team that played with a snarl so opposing clubs came to the rink with trepidation," Burke says. The numbers suggest he achieved that in Hartford. The Whalers established a club record for penalty minutes (2,244) during 1992-93 led by Nick Kypreos (325), Mark Janssens (237), Jim McKenzie (205), Pat Verbeek (197) and Doug Houda (167).

Sean Burke

It was Whalers coach Paul Holmgren who often said "Sean plays bigger than the net" and Hartford fans often saw Burke using his 6-foot-4, 215-pound frame to stonewall rival shooters since his arrival from New Jersey on August 28, 1992. Taken 24th overall in the 1985 Entry Draft, Burke had an immediate impact once joining the Devils after the 1988 Winter Olympics. The rookie went

*Few covered the 24-square foot opening better than goaltender **Sean Burke** and here he extends the pads to rob Pierre Turgeon of St. Louis to enable defenseman **Curtis Leschyshyn** to clear the zone.*

10-1 down the stretch and then 9-8 in the Stanley Cup playoffs before New Jersey finally expired in the Wales Conference Finals against Boston. He continued to be the main puckstopper for the Devils but soon got into a contract squabble with management. Burke left the Devils to play for Team Canada in the 1992 Olympics and eventually forced a trade. Brian Burke, in one of his first deals during his one-year hitch as general manager, shipped Bobby Holik, Hartford's top pick in 1989, along with a second-round draft choice (the Devils took Jay Pandolfo in 1993) for Burke and defenseman Eric Weinrich. Despite missing a handful of games with lower back ailments, Burke never took a night off. He supplied the Whalers a big-play netminder in the mold of Mike Liut who was the king-pin force of Hartford's most succesful teams in the 1980s. The only difference is Burke never had the chance to shine in the playoffs for the Whalers.

*Steady defenseman **Adam Burt** was Hartford's second-round pick in 1987. He played 499 NHL games for the Whalers, the third highest in team annals.*

Adam Burt

A second-round pick by the Whalers (39th overall) in 1987, Burt lasted through a litany of coaching changes and roster shuffles since making his debut in Hartford in 1988-89 and wound up one of a handful of players to appear in over 400 NHL games as a Whaler. Besides moving up the club's all-time team leader board in games played, Burt has shown flashes of offensive pop. His most dramatic goal came in overtime to give the Whalers a 5-4 victory over Minnesota on February 29, 1992. When the franchise shifted to Carolina for 1997-98, Burt and Andrew Cassels were the only ones remaining with the organization who played in the franchise's last playoff series (1992 against Montreal). Cassels has since been dealt to Calgary.

Ron Busniuk

Hartford fans only got to see this hard-hitting blueliner play 66 games in the WHA over two seasons but his aggressiveness to clear the crease of rival forwards was something worth watching. A stay-at-home defenseman, Busniuk played six games with Buffalo of the NHL before jumping to the Minnesota Fighting Saints in 1974-75. In 287 WHA games, he had 9-64-73 with 762 penalty

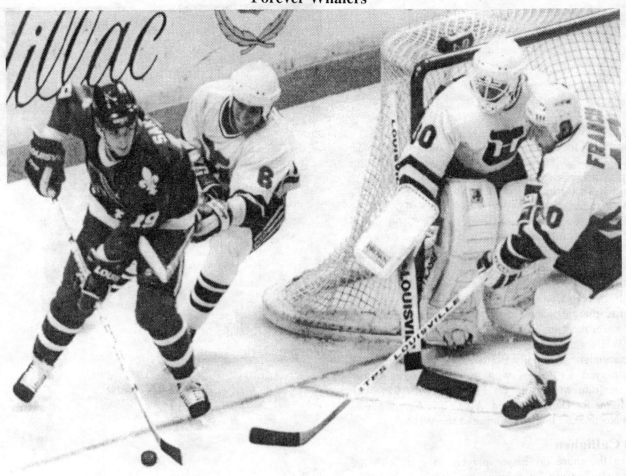

*A look at the attention and defensive coverage by the Whalers against Quebec's Joe Sakic who operates near the net. In goal, **Peter Sidorkiewicz** has cut off the near post while center **Ron Francis** and defenseman **Adam Burt** position themselves to reduce the passing lanes.*

minutes. New England signed Busniuk as a free agent on March 9, 1976 when the Fighting Saints disbanded. He was packaged, along with Brett Callighen, to Edmonton on February 5, 1977 to secure Mike Antonovich and Bill Butters.

Bill Butters

"Captain Crunch" spent two seasons with the Whalers during a five-year tenure in the WHA and NHL. The rugged defenseman signed with the Minnesota Fightning Saints in 1974-75 and also made stops in the WHA in Houston and Edmonton before coming to New England on February 5, 1977 along with Mike Antonovich for Ron Busniuk and Brett Callighen. Roughly a year later, the Whalers, on February 16, 1978, sold Butters to the Minnesota North Stars of the NHL where he eventually finished his career.

Mike Byers

When the Whalers won the first AVCO Trophy in 1973, general manager Jack Kelley made two significant deals during the season. On February 16, 1973, Byers was secured from the Los Angeles Sharks for Mike Hyndman. Much like a swap earlier in the year which brought Brit Selby from Quebec (for Bob Brown), the arrival of another proven veteran gave the Whalers more options and additional two-way players. Byers played 187 games with New England over a three-year stretch and scored two game-winning goals in the 1973 playoff series against Ottawa. He also played 166 NHL games over the years with Toronto, Philadelphia, Los Angeles and Buffalo, leading the Kings in goals (27) in 1970-71. The Whalers sold Byers to Cincinnati on February 24, 1976 where he wrapped up his 266-game WHA career.

C

Terry Caffery

After winning the American League's Rookie of the Year Award in 1971-72 at Cleveland, the Whalers recruited this clever playmaker away from the NHL Minnesota North Stars and wound up with the WHA's first Rookie of the Year. Bookended by Tom Webster and Brit Selby, Caffery tallied 39 goals and 61 assists en route to 100 points (second on the club to Webster's 103). One of seven players to reach the century mark in the first year of the WHA, Caffery sustained a serious knee injury on April 21, 1973 in a 5-4 playoff win over Cleveland which virtually ended his career. After missing 1973-74, Caffery gallantly returned to play 90 games over the next couple of years but was never the same player. Overall, he had 59-111-170 in 164 WHA games, the bulk coming during his rookie WHA campaign. Originally drafted by Chicago in 1966, Caffery played six games with the Blackhawks before getting dealt to Minnesota on February 23, 1971 with Doug Mohns for Danny O'Shea. Also played eight games with the North Stars before jumping to the WHA.

Brett Callighen

One of the more promising players to sign with the Whalers who enjoyed working the corners, Callighen appeared in only 33 games in 1976-77 in the WHA before moving on to Edmonton, along with Ron Busniuk, in a trade on February 9, 1977 for Mike Antonovich and Bill Butters. A speedy skater, Callighen found home in Alberta where he played the bulk of his 213 WHA games and 160 NHL games. His best year was 31 goals in 1978-79 on a line with Wayne Gretzky and Blair MacDonald which helped spark Edmonton to 48 wins and 98 points en route to the last regular season crown in the WHA. Had 23 and 25 goal seasons in the NHL before an eye injury prematurely ended his career.

Wayne Carleton

"Swoop" played in 560 games during his 12-year hockey career in both pro leagues which included 108 games with the Whalers. Earned his nickname from teammates who noticed Carleton's nifty move where he would retreat into his own zone to pick up more speed for a "swoop" down ice. New England obtained the 6-foot-3, 215-pound center from Toronto on September 4, 1974 in a deal for future considerations (Jim Dorey was dealt to the Toros on December 31, 1974) and a second-round draft pick. Led the Whalers in scoring with 35-39-74 in 1974-75. The following year, the Whalers sent Carleton to Edmonton, along with Paul Hurley, for Mike Rogers and Steve Carlyle on January 19, 1976. Carleton broke into pro hockey with the Toronto Maple Leafs in 1965-66 and also played for Boston Bruins and California Golden Seals in the NHL. He was a member of Boston's Stanley Cup

Center **Terry Caffery** was one of seven players to score 100 points in 1972-73 in the WHA's inaugural campaign. He was a sparkplug for the Whalers that winter and earned the league's Rookie of the Year honors. Injured in the playoffs that season, Caffery missed the following campaign but did return in 1974-75 to score 15 goals and 52 points.

champions in 1970. Jumped to the WHA in 1972-73 to help get the new league started and went 42-49-91 for the Ottawa Nationals. Last year was 1976-77 in Birmingham. An avid trackman who owned, trained and drove harness and pace horses in his native Ontario. Had 55-73-128 in 278 NHL games and 132-180-312 in 290 WHA games.

Jack Carlson

One of two Carlson brothers to play for the Whalers, "Big Bopper" rated as one of the heavyweights in WHA annals, amassing 694 penalty minutes in 272 games with stops in Minnesota, Edmonton and New England. Played 136 games for the Whalers, joining the club when the Fighting Saints folded in January of 1977 and exiting on February 1, 1979 when he was sold to the NHL North

*Known as the "Big Bopper" and regarded as one of the all-time best once dropping the gloves, **Jack Carlson** proved capable of scoring at times. He had 18 goals in 136 games with the WHA Whalers including a hat trick against Winnipeg in a 6-3 loss on December 17, 1977.*

*He was called "Swoop" and **Wayne Carleton** goes after the rebound here after Houston's Ron Grahame makes a save in WHA action at the Hartford Civic Center back on March 22, 1975. Carleton, who led the club in scoring that winter with 74 points, would exit to Edmonton midway in the following season in a trade that brought **Mike Rogers** to the Whalers.*

Stars for future considerations. Went on to play 170 NHL games, going 24-7-31 with 309 penalty minutes, in stints with the North Stars and St. Louis Blues before retiring after the 1982-83 season. Most productive year was 1977-78 when he had a hat trick (against Winnipeg's Gary Bromley) among nine goals and 29 points with New England.

Steve Carlson

The more famous of the Carlson brothers to play for the Whalers, Steve was featured (along with Jeff Carlson and Dave Hanson) in Universal Studio's motion picture *Slapshot* which starred Paul Newman. Much like the Charlestown Chiefs, Carlson got his start with Johnstown Jets of the North American Hockey League. Drafted 131st overall by Detroit in 1975, Carlson opted to sign with the WHA and battled his way to play 173 games for the Fighting Saints, Whalers and Oilers and 52 in the NHL with the Kings. Joined New England when the Minnesota club folded in 1977 and remained on board until he was claimed on waivers by Edmonton on September 4, 1978. Had 18 goals and killed penalties in helping the Oilers reach the WHA Finals. Los Angeles reclaimed Carlson from Edmonton where he wound up his career, going 9-12-21. Recently has been featured in a beer commercial in a revival of hockey's Hanson brothers.

Greg Carroll

A center who may have had the longest hair among anyone to play for the Whalers, Carroll had tours of duty in both leagues in Hartford where he wound up his career after the 1979-80 season. Also played for Cincinnati in the WHA, Washington and Detroit in the NHL. Originally drafted by the Capitals as the 15th overall pick in 1976, Carroll opted to sign with the WHA. The Whalers, looking to add some youth, dealt the rights to goaltender Mike Liut to the Stingers on May 26, 1977 to obtain Carroll and defenseman Bryan Maxwell. On February 12, 1978, Carroll was returned to Cincinnati for veteran defenseman Ron Plumb. In 1978-79, Carroll played 24 games with the Capitals before getting claimed by the Red Wings on January 6, 1979. He was back with the Whalers, signing as a free agent on October 30, 1979. Though he once had a 171-point season in juniors at Medicine Hat, Carroll had just 50-100-154 over 281 pro games.

Jimmy Carson

A 55-goal scorer (in 1987-88) who was involved in trades for two of hockey's greatest scorers, Carson completed his NHL career in 1995-96 with the Whalers to wind up with 275 goals and 561 points in 626 NHL games over 10 seasons. Tabbed second overall in 1986 by Los Angeles (Detroit began the Entry Draft by picking Joe Murphy), Carson, a center, went on to score 30 or more goals during five different seasons for three different teams. He was among the players dealt to Edmonton on August 9, 1988 when the Oilers sent Wayne Gretzky to Los Angeles. Carson was soon dealt by the Oilers, moving on to the Red Wings on November 2, 1989 in a multi-player swap that brought Murphy and Adam Graves to Edmonton. Regained the scoring touch in Detroit before going back to Los Angeles in a swap for Paul Coffey on January 29, 1993. Also played for Vancouver before signing with the Whalers as a free agent on July 15, 1994. Played 49 games in Hartford over two seasons before getting released on December 1, 1995. He finished the year playing in Europe. Helped the Detroit Vipers of the International League championship in 1996-97.

*Center **Andrew Cassels** played 438 games for the Whalers over six seasons. He finished with totals of 97-254-351 for Hartford, a steady playermaker who wound up third on the team career board behind **Ron Francis** and **Kevin Dineen**.*

*Though he played just 18 games for the Whalers, defenseman **Steve Chiasson** arrived from Calgary on March 5, 1997 for **Glen Featherstone** and prospect **Hnat Domenichelli** and instantly juiced Hartford special-team unit. Chiasson's first goal as a Whaler came on the power play to clinch a 2-0 win over Montreal on March 7, 1997. It was also the last shutout recorded by the Whalers, a 26-save performance by Sean Burke.*

Lindsay Carson

A journeyman forward who played 373 games over seven years, Carson finished up his stay in the NHL with the Whalers in 1987-88. A product of the Philadelphia system, Carson logged 27 games in Hartford after coming over from the Flyers on January 22, 1987 in a swap for Paul Lawless. Philadelphia selected Carson 56th overall in the 1979 Entry Draft (he and future Whaler draftee Don Nachbaur were teammates on the Billing Bighorns). A member of two Philadelphia teams that reached the Stanley Cup Finals, Carson skated a regular shift on the third or fourth line with the ability to play without the puck. His best year was 1984-85 with 20-19-39 en route to career totals of 66-80-146. Also played 49 postseason games, going 4-10-14. His last goal was tallied in a 7-3 loss to Montreal in Game 2 (April 7) of the 1988 Stanley Cup playoffs.

Andrew Cassels

Drafted 17th overall by Montreal in 1987, the Whalers acquired this playmaker from the Habs on September 17, 1991 for a second round pick (Montreal took Valeri Bure in 1992). Steady but not spectacular, Cassels has led Hartford in scoring twice (1993 and 1995). Has played the bulk of his 400 NHL games with the Whalers. Went 2-4-6 in his first playoff series with Hartford and assisted on Yvon Corriveau's overtime tally in a 2-1 decision over Montreal in Game 6 on April 29, 1992. Also scored one of two goals by the Whalers in a 3-2 double-overtime loss to the Canadiens in Game 7 on May 1, 1992. Like many players in Hartford in recent winters, his play would benefit greatly if the Whalers upgraded the pivot position. The arrival of Keith Primeau from Detroit early in the 1996-97 campaign gave Hartford its best 1-2 punch up the middle in years.

Brian Chapman

Hartford's fourth-round pick (74th overall) in 1986, Chapman played only three games for the Whalers. The defenseman piled up 29 penalty minutes for Hartford, with 20 coming on March 31, 1991 in a 7-3 loss to the Bruins at Boston Garden.

Bob Charlebois

A good special team player for the Whalers, Charlebois spent two seasons with New England in the WHA. After 24-39-63 in 78 games in 1972-73 with Ottawa, the Whalers shipped Brit Selby to the Nationals for Charlebois on June 6, 1973. Originally drafted by Montreal, Charlebois also had a goal in seven NHL games with Minnesota in 1967-68.

Steve Chiasson

Obtained on March 5, 1997 along with a draft pick from the Calgary Flames for defenseman Glen Featherstone, rookie winger Hnat Domenichelli and two draft selections, this veteran blueliner supplied a late-season spark that revived Hartford's special teams and fueled hopes that the Whalers would clinch a playoff berth. In 18 games, Chiasson, who has a booming shot from the left point, went 3-11-14. His arrival helped Hartford close with 14 points (7-9-2)

*Center **Igor Chibirev** holds a couple of distinctions in club annals. Besides being the first Whaler to score a hat trick at the Igloo in Pittsburgh (it came in an 8-4 win on April 5, 1995), Chibirev is the first Russian to play for the Whalers.*

*Forward **Kelly Chase** joined the Whalers on January 18, 1995 in the Waiver Draft, arriving from St. Louis with a truckload of toughness. He led the club in penalty minutes (141) in his first year in Hartford.*

over the final weeks, a bid that finished two points shy of the club qualifying for the Stanley Cup tournament. Originally drafted 50th overall by Detroit in 1985, Chiasson has played in 657 games over 10 NHL seasons with 85-270-355. He delivered a power-play strike in his first game with Hartford to cap a 2-0 win over Montreal on March 7, 1997, the final whitewash in club annals.

Kelly Chase

Claimed from St. Louis in the Waiver Draft on January 18, 1995, Chase led Hartford in penalty minutes (141) during the lockout-shortened campaign, the only time as a regular in the NHL he has failed to reach the 200-penalty minute plateau. The Blues signed the native of Porcupine Plain, Sask. as a free agent on May 24, 1988 after three years with the Saskatoon Blades where he was an on-ice terror. He capped his stay in the Western Hockey League in 1987-88 by leading the junior circuit in penalty minutes (343). One of three in St. Louis annals to collect over 1,000 career penalty minutes, Chase has overcome a series of back, hand and shoulder injuries to average close to 60 games a season. Off the ice, Chase has given many free hours to the community. In St. Louis, for example, he was among the founders of the Gateway Special Hockey Program which enabled children with Down's Syndrome a chance to participate in hockey. Did the same type of charity work in Hartford until traded to Toronto on March 18, 1997 for an eighth-round draft choice.

Igor Chibirev

The first Russian to play for the Whalers, Chibirev made his debut on October 31, 1993 against St. Louis. Selected 266th overall in 1993, Chibirev is one of the few 11th-round picks in the history of the Entry Draft to reach the NHL. A good skater, Chibirev spent five years with the Moscow Red Army before crossing the Atlantic. When he scored a hat trick in an 8-4 win at Pittsburgh on April 5, 1995, Chibirev became the first European to register a three-goal game in Hartford annals.

Shane Churla

When the Whalers picked Churla as the 110th overall player in the 1985 Entry Draft, it was former general manager Emile Francis who remarked, "if you're going to interview this guy, you better wear a mask." Churla began his NHL career with Hartford (22 games over two seasons) before moving on to Calgary with Dana Murzyn on January 3, 1988 in a deal which brought Carey Wilson, Neil Sheehy and Lane MacDonald to the Whalers. Churla, who plays a physical game, has also logged time with Minnesota, Dallas, Los Angeles and the New York Rangers. He led Hartford in penalty minutes (40) despite dressing for just two Stanley Cup games in the 1987 playoff series against Quebec. Despite the rough stuff, Churla has good wheels and has found a niche as a checker. He has played in 400 NHL games and amassed over 2,000 penalty minutes in eight-plus seasons.

Ron Climie

A forward with plenty of speed, Climie played five years in the WHA (249 games) including a 93-game stint over three seasons with the Whalers. One of a handful of players to score five goals in a game (for Edmonton on November 6, 1973 in a 9-0 rout of the New York Raiders), Climie came to New England on February 15, 1975 in a swap for Tim Sheehy. His stay would have been longer but his career prematurely closed because of a back injury.

Paul Coffey

A future Hall of Famer and a member of four Stanley Cup championship clubs, Coffey logged 20 games in Hartford during 1996-97, a brief tenure for the highest-scoring defenseman in NHL history. A big-play maker at the blue line where he once scored 48 goals (in 1985-86) and racked up five 100-point campaigns, Coffey joined the Whalers on October 9, 1996 from Detroit in the swap that also involved Brendan Shanahan and Keith Primeau. Unlike Shanahan who went public with a trade demand, Coffey refused management's offer regarding a future role with the Hartford organization. He also declined to go public with a trade request. Yet Coffey's refusal to play with zest, except for a pair of two-point games (a 3-1 win over Montreal on November 20, 1996 and a 5-2 rout of Florida on December 11, 1996), forced the Whalers to field trade offers from other clubs. The best came from Philadelphia and so on December 15, 1996, Coffey was sent to the Flyers for Kevin Haller and two draft picks (including a first rounder in 1997). In addition to the three

*A last look at two players who both wanted out of Hartford during the 1996-97 season and eventually got their way in hockey's forever-changing landscape. On November 4, 1996, the Whalers played in Detroit, the first meeting between the teams since the blockbuster deal made a month earlier which sent **Brendan Shanahan** to the Red Wings and brought **Paul Coffey** and **Keith Primeau** to Hartford. Detroit rolled to a 5-1 win. The game was one of 20 that Coffey dressed for as a Whaler. By mid-December, his mug was gone to Philadelphia for **Kevin Haller** and a draft choice.*

clubs he wore No. 77 for in 1996-97, Coffey has played for the Los Angeles Kings, Pittsburgh Penguins and Edmonton Oilers. He was taken sixth overall by the Oilers in the 1980 Entry Draft.

Don Conrad

A major decision-maker in the insurance business for close to 20 years in Hartford, Conrad helped get the corporate community involved in bringing major-league hockey to Connecticut when he worked closely in negotiations with Howard Baldwin that resulted in the New England Whalers relocating from Boston in 1974-75. It was Conrad's vision and clout with the Downtown Council that rallied the community to support the team, especially when the Civic Center roof collapsed in 1978 and then into the high-stakes NHL in 1979. Conrad is best remembered for a couple of comments he made to the press during his association with the hockey club. He coined the term "Czar Hunt" in 1982 when the Whalers sought experienced leadership in the general manager's position and hired Emile "Cat" Francis. A few years later, Conrad told the press, "I don't expect to be highly visible, but I won't be invisible," when he and real estate developer Richard Gordon became allies and purchased

the hockey club for $31 million on September 7, 1988. Within a year, Conrad was soon vaporized from the hockey operations when the two partners had a falling out over power and finances. He sold his interests to Gordon who hitched up with Colonial Realty Company, an ill-fated investment group that bilked fund contributors of millions and resulted in jail terms for partners Benjamin Sisti and Jonathan Googel.

Gaye Cooley

When a rash of injuries depleted New England's netminding corps in April of 1976, the Whalers signed Cooley as a free agent when ailments sidelined Christer Abrahamsson and Bruce Landon, leaving Cap Raeder to anchor the position. Cooley, who was named the MVP of the 1966 NCAA hockey tournament when he sparked Michigan State to the title, played one minute in the 1976 AVCO Trophy playoffs, the shortest tenure of any goaltender in team history.

Yvon Corriveau

Washington selected Corriveau 19th overall in the 1985 Entry Draft and the big winger showed flashes of a power game, particularly in 1991-92 when he had a dozen goals in 38 games and was Hartford's most dominant player in the 1992 playoffs against Montreal. Corriveau went 3-2-5 in seven games against the Habs including an overtime game-winner in Game 6. The Whalers obtained Corriveau from the Capitals on March 5, 1990 in a deal for goaltender Mike Liut. Dealt away by Hartford on June 15, 1992 in a transaction which also involved Mark Hunter (it was completed on August 20, 1992 with Nick Kypreos becoming a Whaler), Corriveau wound up being the future considerations in that swap. San Jose took "Ike" in the Waiver Draft on October 9, 1992, played him in 20 games and then shipped him to the Whalers for Michel Picard on January 21, 1993. Overall, in five seasons, Corriveau played 134 games in Hartford. A player who seemed to possess all the tools to be a consistent, offensive force, Corriveau did not reached the expectations of scouts. Wound up playing 280 NHL games with 48 goals and 40 assists. Helped Detroit of the International League to the championship in 1996-97, a team which also included ex-Whalers Brad Shaw, Jeff Reese and Jimmy Carson.

Sylvain Cote

The Whalers selected this quick defenseman as the 11th overall pick in the 1984 Entry Draft. Though he was a regular for five seasons in Hartford and played 382 games, his puck-handling and offensive skills blossomed once getting dealt on September 8, 1991 to Washington for a second-round pick (Hartford took Andrei Nikolishin 47th overall in 1992). His best season with the Whalers was 28 points in 1987-88. With the Caps, "Coco" put up a 21-goal campaign in 1992-93 and also a pair of 50-plus point seasons in a system where the blue line corps were counted on to rush the puck. One of many players drafted by Hartford that went on to enjoy career success elsewhere, Cote reached the 700-game milestone during the 1995-96 season.

Yves Courteau

Despite a couple of 120-point seasons in junior hockey, Courteau had just brief trials in two stops in the NHL, a total of 22 games including four in Hartford in 1986-87. Tabbed by Detroit as the 23rd overall pick in 1982, the Red Wings shipped Courteau to Calgary on December 2, 1982 in a deal for Bobby Francis. Wound up coming to the Whalers on October 7, 1986 when Hartford sent defenseman Mark Paterson to the Flames.

Murray Craven

In a career moving towards 1,000 NHL games, Craven logged 128 over two years in Hartford, proving to be a useful, unselfish player who understood how to play a two-way game. Able to play all three forward slots, Craven has hit 25 or more goals in five seasons including 24 and 25 in his years with the Whalers. Detroit selected the lanky winger as the 17th overall player in 1982 and dealt him to Philadelphia with Joe Paterson for Darryl Sittler on

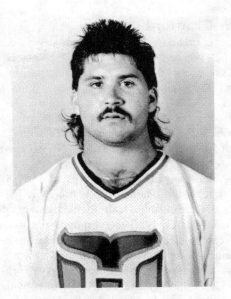

The last Whaler to score a game-winning goal in Stanley Cup annals of Hartford, **Yvon "Ike" Corriveau** *showed flashes of a power game during trials over parts of five seasons. He was one of the most dominant players in the 1992 playoffs against Montreal capped by his OT game-winner just 24 seconds into the extra session for a 2-1 decision in Game 6.*

Sylvain "Coco" Cote *had the ability to rush the puck from the blue line. He was taken 11th overall in 1984 by the Whalers and played 382 games for Hartford.*

October 10, 1984. Played on two Flyer teams that reached the Stanley Cup Finals (1985 and 1987) before arriving in Hartford on November 13, 1991 in a trade for Kevin Dineen. Shipped to Vancouver on March 22, 1993 in a swap that brought Robert Kron, Jim Sandlak and a draft pick (Hartford used it to pick Marek Malik). Craven also reached the Cup Finals with the Canucks in 1994. Has been with the Chicago Blackhawks since 1994-95.

Bobby Crawford

Few teams find a treasure in the annual Waiver Draft. In 1983, the Whalers, thanks to Emile Francis and his awareness of players in the St. Louis chain, plucked Crawford and landed a forward who popped home 36 goals in 1983-84, second behind club-leader Sylvain Turgeon (40) that year. It was the highlight of a 246-game career where the smooth skater finished with 71-71-142. Drafted 65th overall by St. Louis in 1979, Crawford also played for the New York Rangers and Washington Capitals, finishing up with the Caps in 1986-87. Had a penalty shot goal for Hartford on February 1, 1984 against Detroit's Ken Holland. Dealt to the Rangers for Mike McEwen on March 11, 1986. Enjoyed the longest NHL playing career in a hockey family where brothers Marc (176 games with Vancouver and now among the top coaches after leading the Colorado Avalanche to the Stanley Cup in 1996) and Lou (26 games with Boston) also made the grade. Has settled in the Hartford area and is involved with the Whalers Alumni Group and amateur hockey.

Mike Crombeen

Taken fifth overall in the 1977 Entry Draft by the Cleveland Barons, Crombeen proved to be a solid checker and penalty-killer throughout his NHL career, a 475-game tenure which he wrapped up with the Whalers in 1984-85. Teamed with Mike Zuke to improve Hartford's special team play once coming over from St. Louis in the 1983 Waiver Draft. Had only 55-68-123 career totals with a career-best 19 goals in 1981-82 with the Blues.

Doug Crossman

One of several players who was shuttled in and out of Hartford during Eddie Johnston's revolving door years, Crossman was obtained to juice up the power play. Obtained for Ray Ferraro from the New York Islanders on November 13, 1990, the blueliner was soon gone after 41 games, shipped to Detroit for Doug Houda on February 20, 1991. Has played in over 900 NHL games with stops in Chicago, Philadelphia, Los Angeles, Tampa Bay and St. Louis. Had four years of twin numbers in goals, reaching career highs in goals (15), assists (44) and points (59) with the Islanders in 1990-91. Drafted 112th overall by the Blackhawks in 1979.

Ted Crowley

A member of the 1984 U.S. Olympic Team, Crowley joined the Whalers for the final 21 games of the 1993-94 season following his acquisition from Toronto on January 25, 1994. Hartford sent Mark Greig, its top pick in 1990, to the Leafs along with a sixth-round pick. Toronto had drafted Crowley, a defenseman, 69th overall in 1988. Upon his release from the Whalers, the Concord, Mass. native signed with the Bruins on August 9, 1995 but found more ice time when he hooked on with the International League, logging time with Chicago, Houston and Phoenix.

Jim Culhane

Picked 214th overall in the 1984 Entry Draft by Hartford, this blueliner had a short stay with the Whalers in 1989-90, a half-dozen games before getting his release. Notched his only point in a 6-4 win over the New York Rangers on March 3, 1990.

Bob Crawford proved to be a sharp addition for the Whalers in 1983-84. Obtained in the Waiver Draft that season, Crawford popped home a career-high 36 goals including the game-winner in the home opener against Boston, a 4-3 decision on October 8, 1983.

John Cullen

Despite hitting a career milestone of 500 NHL games highlighted by finishing fifth overall (behind Wayne Gretzky, Brett Hull, Adam Oates and Mark Recchi) in the 1990-91 scoring derby with 110 points, Cullen never reached management's expectations in Hartford as part of a multi-player swap with Pittsburgh on March 4, 1991. The Whalers sent Ron Francis, their all-time leader, along with feisty defenseman Ulf Samuelsson and rugged Grant Jennings to the Pens for Cullen, Zarley Zalapski and Jeff Parker. At the time of the trade, it was rated an even swap around the NHL but wound up being one of the worst deals in hockey annals once Francis and Samuelsson played key roles in sparking Pittsburgh to consecutive Stanley Cups in 1991 and 1992. Cullen, more so than Zalapski, soon became the target of boos at the Civic Center Coliseum. Bothered by back and neck problems shortly after getting heavily checked into the boards by Detroit's Vladimir Konstantinov in a 4-0 loss to the Red Wings on February 23, 1992, Cullen was soon on the move, landing in Toronto for a second-round draft choice on November 24, 1992. Despite a strong playoff against Boston in 1991 and leading the Whalers in scoring with 71 points in 1991-92, Cullen will be forever remembered for being in the wrong place at the wrong time in one of hockey's most lopsided trades. One-time Hartford general manager Brian Burke may have assessed it best: "Cullen never had a chance in Hartford. He was traded for two of the most popular players in franchise history." After undergoing surgery to correct a disc problem, Cullen returned to reach the 600-game NHL milestone during 1996-97 with Tampa Bay, his second year with the

*Center **John Cullen**, who had 102 points in 109 games for the Whalers, knocks the puck away from Chicago netminder Dominik Hasek in this battle at the Civic Center. It was one-time Hartford general manager **Brian Burke** who assessed Cullen's tenure with the Whalers by saying "Culley never had a chance. He was traded for two of the most popular players in franchise history." The players involved in that March 4, 1991 swap were **Ron Francis** and **Ulf Samuelsson**.*

Lightning. Cullen's hockey future may come to an end considering he has undergone chemotherapy to fight a form of cancer that poses a threat to his life.

John Cunniff

A member of the AVCO Trophy champs in 1973, this defensive-minded forward played 63 games over New England's first two years in the WHA. An All-American at Boston College, Cunniff was a member of the 1968 U.S. Olympic Team en route to playing pro hockey. Played in the Detroit chain before jumping to the WHA. Later coached the Whalers on an interim basis during the 1982-83 season. Has been a scout or an assistant coach for several organizations, primarily with Boston and New Jersey. In 1996-97, the Devils hired Cunniff to coach Albany of the American League.

Randy Cunneyworth

A forward who enjoyed to forecheck and work the corners, Cunneyworth played 216 games in Hartford of a career which has now reached over 700 games. Drafted by Buffalo as the 167th overall pick in 1980, Cunneyworth eventually joined the Sabres, then moved on to the Penguins where he had goal seasons of 26, 35 and 25. Became a Whaler on December 13, 1989 in a swap for Paul MacDermid. Had the game-winning goal in Game 4 of the 1992 Stanley Cup playoffs against Montreal, a 3-1 win on April 25. Dealt to Chicago with Gary Suter on

March 11, 1994 in a trade which brought Jocelyn Lemieux and Frantisek Kucera to Hartford. Signed as a free agent by Ottawa, Cunneyworth has captained the Senators since 1995-96.

Tony Currie

A native of Sydney Mines, Nova Scotia, Currie played 45 games for the Whalers, a productive 12-16-28 in 32 games after signing as a free agent on January 21, 1984. Drafted by St. Louis as the 63rd overall pick in 1977, Currie also played for Vancouver in 1984-85 where he ended up his career. Appeared in 290 NHL games with his highlight in Hartford being a hat trick in a 6-6 tie against Quebec on March 25, 1984. Finished playing in the American League at Nova Scotia and Fredericton.

Paul Cyr

The ninth overall pick in the 1982 Entry Draft by Buffalo, Cyr never became the high-scoring winger he was in the Western Hockey League but did play a good two-way style to last 470 NHL games. He finished his career with the Whalers in 1990-91, a remarkable story considering he missed all of 1989-90 with serious knee injuries. Underwent two operations and months of rehab to make a comeback and docked in with career totals of 101-140-241. Also survived a robbery while on vacation in the Carribean in 1987 when he was shot in the stomach when a cab he was riding in was the target of a theft in the Dominican Republic.

*Action here against the Boston Bruins involves **Randy Cunneyworth** and Bob Sweeney. Cunneyworth worked the corners for the Whalers for 216 games once arriving in a deal for **Paul MacDermid** on December 13, 1989 from Winnipeg.*

D

Corrie D'Alessio

Obtained from Vancouver on October 1, 1992 for goalie Kay Whitmore, D'Alessio mopped up 11 minutes of a 9-3 loss at Buffalo on December 11, 1993 in his only appearance between the pipes for Hartford. Played collegiately at Cornell and was drafted 107th overall by the Canucks in 1988.

Jake Danby

One of the better special-team players in club annals, Danby was a two-year regular for New England and a member of the 1973 AVCO Trophy champions. A prolific scorer at Boston University who played on two NCAA titlists, Danby followed BU coach Jack Kelley to the WHA and anchored New England's checking unit. Went 14-23-37 with four game-winning goals in 1972-73, the highlight of a 150-game career over three seasons with the Whalers. His only postseason goal was a short-handed strike in Game 1 of the 1974 AVCO Trophy playoffs, a 6-4 win over Chicago on April 6.

Jeff Daniels

Signed by the Whalers as a free agent on July 18, 1996, Daniels reached Hartford for his third tour in the NHL in 1996-97. Collected two helpers in a 6-4 win over Buffalo on December 7, 1996 during a stretch of nine games where he performed well in a checking role. Suffered a knee sprain in a 4-1 loss to Dallas on December 20, 1996, an injury which limited his chances for the rest of the year. Eventually returned to the lineup for his final game as a Whaler, a 5-3 loss at Washington on March 16, 1997. Originally selected 109th overall by Pittsburgh in 1986, Daniels has been cast into the role of a journeyman, a winger with the smarts to play in the NHL (154 career games entering 1997-98) but never given a serious opportunity to play for an extended period of time. Besides the Penguins and Whalers, Daniels has also played for the Florida Panthers.

Scott Daniels

Just when he finally arrived at 25 to get regular work in Hartford, the "Chief" left to sign with Philadelphia as a free agent on June 17, 1996. Hartford's 136th overall pick in 1989 had brief trials over the years before he made an impact on a checking line in 1995-96 when he led the Whalers in penalty minutes (254). Able to quickly drop the gloves as well as use his 6-foot-3, 200-pound frame to clog sightlines for opposing netminders. Did exactly that for the Flyers in 1996-97, leading the team in penalty minutes (237) with 5-3-8 in 56 games.

Joe Day

The Whalers tabbed this pesky forward with the 186th overall pick in 1987 out of St. Lawrence University. Day played 48 games with Hartford over two seasons. His only NHL goal came in a 7-6 overtime loss to the New York

*Journeyman **Jeff Daniels** signed with Hartford for 1996-97 and managed two assists in 10 games. He seemed to find a niche as a checker and likely would have played in more games but a knee sprain sustained in a 4-1 loss to Dallas on December 20, 1996 virtually ended his chance.*

Islanders at the Civic Center Coliseum on February 28, 1993. Also played 24 games with the Islanders who signed him as a free agent on August 24, 1993.

Dave Debol

Drafted by the Whalers in 1976, Debol played four years at Michigan where he reached All-American status after making the Wolverines as a walk-on. Once he established school records with 43-56-99 in 45 games 1976-77, Debol became a sought-after commodity. The Chicago Blackhawks of the NHL had picked Debol 63rd overall in 1976 but the WHA plotted to keep the rising star by shuttling his collegiate draft rights around. Cincinnati eventually signed Debol where he played 68 games over two seasons before suffering an eye injury in 1978-79 which resulted in blurred vision. Hartford took a chance by claiming Debol in the dispersal draft of WHA players and the center eventually made it back after two operations. Played 92 games over two years with the Whalers, winding up in 1980-81. That season, Debol played in 44 NHL games and 18 in the minors and was never called for a penalty. The only player that year to play more games in professional hockey without a penalty was Butch Goring (78) of the New York Islanders. Signed on with Toronto after getting released by the Whalers but never made it back to the NHL.

Gerald Diduck

A journeyman closing in on 800 NHL games played, Diduck, taken 16th overall in the 1983 Entry Draft by the New York Islanders, has been a regular at the blue line for several teams, making stops in Montreal, Vancouver and Chicago prior to signing as a free agent with the Whalers on August 24, 1995. A steady defenseman who played a key role during Vancouver's ride to the Stanley Cup Finals in 1994, Diduck logged 135 games in Hartford over parts of two seasons but never seemed to play with fire. Tallied two goals as a Whaler and both came in victories, a 5-4 win at Pittsburgh on April 8, 1996 and in a 5-2 decision over Florida on December 11, 1996. A regular for most of 1996-97 with the Whalers, Diduck was swapped on March 18, 1997 to Phoenix in a deal for winger Chris Murray. With the Whalers and Coyotes, Diduck finished with a combined 2-12-14 in 67 games.

Bill Dineen

Bill Dineen joined the New England Whalers as their bench boss in 1978-79 when the Houston Aeros, one of the best clubs in WHA history, ceased operations after the 1977-78 season. Dineen coached the Whalers until the closing weeks of the campaign when New England reshuffled its management ranks. Remained with the organization in its early NHL days and was responsible for Hartford's dynamic talent infusion in 1982. Dineen had plenty of input when the Whalers drafted several future NHLers (particularly Kevin Dineen, Ulf Samuelsson and Ray Ferraro) that formed the nucleus of the franchise's better clubs during the mid-1980's. Dineen left the organization shortly after Emile Francis was hired as general manager. Returned to coaching and had plenty of success in the minor leagues, most notably with the Adirondack Red Wings of the American League. Piloted Detroit's farmhands to Calder Cups in 1986 and 1989, a springboard he used to resurface in the NHL. Philadelphia hired Dineen as a scout in 1990 and he wound up coaching the the club to a 60-60-20 mark over a two-year period (1991-92). As a player, "Foxy" played in the NHL when the Red Wings ruled the six-team league and was a member of Stanley Cup winners in 1954 and 1955. After serving as the player-coach for several clubs in the Western Hockey League for many years, Dineen, in 1972-73, was hired as the first general manager of the WHA Aeros. After procuring the draft rights to Boston's Bobby Orr and Phil Esposito as a major publicity stunt at the new league's first draft, Dineen stunned the NHL a year later when he successfully recruited a former Detroit teammate to make a comeback. The Aeros made hockey history when they signed Gordie Howe along with his sons, Mark and Marty. The Howes, besides bringing Houston into national prominence, turned the Aeros into a powerhouse. Dineen's club won four division titles in six seasons, including back-to-back AVCO Trophy titles in 1974 and 1975.

*Defenseman **Gerald Diduck** played 135 games with the Whalers over two seasons before getting swapped to Phoenix on March 18, 1997 in a deal which brought rugged forward Chris Murray to Hartford. Diduck scored two goals for the Whalers and both came in victories.*

***Bill Dineen** had his greatest success behind the bench with the Houston Aeros, twice winning the AVCO Trophy. In 1978-79, Dineen took over the coaching duties of the Whalers. He later scouted for Hartford in the NHL.*

*The goalmouth is where **Kevin Dineen** often made his presence felt and here he tumbles into Montreal goaltender Patrick Roy. Dineen, who scored a career-high 45 goals in 1988-89, was the last captain of the Whalers. He also scored the final goal in team annals.*

***Kevin Dineen** grew up in Glastonbury when his father, Bill, coached the Whalers in the WHA in 1978-79. Eventually drafted by Hartford in 1982, Dineen went on to play over 500 games for the Whalers.*

Kevin Dineen

The last captain of the Whalers and one of two players to play over 500 NHL games for the franchise (Ron Francis is the other), Dineen ranks high on the list of many career statistics in team annals and arguably might be the franchise's most popular and most identifiable player. Drafted 56th overall in 1982 out of the University of Denver, Dineen was a sparkplug on Hartford's best clubs in the 1980s. Played with grit and determination every shift. Nicknamed "John Wayne" because the bigger the game, the bigger he played. Holds several team playoff records including four game-winning goals (with two coming in overtime). Dealt away on November 13, 1991 to Philadelphia for Murray Craven, Dineen eventually captained the Flyers. Returned to the Whalers on December 28, 1995 and played 20 games before a wrist injury put him on the sidelines. Wore the "C" in Hartford for 1996-97, getting a stirring ovation from the faithful on Opening Night. It was only fitting he scored the last goal by a Hartford player, the game-winner in a 2-1 decision over Tampa Bay on April 13, 1997. One of three sons of

*The Whalers traded a number of young players away over the years. Some like **Bobby Holik** and **Jody Hull** blossomed elsewhere. Charting the future of **Hnat Domenichelli** will be worth watching. A two-time 50-goal scorer in the Western League at Kamloops, Domenichelli looked like the real deal in training camp en route to the 1996-97 season. Given little ice time, the clever winger went just 2-1-3 in 13 games before getting dealt to Calgary on March 5, 1997 in a multi-player move.*

former NHLer and Whaler coach Bill Dineen to reach the NHL (Gord and Peter are the others). Played for the Canadian Olympic Team in the 1984 Winter Games, a squad that included eventual Whalers Dave Tippett and Carey Wilson. Represented the Whalers in RendezVous '87 and the the 1987 Canada Cup Tournament. Best offensive season was 45-44-89 in 1988-89, one of seven campaigns in the NHL where he scored at least 25 goals.

Hnat Domenichelli

Taken 83rd overall by the Whalers in the 1994 Entry Draft, it will be worth charting to see if Domenichelli will join the list of young forwards Hartford traded over the winters that blossomed elsewhere. It might happen in coming years in Alberta where Domenichelli, now a member of the Calgary Flames, has been reunited with junior hockey linemate Jarome Iginla who led all NHL rookies in scoring in 1996-97. Throughout Domenichelli's junior career with Kamloops in the Western League, he tallied impressive numbers, capping his stay in 1995-96 with a 59-89-148 spree in 62 games to garner All-Star status. Given a chance to crack the Hartford lineup in training camp, Domenichelli rated among the best performers in preseason games and opened the 1996-97 season with the Whalers. Though he was shipped to Springfield of the American League for most of the first half of the campaign, Domenichelli returned to the Whalers and notched his first NHL goal in a 5-1 rout of Buffalo on January 25, 1997. Also connected in a 5-2 loss to the New York Rangers on February 5, 1997. That tally

...ned out to be his last goal as a Whaler. After many weeks of a rumored deal between Hartford and Calgary, the clubs worked out a swap on March 5, 1997 where Domenichelli and defenseman Glen Featherstone were sent to the Flames along with a second-round pick in 1997 and a third-round choice in 1998 for Steve Chiasson and Calgary's third-round selection in 1997. In 23 total games, Domenichelli went 3-3-6.

Jim Dorey

A teammate of Rick Ley and Brad Selwood in Toronto, "Flipper" became the third former Maple Leaf to jump to the Whalers to help launch the franchise in the WHA. Add Ted Green to the mix, New England easily had one of the better blue line corps in hockey in the early 1970s, a major reason why the Whalers won the first AVCO Trophy in 1973 and three straight division titles (1973-1975). Dorey was a member of two of those teams and appeared in 183 games before getting shuttled to the Toronto Toros for Wayne Carleton on December 31, 1974. Eventually was dealt to Quebec where he played on a second AVCO Trophy winner in 1977 with the Nordiques. Played 431 WHA games with 52-232-284 and 617 penalty minutes. Came up as a brawler and collected 553 penalty minutes in 232 NHL games over five seasons with the Maple Leafs and New York Rangers. On October 16, 1968 at Maple Leaf Gardens against Pittsburgh, Dorey became the first player in NHL annals to be assessed nine penalties (four minors, two majors, two 10-minute misconducts, one game misconduct) in a game where he set a record with 48 penalty minutes. The record lasted for many years but is now shared by Boston's Chris Nilan (10 penalties vs. Hartford, March 3, 1991) and Randy Holt of the Los Angeles Kings (67 PM vs. Philadelphia, March 11, 1979).

Jordy Douglas

When this big winger scored 33 goals in 1979-80, the Whalers believed they had found a scoring forward for many years to come. Injuries often take that special label away from an athlete and a bad shoulder late in the season began a litany of other ailments for Douglas who was never the same player. Douglas, who once tallied 60 goals at Flin-Flon of the Western Hockey League in 1977-78, signed with the Whalers in the WHA on May 25, 1978 rather than Toronto which selected him as the 81st overall pick in the 1978 NHL Entry Draft. Played 51 games as a rookie in the WHA, going 6-10-16 before adding four goals in the playoffs. Hartford made Douglas one of its priority picks when the Whalers joined the NHL. Played 152 NHL games with Hartford before moving on to Minnesota on October 1, 1982 in a swap which brought Mark Johnson to the Whalers. Douglas lasted with the North Stars until January 12, 1984 when he was obtained by his hometown Winnipeg Jets. He finished in 1984-85 with career totals of 76-62-138 in 268 NHL games.

Ted Drury

The first Connecticut-produced player to wear the Whalers sweater, this two-time U.S. Olympian logged 50 games in Hartford over two seasons before getting claimed by Ottawa on October 6, 1995 in the Waiver Draft. A Trumbull native, Drury played scholastically at Fairfield Prep and reached All-American status at Harvard where his skills were spotted by Calgary scouts who made the versatile forward the 42nd overall pick in 1989. Drury played 34 games with the Flames in 1993-94 before coming to Hartford on March 10, 1994 in a six-player deal which involved Gary Suter and Paul Ranheim heading east and Zarley Zalapski, James Patrick and Michael Nylander heading to Alberta. Played for Anaheim in 1996-97, going 9-9-18 in 73 games. Hails from a sports family which includes a brother named Chris Drury who was selected by Quebec (72nd overall in 1994) and reached international prominence as a 12-year old when he pitched Trumbull to the 1989 Little League World Series Championship.

Richie Dunn

A Boston native who played 483 NHL games and 36 Stanley Cup games with Buffalo, Calgary and Hartford, Dunn came to the Whalers on July 5, 1983 along with Joel Quenneville from the Flames for Mickey Volcan. Though overlooked in the 1977 Entry Draft, Dunn signed with the Sabres as a free agent and emerged as a regular at the blue line for three seasons before going to Calgary with goaltender Don Edwards in a swap that involved draft choices. Had 36-140-176 in his career including 6-24-30 in 76 games with the Whalers.

Norm Dupont

Taken 18th overall by Montreal in 1977, Dupont played 256 games over five winters with the Canadiens and Winnipeg Jets before winding up with the Whalers in 1983-84, his last stop in the NHL. A prolific scorer in

*A priority pick when the Whalers were able to protect two skaters and two goaltenders once joining the NHL, winger **Jordy Douglas** enjoyed his best campaign in 1979-80, notching 33 goals including four in a 5-3 triumph on February 3, 1980, Hartford's first-ever win over the New York Islanders.*

*Forward **Ted Drury** became the first Connecticut scholastic prep product to play for the Whalers. The Trumbull native played 50 games for Hartford before being selected by Ottawa in the 1995 Waiver Draft.*

Canadian juniors where he had years of 70 and 69 goals, Dupont was named the top rookie in the American League in 1978. His best NHL campaign was 1980-81 when he tallied 27 goals as a regular for the Jets. Hartford shipped a fourth-round pick to Winnipeg for Dupont on July 4, 1983 where he added 7-15-22 in 40 games to push his career totals to 55-85-140.

Steve Dykstra

A free agent defenseman who signed with Buffalo in 1982, Dykstra played 217 NHL games with Buffalo, Edmonton, Pittsburgh and Hartford. In 1989-90, Dykstra went scoreless in nine games with the Whalers who shipped him to Boston on March 3, 1990 for blueliner Jeff Sirrka.

E

Tommy Earl

One of the best penalty-killers to ever play for the Whalers, Earl spent five years with New England of the WHA, never scoring more than 23 points in a season yet a valuable component in the overall team concept. A good worker along the wall, Earl combined anticipation with smarts, often paired with Jake Danby or Garry Swain to give the Whalers reliable special-team play. Earl played for future New England coach Ron Ryan at Colgate and jumped to the Whalers after a couple of years in the St. Louis chain. A member of the AVCO Trophy club in 1973, Earl is among many former Whalers to settle in Connecticut and currently works in administration at Westminster Prep in Simsbury.

Jack "Tex" Evans

When he died on November 10, 1996, "Tex" was kindly remembered as a coach who stood behind his players, through good times and lean times. In the four-plus seasons that Evans was behind the Hartford bench, he established coaching marks for most games (374), most wins (163), most losses (171) and most ties (37) in club annals. He also coached the most Stanley Cup playoff games in team history (16). Often booed by the Civic Center faithful because of his stoic style, Evans turned the catcalls into loud cheers while calling the line changes during two successful winters in the 1980's. The fans roared with delight in 1986-87 when the Whalers reached their pinnacle in Hartford, capturing the Adams Division crown on the strength of a 43-win, 93-point year, a significant milestone considering the Whalers finished ahead of both Montreal and Boston. Evans admitted he had no secret formula for winning hockey games. "Just hard work and repetition," he would often say. That was Jack. Quiet and reserved. He never said much into a microphone or to help a reporter fill a notebook. His most memorable line came after an 8-1 loss to Los Angeles on home ice on November 2, 1985 when he quipped, "I'm going home and pour me a scotch." Another gem from

*Tommy Earl logged five years in the WHA with the Whalers. Known for his ability to play without the puck, Earl was often paired with either **Jake Danby** or **Garry Swain** in killing penalties. He twice scored 10 goals in a season.*

*The date is July 7, 1983 and a new chapter in Hartford hockey begins with general manager **Emile Francis** (center) flanked by coach **Jack Evans** and assistant coach **Claude Larose**. In 1986-87, Evans piloted the Whalers to their best finish in team annals, 43 wins and 93 points and the top perch in the Adams Division.*

Evans came on April 16, 1987 when Quebec's Peter Stastny ended another Hartford season early by scoring in overtime at 6:05, the clincher in a 5-4 loss in the Stanley Cup playoffs. "It's terrifying to see Stastny on a breakaway." For the most part, Evans low-keyed his role. He preferred to let others speak. One of the last times he addressed the media led to his dismissal as bench boss of Hartford. Evans, nearing 60, had spent his life in hockey. He was a fierce competitor at the blue line and played for Chicago's last Cup winner in 1961. He had also coached the California Golden Seals and Cleveland Barons in the NHL before taking the post with the Whalers on July 7, 1983. When the subject of his future plans came up during the 1987-88 season and general manager Emile Francis sensed Evans was hinting about early retirement, the "Cat" made it official. After a 5-4 loss at Pittsburgh on February 6, 1988, the Whalers had a new bench boss 24 hours later when Francis dismissed Evans and elevated Larry Pleau from Binghamton.

Nelson Emerson

A small but determined winger, Emerson came to the Whalers on October 6, 1995 in a swap with Winnipeg for Darren Turcotte and wound up third on the club in goals (29) and fourth in scoring (58 points) in his first year with Hartford. A good pointman on the power play, Emerson possesses a good combination of speed and cleverness to find open ice or make a pass to spring a teammate free. At Bowling Green where the Wave retired his jersey on February 24, 1996, Emerson established several school records en route to ranking among the Top 10 scorers in

*Nelson Emerson was at his best on the power play where he scored 14 of his 38 career goals for Hartford. The speedy winger played 147 games over two years with the Whalers following his arrival from Winnipeg on October 5, 1995 for **Darren Turcotte**.*

college hockey history. Drafted 44th overall in 1985 by St. Louis, Emerson twice cracked the 20-goal plateau with the Blues before moving on to Winnipeg, along with defenseman Stephane Quintal, for blueliner Phil Housley.

With the Jets, Emerson notched a career-high 33 lamplighters in 1993-94. Having now played over 400 games in the NHL, Emerson is a testament that there is room in the league for small players. One of several Whalers who had a subpar winter in 1996-97, a year where an ankle injury sustained in training camp limited his effectiveness and dipped his production to 9-29-38 in 66 games.

Dean Evason

A compact center who could put a charge in a slapshot, Evason belongs to a small club of Whalers to play over 400 NHL games in Hartford. Drafted by Washington as the 89th overall pick in 1982, the Whalers secured Evason, along with goalie Peter Sidorkiewicz, on March 12, 1985 in a trade with the Capitals for David A. Jensen (one of Hartford's first-round picks in 1983). The swap tilted in favor of the Whalers. Sidorkiewicz eventually handled the netminding chores in the late 1980's and Evason proved to be a valuable player in the pivot, twice scoring 20 or more goals including a career-high 22 in 1986-87 when Hartford won the Adams Division flag. He centered the "LEG" line with Paul Lawless on the left flank and Stewart Gavin on the right wall, a dependable two-way trio. Moved on to San Jose on October 2, 1991 in a deal for defenseman Dan Keczmer. Has also played for Dallas and Calgary, the last stop in 1995-96 where he played in his 800th NHL game.

*Center **Dean Evason** anchored a second line during his years in Hartford from 1984 to 1991. His best year was 22-37-59 in 1986-87 to help the Whalers to the Adams Division flag.*

Glen Featherstone

After several seasons battling injuries, Featherstone put together his best year in 1995-96 with Hartford. Drafted 73rd overall by St. Louis in 1986, Featherstone spent three years with the Blues before signing with Boston as a free agent on July 25, 1991. Back injuries curtailed his appearances with the Bruins and eventually led to a deal with the New York Rangers for Daniel Lacroix on August 19, 1994. A spare with the Blueshirts, the Whalers secured the big blueliner in a 4-for-1 swap with New York involving Pat Verbeek on March 23, 1995. It took 20 seconds for Featherstone to score his first goal for the Whalers which sparked a 5-1 verdict over the Islanders on March 25, 1995. Dealt to Calgary on March 5, 1997 with prospect Hnat Domenichelli for veteran defenseman Steve Chiasson.

***Glen Featherstone** battled injuries and opposing wingers in the corners during his tenure with the Whalers, a total of 122 games over three seasons with 5-16-21 and 257 penalty minutes. The defenseman came to Hartford in a 4-for-1 swap with the Rangers for **Pat Verbeek** on March 23, 1995. He departed on March 5, 1997 in a deal with Calgary that brought **Steve Chiasson** to the Whalers.*

Paul Fenton

A free agent who signed with Hartford in 1984-85, Fenton played 411 NHL games in his career, lasting until 1991-92 with stints with the New York Rangers, Los Angeles Kings, Winnipeg Jets, Toronto Maple Leafs, Calgary Flames and San Jose Sharks. The Springfield, Mass. native played at Boston University and after a 60-goal year in the International League in 1982-83 at Peoria,

Fenton followed it up with a good training camp in Hartford, earning a contract on October 6, 1983. Played only 34 games with the Whalers over two seasons at a time when Hartford shuttled in plenty of wingers trying to find the right combination en route to assembling the Adams Division champion in 1987. Moved on to the Kings on October 5, 1987 in the Waiver Draft which led to a career-high 20 goals in 1987-88.

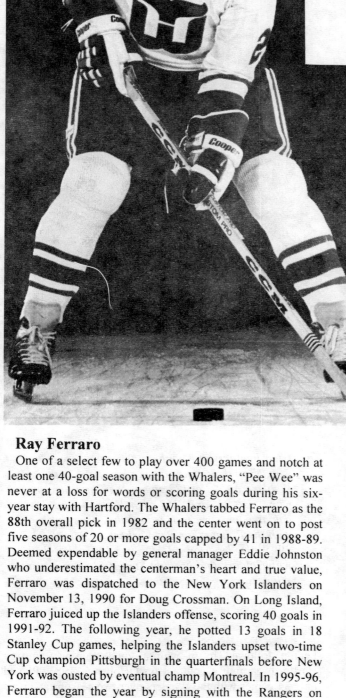

*Rugged winger **Nick Fotiu** played for the Whalers in both leagues and was considered the team's on-ice policeman. In 1979-80 when he was tutored by **Gordie Howe,** "Big Nick" developed a wrist shot which led to 10 goals. A big fan favorite, Fotiu was among the first in hockey to toss pucks into the stands following the pregame skate.*

*Ray Ferraro was at his best in power-play situations during his tenure with the Whalers, able to deflect home shots or find the open man. In 1988-89, the centerman tallied 41 goals for Hartford. He was dealt to the New York Islanders on November 13, 1990 for defenseman **Doug Crossman.***

Ray Ferraro

One of a select few to play over 400 games and notch at least one 40-goal season with the Whalers, "Pee Wee" was never at a loss for words or scoring goals during his six-year stay with Hartford. The Whalers tabbed Ferraro as the 88th overall pick in 1982 and the center went on to post five seasons of 20 or more goals capped by 41 in 1988-89. Deemed expendable by general manager Eddie Johnston who underestimated the centerman's heart and true value, Ferraro was dispatched to the New York Islanders on November 13, 1990 for Doug Crossman. On Long Island, Ferraro juiced up the Islanders offense, scoring 40 goals in 1991-92. The following year, he potted 13 goals in 18 Stanley Cup games, helping the Islanders upset two-time Cup champion Pittsburgh in the quarterfinals before New York was ousted by eventual champ Montreal. In 1995-96, Ferraro began the year by signing with the Rangers on August 9, 1995 and wound up getting dealt to Los Angeles in a swap for Jari Kurri on March 21, 1996. In between, Ferraro played in his 800th NHL game and crossed the 600-point career milestone. Led the Kings in goals (25) in 1996-97 including 11 on the power play.

Mike Fidler

A hard-nosed winger who played 271 NHL games, Fidler played for the Cleveland Barons, Minnesota North Stars, Chicago Blackhawks and Hartford Whalers. Originally drafted by the California Golden Seals as the 41st overall pick in 1976 (New England of the WHA also picked him that year), Fidler opted for the NHL but remained a player who had caught Jack Kelley's fancy. Though shoulder and leg injuries had limited Fidler's contributions, Kelley seemed convinced a change of scenery would recharge the forward and on December 16, 1980, Hartford shipped Gordie Roberts to Minnesota for Fidler. The Everett, Mass. native, who also played at Boston University, logged 40 games over two years with the Whalers, before getting released. Roberts, meanwhile, went on to play through 1993-94, a span of 1,097 games, a milestone for American-born players.

*The Whalers had a number of rugged brawlers during their WHA days and captain **Rick Ley** is flanked here by **Nick Fotiu** (left) and **Jim Troy** (right) during a break in practice. Fotiu had a couple of memorable altercations over the years with Minnesota's Jack Carlson. Troy, who eventually moved into professional wrestling as a promoter, had a handful of bouts for New England including a wild one against Curt Brackenbury of Quebec during the 1976-77 campaign.*

Nick Fotiu

"Big Nick" had tours of duty with the Whalers in both leagues, signing as a free agent in 1974-75 and going on to play 110 games in the WHA and later coming back for 106 NHL games when Hartford obtained him from the New York Rangers in the 1979 Expansion Draft. Besides throwing pucks into the stands for fans following the pregame skate, Fotiu could also throw punches. WHA fans recall some memorable bouts with Calgary's Rick Jodzio and Minnesota's Jack Carlson. Fotiu had nine points and 238 penalty minutes in the WHA, honing his skills to become the first native New Yorker to play for the Blueshirts when he signed a contract on June 23, 1976. Fotiu was a regular and Madison Square Garden favorite for three years, then returned to the Whalers for a second hitch. In 1979-80, Fotiu, schooled by Hall of Famer Gordie Howe, worked on his wrist shot and produced a career-high 10 goals. Midway in his second NHL season in Hartford, Fotiu was upset with his lack of playing time and wound up back with the Rangers on January 15, 1981 for a fifth-round draft pick (Hartford took Bill Maguire in 1981). Fotiu also played for Calgary, Philadelphia and Edmonton, winding up a 646-game career in 1988-89 with 60 goals and 1,362 penalty minutes, a remarkable achievement for someone who didn't begin skating until he was 15. He is currently coaching in the minor leagues.

Emile "Cat" Francis

One comment made by a rival executive shortly after Hartford hired Emile Francis on May 2, 1983 as president and general manager of the hockey club suggested a new ice era had begun in southern New England. "Now I have to worry about the Whalers," Boston general manager Harry Sinden said. Few ever commanded the respect or possessed the network contacts in hockey that the "Cat" developed during 47 years in the game as a player, coach and organization decision-maker. Following the same strategy that reshaped the New York Rangers and St. Louis Blues in previous appointments, the former netminder retooled a dormant franchise in Connecticut's capital city just like he did in previous NHL stops. From the depths of inheriting a club that had lost 54 games, Francis revived the fortunes of the Whalers. Though

Larry Pleau gets the baton from general manager Emile "Cat" Francis on February 7, 1988 to take over the bench duties of the Whalers. Francis opted to make a change when Jack Evans announced he would "like to retire" after one more season as coach. Francis quickly accommodated the request.

Hartford fell short of what Francis called the "Holy Grail," the Whalers, capped by an Adams Division title in 1987 and an emerging box office draw, were a competitive outfit throughout his stewardship. In 1986, Hartford won its only post-season series in the NHL annals and took eventual Stanley Cup champion Montreal to an epic duel that ended at 5:55 of overtime in Game 7 at the Montreal Forum in the quarterfinals. The following winter, the Whalers racked up 43 wins and 93 points to win the division crown, a nifty achievement considering that Hartford had bested both Montreal and Boston over the 80-game grind. In subsequent years, the Whalers never matched that pinnacle on ice or at the gate. In 1987-88, for example, the Whalers played before 24 sellout crowds and averaged a club-high 14,616 per game en route to a total attendance of 628,476. Armed with a long-term contract, Francis remained with the organization during the turbulent ownership of Richard Gordon until June 30, 1993 when he exited the game. In honestly looking at the history of the franchise, it was Francis who literally "saved the Whalers." Critics may suggest the chronic roster moves or miscalculations at the draft table (pointing to Sylvain Turgeon over Pat LaFontaine in 1983 or Scott Young over Craig Janney in 1986), but the simple fact is the Whalers were no longer pushovers throughout Francis' stewardship. Cat's signature trade was made on February 22, 1985 when the Whalers obtained goaltender Mike Liut from St. Louis for netminder Greg Millen and center Mark Johnson. That transaction solidified the Whalers in hockey's key position. Coupled with moves such as Dave Babych from Winnipeg (for Ray Neufeld), Dean Evason and Peter Sidorkiewicz from Washington (for David A. Jensen), Doug Jarvis from Washington (for Jorgen Pettersson), Stewart Gavin from Toronto (for Chris Kotsopoulos) and John Anderson from Quebec (for Risto Siltanen), Hartford finally had assembled the parts to go with its identifiable cadre of players such as Ron Francis,

Kevin Dineen, Ray Ferraro, Joel Quenneville and Ulf Samuelsson. Media types often chuckled when the Cat said "Dineen has more guts than a slaughterhouse" or "you never have enough goaltending" as to why he pulled the trigger to obtain yet another puckstopper." In retrospect, it was Emile Francis who brought respectability and stability to the Hartford Whalers.

Ron Francis

Arguably the greatest player to play for the Hartford Whalers, Ron Francis holds several longevity records in club history including most games (714), goals (264), assists (557) and points (821). His selection as the club's first-round pick in 1981 is a testament that even the experts can look foolish. In days leading up to the draft, the Whalers, who had the fourth overall pick, made their intentions known that they would select Massachusetts scholastic sensation Bobby Carpenter. Stonewalled when Colorado and Washington maneuvered on draft day and flipflopped draft positions in a trade, the Whalers were stunned when the Capitals picked Carpenter after Winnipeg took Dale Hawerchuk and Los Angeles tabbed Doug Smith. Carpenter's dad, who was sitting with the Hartford delegation, left the Montreal Forum in a huff. Somewhat redfaced, the Whalers opted for Francis, a two-way center who was a budding standout on Terry Crisp's Sault Ste. Marie Greyhounds. Francis put up 21 or more goals in each of his 10 seasons in Hartford capped by a 32-goal, 101-point campaign in 1989-90. The following year, the Whalers, in a move by general manager Eddie Johnston with the total blessing of then-owner Richard Gordon, decided to swing a trade with Pittsburgh. On March 4, 1991, Hartford sent Francis, Ulf Samuelsson and Grant Jennings to the Penguins for John Cullen, Zarley Zalapski and Jeff Parker. The transaction seemed even at the time of the trade but the pendulum swung heavily towards Pittsburgh when the ex-Whalers played pivotal

*A look at Hartford's all-time favorite **Ron Francis** in what seemed to be an annual battle against Montreal's defensive style against the Whalers over the years. No matter where he skated, with or without the puck, Francis was often shadowed in each zone by Guy Carbonneau.*

roles in helping the Penguins to Stanley Cups in 1991 and 1992. Francis finally received the kudos for what he meant to a hockey team and epitomized hockey's second-line center. The problem was no longer finding players to build a team around Francis as he became the missing link behind Mario Lemieux, giving Pittsburgh another top-draw center who could anchor another potent line. One can only wonder if the Whalers had any good fortune in the Entry Draft over the years to secure another offensive force to go with Francis what type of tradition would have been achieved in Hartford. In 1982, for example, the Whalers passed on Dave Andreychuk and took Paul Lawless. In 1983, Hartford, picking second overall, could have taken Pat LaFontaine or Cam Neely yet tabbed

GM-Coach *Larry Pleau* (center) welcomes *Rick MacLeish* and rookie *Ron Francis* to Hartford during the summer of 1981. *MacLeish came over in a deal from the Flyers. Francis was taken fourth overall by the Whalers in the Entry Draft that year. MacLeish played 34 games with Hartford before moving on to Pittsburgh. Francis began the journey to a club-record 714 games in 1981-82 with 25-43-68 in 59 games.*

Sylvain Turgeon, a gifted scorer who had seasons of 40 and 45 goals before a stomach injury zapped his quick first step. In 1984, the Whalers took Sylvain Cote (when Gary Roberts was still available). In 1986, Hartford overlooked Enfield native Craig Janney in favor of Scott Young. In 1990, Hartford passed on Keith Tkachuk and took Mark Greig. Francis, who has played in over 1,000 NHL games, continues to skate towards the Hall of Fame. In 1994-95, he copped the Frank Selke Trophy as the league's top defensive forward, the Lady Byng Trophy for sportsmanship and the Alka-Seltzer Plus Award for having the best plus-minus ratio in the NHL. Led the NHL in assists in 1994-95 and 1995-96. Went 237-63-90 in 1996-97 to finished ninth in the NHL scoring race.

Ron Francis, *the eternal captain of the Whalers*

Hartford's all-time favorite **Ron Francis** *battles here with Montreal's Robert Picard. Francis, who played in more games than any Whaler in club history, is wearing long pants rather than shorts and stockings, a style the hockey club used during 1982-83 before returning to hockey's traditional look.*

*Winger **John French** spent three seasons in the WHA with the Whalers, going 60-124-184 in 226 games. He led the squad in scoring in 1973-74 with 24-48-72 in 77 games.*

John French

Originally drafted by Montreal, French joined the Whalers in the WHA and scored 24 goals and 63 points to help New England to the AVCO Trophy in 1973. "Frenchy" added 24 goals and led the club in scoring with 72 points in 1973-74. Slipped to a dozen goals in 1974-75 and was dealt to San Diego for a second round pick before the start of 1975-76. Played two years with the Mariners before winding up with Indianapolis in 1977-78. Had 108-192-300 in 420 games.

Dan Fridgen

A strong scorer at Colgate who led the ECAC in goals scored in consecutive years (38 in 1980-81 and 37 in 1981-82), the Whalers signed this quick forward on April 1, 1982. Had brief trials in two seasons in Hartford, scoring two goals in 13 NHL games. Currently the head coach at his alma-mater.

Mark Fusco

Hartford signed the 1983 Hobey Baker Award winner following the Winter Games on February 25, 1984. The diminutive defenseman played 80 games in Hartford over two seasons before a head injury forced Fusco to prematurely retire. During his collegiate career at Harvard, he was named to the Ivy League''s All-Star team each year (1980-1983). Played his first two NHL games against the Bruins, getting an assist in a 3-3 tie on March 3, 1984 and two helpers in a 6-4 win on March 4, 1984. The Whalers also obtained the draft rights to Scott Fusco, Mark's younger brother, who opted not to play pro hockey.

Mark Fusco, the 1983 Hobey Baker winner as college hockey's top player, joined the Whalers following the 1984 Winter Games. Fusco took a regular turn on defense in 1984-85, going 3-8-11 in 63 games.

Michel Galarneau

Hartford's second overall pick in 1980 (29th overall) played 78 games with the Whalers over parts of three seasons but the speedy center never developed into the player club officials projected after a 39-goal, 103-point year with Hull of the Quebec League in 1979-80. The Montreal native finished his NHL career in 1982-83, going 5-4-9 in 38 games in Hartford.

Bill Gardner

The Whalers acquired this centerman as postseason insurance on February 3, 1986 from Chicago for a third-round draft pick. The veteran played in 26 games over two winters in Hartford before getting his release. Tabbed by the Blackhawks as the 49th overall pick in 1979, Gardner played the bulk of his 380 NHL games in Chicago, winding up with 73-115-188 career totals. Joined the Hartford TV broadcast booth in 1996-97.

John Garrett

"Cheech" played for the Whalers in both leagues. He also made NHL stops in Quebec and Vancouver, ending a 206-game stint in 1984-85 with career totals of 68-91-7 with a 4.27 GAA. He also played in nine Stanley Cup games, going 4-3 in 1982 as he helped the Nordiques to the Wales Conference Finals. In the WHA, Garrett played for Minnesota, Toronto and Birmingham before joining New England on September 18, 1978. A workhorse, Garrett usually played 50 or more games with the most being 65 with the Bulls in 1976-77. Originally drafted by the Blues as the 38th overall pick in 1971, St. Louis shipped the netminder to Chicago on September 18, 1972. One year in the Blackhawk chain quickly made Garrett realize that it "would take a miracle to unseat Tony Esposito," so he elected to jump to the Fighting Saints in 1973-74, the first of six years in the WHA. Among his 323 appearances, Garrett was between the pipes on December 7, 1977 when Gordie Howe scored his 1,000th career goal against Birmingham. In 1979, the Whalers made Garrett one of their priority selections upon joining the NHL. Shared net chores with Al Smith in Hartford's first campaign. Found security with the Whalers but not the playing time he was used to on other clubs. On January 12, 1982, he was sent to Quebec in a swap for goalie Michel Plasse.

Marty Gateman

A defenseman who spent most of his career in the minors with the Providence Reds, Gateman had one helper while playing a dozen games at the blue line for the Whalers in 1975-76, his only shot in the majors.

John "Cheech" Garrett played for the Whalers in both the WHA and NHL. On December 7, 1977 he was in goal for the opposition when Gordie Howe set history in Alabama by scoring his 1,000th career goal against Birmingham in a 6-3 win by New England.

Stewart Gavin played a strong two-way game at forward for the Whalers. His most productive year was 1985-86, going 26-29-55 in 76 games. Left unprotected in the 1988 Waiver Draft, Gavin was snared by the Minnesota North Stars.

Dallas Gaume

A playmaker who had 32-67-99 in 47 games with Denver Unversity in 1985-86, the Whalers signed the 5-foot-10, 180-pound center as a free agent on July 10, 1986. Spent three years in the Hartford chain. His only trial with the Whalers came in 1988-89. During a four-game stint, Gaume notched his only NHL goal in a 5-2 loss to Buffalo on February 9, 1989.

Stewart Gavin

A durable two-way winger who played 780 games over 13 NHL seasons, Gavin proved to be a very useful player without the puck during his stay with the Whalers. Drafted 75th overall in 1980 by Toronto, the Whalers, on October 7, 1985, dealt disgruntled defenseman Chris Kotsopoulos to the Maple Leafs to secure Gavin. His forechecking skills and speed fit perfectly with center Dean Evason and winger Paul Lawless, a combination which was dubbed the "LEG" line, a strong supporting trio to the club's number one line of Ron Francis flanked by Kevin Dineen and John Anderson. Deemed expendable with no concrete evidence that his skills had diminished, Gavin was plucked by Minnesota in the 1989 Waiver Draft. He contributed to the North Stars who staged a dramatic run to the Stanley Cup Finals in 1991.

Jean-Sebastien Giguere

When injuries thinned Hartford's netminding ranks during 1996-97, the Civic Center faithful had an opportunity to gauge the progress of the franchise's future goaltender, a Montreal native who was taken 13th overall in the 1995 Entry Draft. In a five-game trial, Giguere became the youngest ever in club history to start a game between the pipes, making his debut in a 3-2 loss at Philadelphia on December 12, 1996. Notched his first NHL win on December 28, 1996, a 4-3 decision over Ottawa. Giguere developed his skills in the Quebec League, the proving ground for many goaltenders in recent years such as Patrick Roy, Felix Potvin, Stephane Fiset and Eric Fichaud.

Hartford tapped the Quebec Junior League in 1995 to pluck promising goalie J.S. Giguere with the 13th overall pick on the first round. When Giguere, just 19, made his NHL debut in a 3-2 loss at Philadelphia on December 12, 1996, he was the youngest goalie to ever start a game in club history. He played in eight games and notched his first win on December 28, 1996, a 3-2 decision over Ottawa.

Randy Gilhen

Taken 109th overall by Hartford in the 1982 Entry Draft, Gilhen played two games with the Whalers in 1982-83, the start on the compass of a career which reached 457 games through the 1995-96 campaign. A dependable checker, Gilhen played for Winnipeg, Pittsburgh, Los Angeles, Tampa Bay, Florida and the New York Rangers. Best year was 15 goals (1990-91 with the Penguins) but it is how Gilhen played without the puck that kept a number of clubs interested in his availability.

Don Gillen

Tabbed 77th overall by Philadelphia in 1979, Gillen once rolled up 372 penalty minutes for the Brandon Wheat Kings of the Western Hockey League. Made his NHL debut with the Flyers in 1979-80, scoring a goal in his only game. Joined the Whalers in a trade on July 3, 1981 which involved Rick MacLeish and a flip-flop of draft picks. Logged 34 games with Hartford in 1981-82, going 1-4-5 with 22 penalty minutes, his last crack at the NHL.

Paul Gillis

A fearless pest and checker who played 11 seasons in the NHL, Gillis played the final 33 games of a 624-game hitch in 1992-93 in Hartford with the same robust he displayed for Quebec and Chicago. Drafted 34th overall by the Nordiques in 1982, Gillis amassed plenty of bruises and penalties as he racked up 1,351 minutes, second only in club annals to Dale Hunter's 1,545. Finished with career numbers of 88-154-242 and 1,498 penalty minutes. Obtained from the Blackhawks by Hartford on February 27, 1992.

*A defenseman who keyed the playoff drive in 1980, **Larry Giroux** anchored the blue line and at his best clearing the crease of rival forwards seeking to pitch the proverbial campsite in the low slot.*

Larry Giroux

What turned out to be Giroux's last of seven seasons in the NHL also turned out lucky for this burly defenseman who supplied a positive influence in 1979-80 at the blue line once obtained on waivers from St. Louis. Giroux went 2-5-7 with 44 penalty minutes in 47 games but it was his blue-collar approach and ability to clear the crease which helped Hartford gain a playoff berth. Played 274 career games with hitches in St. Louis, Kansas City and Detroit. Signed with the Whalers on December 13, 1979, one of two shrewd moves by general manager Jack Kelley that season. The other was obtaining winger Pat Boutette from Toronto.

Brian Glynn

Originally taken 37th overall by Calgary in 1986, Glynn reached the 400-game career milestone during 1995-96 with the Whalers. Hartford snared the blueliner in the 1995 Waiver Draft and he played 98 games over parts of three seasons. Glynn notched a goal in his last game as a Whaler, a 7-3 win over Pittsburgh on October 8, 1996. The next day, he was dealt with Brendan Shanahan to Detroit for Keith Primeau, Paul Coffey and a first round draft choice. A defenseman with a big reach, Glynn has also made stops in Minnesota, Edmonton, Ottawa and Vancouver, the highlight being a trip to the Stanley Cup Finals with the Canucks in 1994.

*Alexander Godynyuk came to the Whalers in a swap with Florida for **Jim McKenzie**. He logged 115 games at the blue line for Hartford over parts of three seasons. He scored the game-winner in the 1996-97 lidlifter on October 5 to give the Whalers a 1-0 win over the Phoenix Coyotes.*

Alexander Godynyuk

Entering 1997-98, this Ukraine-born defenseman has played six seasons in the NHL, a span of 223 games including 115 for the Whalers. After an impressive debut in 1993-94 (acquired from Florida by Whalers general manager Paul Holmgren for Jim McKenzie on December 16, 1993), Godynyuk fell in disfavor, playing just 15 games over the next two years. In 1995-96, Godynyuk spent most of the winter with the Detroit Vipers of the International League, then worked himself back into the mix for 1996-97 to play 55 games. Made his only goal decisive, the game-winner in a 1-0 decision on October 5, 1996 to welcome the Phoenix Coyotes (nee Winnipeg Jets) into the NHL record book. Originally tabbed by Toronto as the 115th overall pick in 1990, Godynyuk has also played for Calgary and Florida. Was part of a nine-player blockbuster between the Maple Leafs and Flames on January 2, 1992, a trade that involved Doug Gilmour.

*Hartford goalie **Mario Gosselin** manages to stop Steve Thomas of the New York Islanders from close range here before rugged defenseman **Bryan Marchment** arrives. Marchment was awarded as compensation to Edmonton on August 30, 1994 when the Whalers inked free agent forward Steven Rice.*

Mario Gosselin

When injuries crippled the Hartford netminding corps in 1992-93, the "Goose" stepped in and went 5-8-1 with a 3.86 GAA while making 14 straight starts, a highlight in one of Hartford's darker seasons. Gosselin also played seven games in 1993-94 to conclude a 241-game hitch in the NHL that began in 1984-85 with Quebec. Drafted 55th overall by the Nordiques in 1982, Gosselin vaulted into prominence by winning the Jacques Plante Trophy as the Quebec League's top netminder on the strength of a 3.12 GAA in 46 games with Shawinigan. Gosselin played for the Canadian Olympic Team in 1984 and posted a shutout in his NHL debut, a 5-0 decision over St. Louis on February 26, 1984. Shared net duties in Quebec until moving on to Los Angeles in 1989-90. Joined Hartford as a free agent on September 4, 1991, an organization where he gained a last hurrah in the NHL.

Chris Govedaris

Taken by the Whalers as the 11th overall pick in the 1988 Entry Draft, this Toronto native had three brief cracks at trying to make the grade in Hartford where he played a total of 33 of his 45 career NHL games. Had 2-4-6 with 10 penalty minutes for the Whalers. Notch his final goal with Hartford in a 6-3 win over Washington on December 16, 1992. Released following the 1992-93 season, Govedaris signed with the Maple Leafs and went 2-2-4 in a dozen games in 1993-94 following a 35-goal year in the American League at St. John's. Another of the many first-rounders taken by the Whalers that easily could have been taken on a latter round.

Ted Green

Once know as "Terrible Ted" when he was voted in one poll as the meanest player in professional hockey, Green went out of hockey as a pacifist in his latter years, advocating that league officials curb violence on ice. Among the established NHL players who helped launch the WHA in 1972, the gung-ho defenseman made a remarkable comeback to get back on skates. An intimidating member of the powerhouse Boston Bruins, Green sustained a severe head injury in a stick fight with Wayne Maki of the St. Louis Blues during an exhibition game on September 21, 1969. As he had done routinely dozens of times, Green went to clear the defensive zone and complete a hit on a rival forechecker. Green advanced the puck and then leveled Maki who fell to the ice. Maki took exception to a spirited check. While getting up, Maki speared Green who responded by swinging his stick, again knocking Maki down. Maki, who tragically died in 1974 from a brain tumor, retaliated by swinging his stick, hitting Green squarely on the head. Onlookers gasped as Green collapsed to the ice from the force of the blow which fractured his skull and embedded bone fragments into his brain. Two operations were needed to save Green's life. For several weeks, Green's left side was paralyzed, raising thoughts if he would live a normal life again. Time is a great healer and Green recovered, eventually getting medical clearance to resume playing hockey. Though he sat out 1969-70, a season that Boston ended the franchise's long drought for the Stanley Cup, Green

*The first captain of the Whalers was **Ted Green**. He enjoyed his best offensive season as a pro in 1972-73 with 46 points in helping New England to the AVCO Trophy.*

rejoined the Bruins to reach the 600-game career milestone and contributed to the club's Cup success in 1972. A few months later, Green moved over to the WHA where he captained the Whalers and played a key role in sparking New England to the championship that first season, becoming the first player in hockey annals to play on Stanley Cup winner and AVCO Trophy winner in consecutive years. Green's debut with the Whalers, which produced career highs with 16-30-46, launched a 452-game tenure in the new league where the veteran defenseman also played for a championship club in Winnipeg. Green is among a list of former Whalers to coach in the NHL, serving in that capacity for the Edmonton Oilers.

*The "Grim Reaper" came to the Whalers from Detroit via waivers on October 12, 1996 and **Stu Grimson** quickly became a fan favorite. He went 2-2-4 in 76 games while leading the club in penalty minutes (218). Few in hockey annals were as rough or as willing to drops the gloves as Grimson. Among the heavyweights to play for the Whalers, it's a toss-up as to who was the best fighter. In the WHA, **Jack Carlson** and **Nick Fotiu** hold that honor. In the NHL, **Torrie Robertson**, who racked up a record 358 with Hartford in 1985-86, rated among the league's most feared sluggers.*

Stu Grimson

The "Grim Reaper" came to the Whalers on October 12, 1996 off waivers from the Detroit Red Wings and took over as Hartford's main enforcer, leading the club in penalty minutes (218) and becoming a fan favorite. Grimson played in 76 games with the Whalers and pushed his career totals to 422 NHL games over nine seasons with 6-13-19 and 1,324 penalty minutes. Taken 186th overall by Detroit in the 1983 Entry Draft, Grimson has battled every challenger throughout his career. He remains one of the toughest and strongest heavyweights in NHL annals. Grimson has also played for Calgary, Chicago and Anaheim.

Richard Gordon

Few ever made money bankrolling the Hartford Whalers. Richard Gordon was the exception. He turned a profit of $10 million after owning the franchise for six years, a segment in club annals where it should be noted that he kept the hockey club docked in the capital city rather than peddle it to the highest bidder. As an entrepreneur, Gordon made his fortune in real estate development. He invested heavily in land and office buildings in the Farmington Valley and downtown Hartford. As a businessman, he was sharp with a pencil figuring out the bottom line. Besides an avid tennis player, Gordon owned season tickets to the Whalers for many years. That led to a $31 million transaction on June 28, 1988 when Gordon and Donald Conrad agreed to become majority owners of the Hartford franchise. What had worked so easily for Gordon in terms of mortgaging and leasing structures did not in the world of sports. His friendship with Conrad, a former insurance mogul and head of the Hartford Downtown Council, soured within a year. Soon allied with the Colonial Realty Company, Gordon had the last word on all deals but the arrangement with CRC quickly became a nightmare. The group, headed by Benjamin Sisti and Jonathan Googel, filed for bankrupcty. Within a short time, federal authorities indicted Sisti and Googel for siphoning $350 million dollars from clients in a scheme which defrauded hundreds of real estate investors. Meanwhile, Gordon's choices with the hockey club were equally disturbing. An admitted novice, Gordon made a series of bad front office hirings. Refusing to let a proven hand like Emile Francis

*Mark Greig was Hartford's top draft pick in 1990. Like many over the years, the winger never reached expectations of scouts. Greig had 7-17-22 in 74 games over parts of four seasons before getting shipped out to Toronto for **Ted Crowley** on January 25, 1994.*

Mark Greig

Hartford's top pick and 15th overall in 1990, Greig had a handful of chances over four different seasons to crack the lineup but lacked the confidence and zest that produced a 55-goal season in junior hockey. Greig is one of numerous examples of just how inexact a science the NHL Entry Draft is. Deemed to be among the better amateur prospects in his draft year, Greig would play his best hockey in the minor leagues, the same fate of many top picks that year. Notables who were taken on later rounds were Geoff Sanderson, Hartford's second pick that year and 36th overall, and Peter Bondra, tabbed 156th overall by Washington. Greig went 5-17-22 in 74 games with the Whalers before getting dealt to Toronto in a swap for U.S. Olympian Ted Crowley on January 25, 1994. He also played for Calgary in 1994-95. Had 13-32-45 in 64 games with Houston of the International League in 1996-97.

*The date is September 1, 1993 and Connecticut Governor **Lowell P. Weicker Jr.** signs an agreement with the state, the City of Hartford and the Whalers that was to keep the NHL team docked at the Trumbull Street mall complex for 20 years. Looking on are Hartford mayor **Carrie Saxon Perry** and Whalers owner **Richard Gordon** who share a laugh. Within a year, Gordon sold the hockey club to **Peter Karmanos**.*

*Real estate developer **Richard Gordon** purchased the Whalers in 1988 when the franchise made the transition from corporate to individual ownership. Gordon stayed at the helm six seasons, a checkered tenure where his club seemed to make more news off the ice than on it. An after-hours brawl at a restaurant in Buffalo in 1994 helped to sway Gordon to sell the hockey club. The result was the arrival of a Michigan computer entrepreneur, the acerbic **Peter Karmanos**.*

run the hockey operations, Gordon, on the advice of his sports committee, hired Eddie Johnston as general manager. In the same fashion a demolition worker goes about an assignment and sticks dynamite in strategic points to raze a building, Johnston had the green light to blowup whatever ties the hockey team had developed over the years. With Gordon's approval, Johnston curtailed community relations and began trading players. In a three-year period, the Whalers were constantly on the transaction wire with Johnston making 32 deals. Many prominent players in team history (particularly Ron Francis, Kevin Dineen, Ulf Samuelsson, Mike Liut, Ray Ferraro and Dave Tippett to name a half-dozen) are cut adrift. The changes gut the club of players and names familiar to the Civic Center faithful. Interest wanes. Attendance drops close to 4,000 per game. Sailing into darkness, Gordon fires Johnston, a decision cheered by anyone whoever bought a ticket to see the Whalers. Brian Burke is hired but opts out after a 52-loss season for a post with the National Hockey League. Paul Holmgren, a blue-collar player who feared no one during his playing days with the Flyers, follows for 1993-94. On the ice, the Whalers stumble early and Holmgren inserts Pierre McGuire as coach. Though the team plays better for a stretch, the players are keenly aware of McGuire's shortcomings and the likeliness of eventual changes. Off the ice, they rumble. Deeply stung publicly when several players and members of McGuire's coaching staff are involved in an after-hours brawl at a Buffalo nightclub on March 24, 1994, Gordon is betrayed for the last time. With his attention more on the health of his real estate empire now caught in the web of a tougher business climate, Gordon decides to sell the hockey club. The Connecticut Development Authority (CDA), under terms of negotiated arrangements gained in a quasi-partnership with the hockey club to giving Gordon economic relief over the years, buys the Whalers on June 1, 1994 for $45 million. Within a month and despite better offers from suitors on the table, the CDA selects a Michigan-based group headed by Peter Karmanos to take control of the franchise. Despite optimism at the outset and a season-ticket drive in 1996-97 that boosted attendance, the on-ice results for the Whalers didn't change. . . only the address. Able to cut a deal to escape from a four-year lease agreement a year early, Team Karmanos sets sail for North Carolina to begin play as the Carolina Hurricanes in 1997-98.

Alan Hangsleben ranked among the most versatile of Whalers, able to play both defense and forward. "Hank" played 334 games with the franchise, starting in the WHA and 37 games in 1979-80 when he was shuttled to Washington for winger Tom Rowe.

Kevin Haller

Drafted by Buffalo as the 14th overall choice in 1989, Haller arrived in Hartford on December 15, 1996 from the Philadelphia Flyers in a multi-player deal that involved Paul Coffey. A regular since 1991-92 in the NHL which included a stop in Montreal, Haller played 35 of his 397 career games as a Whaler. Haller's most notable games with Hartford involved the New York Rangers where he was a pest. It was Haller's check against Wayne Gretzky in the neutral zone which resulted in controversy during a 5-2 loss to the Blueshirts on February 5, 1997. In a return meeting in Hartford on February 21, 1997, Haller had a goal and an assist in a 7-2 rout by the Whalers.

Alan Hangsleben

Drafted by Montreal as the 56th overall pick in 1973, "Hank" elected to sign with the New England Whalers where he played 334 games in the WHA and 37 in the NHL before getting dealt to Washington on January 17, 1980 in a trade for winger Tom Rowe. Able to play forward and defense with a touch of gusto, Hangsleben used his 6-foot-1, 195-pound frame to overpower people at times. It was coach Harry Neale who once said, "Alan is at his best when I can hear him forechecking." At North Dakota, Hangsleben reached All-American status. Notched 10 or more goals three times in the WHA. Also played for Los Angeles, capping a 185-game NHL career with the Kings in 1981-82. Went 21-48-69 with 396 penalty minutes in the NHL.

Dave Hanson

Though his stay with the Whalers consisted of just one game in 1976-77, Hanson gained famed in *Slapshot*, the Hollywood version of hockey when he was cast as one of the Hanson brothers with Steve and Jeff Carlson on the Charlestown Chiefs for coach Reggie Dunlop (Paul Newman). Hanson played for the Minnesota Fighting Saints and Birmingham Bulls in the WHA as well as for Detroit and Minnesota of the NHL. In 103 WHA games, Hanson racked up 497 penalty minutes.

Todd Harkins

The Whalers dealt forward Scott Morrow to Calgary on January 24, 1994 for Harkins who logged the final 28 games of his 48-game NHL career in Hartford. Taken 42nd overall in the 1988 draft by the Flames, Harkins had great size at 6-foot-3 and 210 pounds but not the polish to stick as a regular in the NHL. Went 3-3-6 with 78 penalty minutes.

Hugh Harris

Small but effective, Harris signed with New England for 1973-74 after playing 60 games with the Buffalo Sabres of the NHL the previous winter. A consistent performer in his one year with the Whalers, Harris went 24-28-52 before he was shuttled to the Phoenix Roadrunners in a cash swap on July 31, 1974. Logged 336 WHA games, with additional stops in Vancouver, Calgary, Indianapolis and Cincinnati. Had explosive speed and a fair touch around the net to score 107-173-280. In 1976-77, Harris had 21-35-56 in 46 games for the Racers before suffering a knee injury which forced him out of hockey prematurely.

Archie Henderson

One of the most feared fighters who possessed a great reach at 6-foot-6 and 220 pounds, Henderson played 15 games for the Whalers in 1982-83, collecting 64 penalty minutes with two goals and one assist. Netted one of his goals in a 5-2 win at Boston Garden on November 4, 1982. Drafted by Washington in 1977 as the 156th overall pick, Henderson racked up 419 penalty minutes in the International League in 1977-78 and then 337 with Hershey in the American League in 1978-79. In 1981-82, Henderson amassed 320 minutes at Nashville to lead the league. Played seven games with the Capitals and one with Minnesota in 1981-82 before coming to the Whalers. After his stint in Hartford, Henderson continued to throw punches back in the minor leagues for several years before eventually coaching in the International League.

Bob Hess

A defenseman who played 329 NHL games, Hess ended his career with a three-game trial in Hartford in 1983-84. Drafted by St. Louis as the 26th pick in 1974, Hess had 27-95-122 for his career, the bulk as a regular with the Blues from 1974 to 1976 when he was considered one of the more promising blueliners in the league. Also played for Buffalo.

Bobby Holik, *Hartford's top pick in 1989, goes to the net here against Boston's Andy Moog. Holik had back-to-back 21-goal campaigns before exiting to New Jersey in a trade for goalie* **Sean Burke**.

Brian Hill

The Whalers claimed this lanky forward from the Atlanta Flames in the Expansion Draft on June 13, 1979. The right winger played 19 games in 1979-80 in Hartford, notching his only goal to produce a 4-4 tie against Philadelphia on December 26, 1979. Hill was taken 31st overall by the Flames in 1977.

Rick Hodgson

Taken by Atlanta as the 46th overall pick in 1976, this defenseman was plucked by the Whalers in the 1979 Expansion Draft. Played seven total games with Hartford including one in the 1980 Stanley Cup playoffs, his only appearance in the NHL.

Mike Hoffman

Drafted 67th overall by Hartford in 1981, Hoffman played nine games with the Whalers over three different seasons including a six-game trial in 1985-86. The left winger notched his only NHL goal in a 7-2 loss to Winnipeg on January 25, 1986.

Paul Hoganson

The Whalers signed Hoganson as a free agent when the Baltimore Blades (nee Michigan Stags) were having financial woes in 1974-75. Hoganson played four games with New England before getting his release. Future opportunities followed for Hoganson in the WHA with Cincinnati and Indianapolis which resulted in a 143-game career including 32 in a row for the Stingers. Also played two games with Pittsburgh in the NHL in 1970-71. Had a 44-71-4 record with a 4.11 GAA in the WHA.

Bobby Holik

When taken 10th overall in 1989, the Whalers felt they had secured the big-play center that would help them take a run at the Stanley Cup. The 6-foot-3, 220-pound Czech pivot did put up back-to-back 21-goal campaigns in his only two years in Hartford but was soon gone, dispatched to New Jersey in a deal that brought netminder Sean Burke to the Whalers on August 28, 1992. Holik had his best offensive years in Hartford but gained his championship ring with the Devils in 1995 when New Jersey won the Cup.

Ken Holland

Drafted by Toronto 188th overall in 1977, Holland signed with the Whalers in September, 1979. Played his only game in Hartford in 1980-81, a 7-3 loss to the New York Rangers on November 16, 1980. Holland also played three games with Detroit in 1983-84. Soon made the move into management and is currently working in the front office of the Detroit Red Wings.

Paul Holmgren

"Homer" served terms as coach and as general manager of the Whalers from 1992 to 1995. Though his motto was simply "have pride in yourself, work hard and play for the sweater," Holmgren had little chance for success in Hartford given a club shy of talent and depth. A good man and a warrior as a player for the Philadelphia Flyers, Holmgren deserved better. Holmgren's three-year relationship with the Hartford Whalers symbolized bad hockey, both on and off the ice. Hired to coach the team by Brian Burke, "Homer" had a bigger demon than the general manager who soon realized the biggest stumbling block in Hartford was meddling owner Richard Gordon. Burke, after one season, wound up taking a job with the

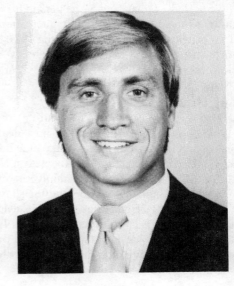

Paul "Homer" Holmgren *served as coach and general manager of the Whalers in the 1990s. Originally signed by the WHA Minnesota Fighting Saints, Holmgren was a blue-collar forward for the Philadelphia Flyers as a player. His teams in Hartford were short of talent and depth, a combination that does not often result in success.*

league in New York. Holmgren, meanwhile, moved into the dual capacity of coach and general manager during training camp in 1994. The Whalers stumbled on the ice under Holmgren who decided to concentrate on managerial duties and gave the coaching reins to Pierre McGuire 17 games into the campaign. The Whalers fell through on the ice throughout 1993-94 and it was Holmgren who eventually stumbled off of it. A late afternoon car wreck in suburban Simsbury resulted in a DWI offense. To Gordon's credit, he assisted Holmgren who checked into a detoxification clinic in California to begin recovery. Upon his release, Holmgren went public with a memorable press conference, admitting he had a drinking problem since his teen years and vowed to change his life. When Gordon sold the franchise, there was growing speculation that Holmgren would be out of the organization. Peter Karmanos and Jim Rutherford may have coveted another coach but most of the established ones were under contract to other teams. To keep some stability with the organization, the new management team gave "Homer" a second chance rather than immediately hire someone it had groomed for several years with Compuware's junior hockey programs. After failing to make the postseason in the 48-game season in 1995, Holmgren had the Whalers flying out of the gate to open 1995-96. Hartford won its first four games, then sank like a rock. Though a six-game winless slide ended on November 4, 1995 when a late goal by Nelson Emerson gave the Whalers a 5-4 win over Ottawa, Karmanos viewed a listless effort the following night in Philadelphia, a 6-1 spanking at the hands of the Flyers. The volatile Karmanos had seen enough. On November 6, 1995, Holmgren was out and Paul Maurice of Team Compuware was in.

Ed Hospodar

Taken 34th overall by the Rangers in 1979, "Boxcar" logged 450 games over nine NHL seasons with New York, Hartford, Philadelphia, Minnesota and Buffalo before winding it up in 1988-89. The Whalers obtained Hospodar, a rugged defenseman who amassed 1,314 penalty minutes during his career, by sending Kent-Erik Andersson to the Blueshirts on October 1, 1982. Had only one goal in 131 games with Hartford before moving on to the Flyers. Had 17 goals and 51 helpers in his career but is often remembered for instigating altercations. Led the NHL in penalty minutes (93) during the 1981 Stanley Cup tournament with the Rangers. Hospodar lost a bit of his fearlessness on December 30, 1981 when Clarke Gillies of the Islanders broke his jaw in a fight.

Doug Houda

The 28th overall pick in 1984 by Detroit, Houda came to the Whalers on February 20, 1991 in a deal for Doug Crossman and logged 142 games at the blue line over four seasons in Hartford. Dealt to Los Angeles on November 3, 1993 for Marc Potvin, it was simply an exchange of players who had no trouble dropping the gloves. Blessed with good mobility, Houda has played in over 400 NHL games, reaching the milestone during 1995-96 with Buffalo. Most memorable night as a Whaler was February 11, 1992 when he was called for three minors, a major and a game misconduct in a 5-1 win over Buffalo.

*Defenseman **Ed "Boxcar" Hospodar** amassed 362 penalty minues in 131 games over two seasons with the Whalers. He came to Hartford in a deal with the New York Rangers for Kent-Erik Andersson.*

Garry Howatt

A veteran of 12 NHL seasons, Howatt spent 1981-82 with the Whalers after coming to Hartford on October 2, 1981 in a swap with the New York Islanders for a fifth-round draft choice. That year, Howatt led the squad in penalty minutes (242) and chipped in with 18-32-50 in 80 games. The only two Whalers to play in every game that winter were Howatt and Blaine Stoughton. Originally a 10th round draft pick by the Islanders in 1972, Howatt overcame epilepsy to carve out a lengthy career that ended following the 1983-84 campaign with New Jersey. The Whalers dealt the gritty forward to the Devils on October 15, 1982 with center Rick Meagher for center Merlin Malinowski and the draft rights to Harvard's Scott Fusco.

Gordie Howe

The greatest name in hockey played in Hartford for two seasons in the WHA and then closed out his 32-year hockey career with one last hurrah in 1979-80 when the Whalers joined the NHL. Often called "Mr. Hockey," "Number 9," "Blinky" or just "Gordie," Howe played in 2,186 games during a tenure which began in 1946-47 with the Detroit Red Wings, stalled in retirement for two winters in the early 1970s and flourished again starting in 1973-74 when the Houston Aeros cut a deal with Howe and sons Mark and Marty to play in the WHA. Howe played on four Cup winners in Detroit and won six Hart Trophies as the NHL's top player. He may have been a step slower in the WHA but his skills and popularity never waned. He led the Aeros to AVCO Trophies in 1974 and 1975 and a trip to the Finals in 1976. When Houston had fiscal difficulties in 1976-77, the Howes were declared free agents and free to negotiate with teams in both leagues. Boston owned the draft rights to Mark and Detroit had obtained the rights to Marty in a deal with Montreal. At 48, Gordie, meanwhile, was still keen on playing. When opportunities with the Bruins and Red Wings failed to materialize, the Howes landed in Hartford with the Whalers in a blockbuster announcement on May 23, 1977.

Gordie Howe, *often considered hockey's greatest player, wore the jersey of the Whalers proudly during three seasons in Hartford, two winters in the WHA and his last NHL campaign in 1979-80.*

Gordie Howe, a gallant warrior over five decades of professional hockey, takes a deep breath before a face-off in his final NHL season. Howe proved he was still capable of scoring as well as selling tickets, notching 15 goals while playing in all 80 games in 1979-80 Howe's Whalers were among the league's best draws on the road. Fans across North America got a final look at Howe trying to match strides with Father Time. The most vocal gathering turned out in Detroit where Howe played for 25 seasons including 21,002 for the All-Star Game when Gordie represented the Whalers.

When the WHA Houston Aeros deemed the Howe family free agents and various deals with NHL clubs failed to materialize, **Howard Baldwin** engineered the signing of **Gordie, Marty and Mark** that was announced on May 23, 1977. The transaction enhanced Hartford's position to join the NHL in 1979. One of the few teams to vote against the Whalers entering the league was Boston. Harry Sinden, the general manager of the Bruins, held a grudge against Hartford for many years based solely on losing Mark Howe to a team in southern New England.

It's December 7, 1977 in Birmingham and there is little goalie John Garrett of the Bulls can do but watch No. 1,000 off the stick of New England's **Gordie Howe** *roll into net. Howe became the first in hockey annals to reach a grand, notching the historic milestone in a 6-3 win.*

The additions, along with the recent acquisitions of Dave Keon and John McKenzie from the disbanded Minnesota franchise, revived the fortunes of the New England franchise. In 1977-78, Howe led the Whalers in scoring with 34-62-96 and gallantly sparked New England to the AVCO Finals. During that year, Gordie became the first player in hockey annals to reach 1,000 career goals, reaching the milestone on December 7, 1977 in a 6-3 win at Birmingham when he tallied against Bulls goalie John Garrett. When the WHA ceased operations, Howe helped usher the Whalers through their first NHL season, playing in 80 games and going 15-26-41. Hartford surprised the established league by making the playoffs and setting a handful of expansion team records. Gordie's final NHL goal was registered in a 8-4 loss in Game 2 of the Stanley Cup playoffs at the Montreal Forum on April 9, 1980. Mark Howe assisted on the goal at 13:59 of the third period. The crowd of 15,242, sensing it was Howe's last game in Montreal, gave a huge standing ovation. Howe laced the skates for the final time in Hartford on April 11, 1980, a 4-3 loss to the Habs which eliminated the Whalers from the playoffs. On June 4, 1980, before a throng of international media which huddled in Hartford, Gordie announced his retirement. "It's not an easy task to retire," Howe said. "No one teaches you how. I felt that the first time. I'd hate to go out after 32 years and find out in the middle of the winter I'd run short. When I started playing

hockey, my goal was to provide a proper house and a proper heater for my parents. Heck, I just wanted to be lucky enough to have two suits. I've decided to hang up the blades. . . but I still think I'm damn good enough to play."

For five decades, Howe's name was synonymous with hockey. His numbers stood for years such as games played (1,767), goals (801), assists (1,049) and points (1,850) until Wayne Gretzky eventually revised the NHL record book. Howe's occasional penalties added up as well (1,685 PM). In postseason play, Gordie played in 157 Stanley Cup games, going 68-92-160 with 220 PM. In the WHA, Howe played in 419 games with 174 goals, 334 assists, 508 points and 399 PM. In 78 playoff games, Howe went 28-43-71 with 115 PM. Howe's major league totals are 2,186 games, 1071 goals, 1,518 assists and 2,589 points (only Gretzky has scored more goals and points). Howe's career numbers and longevity dwarf all who have played professional hockey. Most of his points were recorded when the NHL consisted of just six teams and he was the most feared player in the game. Dave Keon, a Hall of Famer who was a rival and a later teammate of Howe with the Whalers, often summed up the Original Six this way: "There are two weak teams in the league and four strong ones. The weak ones are New York and Boston. The strong ones are Toronto, Montreal, Chicago and Gordie Howe."

Mark Howe, seen here with the Whalers in the early 1980s, recalls the decision to turn pro with the WHA was not easy. "I had severe migraine headaches for days," Mark states about his first training camp with Houston in 1973. "I put so much pressure on myself to do well. I felt I had to prove to people I could play hockey because of my abilities, not because I was Gordie Howe's son.."

Gordie often said he was fortunate to avoid serious injuries to last as long as he did in hockey. The biggest injury of Howe's career happened in his fourth pro season. On March 28, 1950 in a playoff game against Toronto, Howe crashed into the boards at the Olympia in Detroit when he missed a check on Leafs captain Ted Kennedy. The prayers of hockey fans were heard when Howe survived a series of operations to ease a head injury which put Howe's life in peril. Gordie recovered to become the game's icon for many years until he was finally stopped by Father Time.

Mark Howe

When Mark Howe announced his retirement following the 1994-95 season, it ended a chapter in the game where the Howe name would no longer be on a pro hockey uniform, the first time in close to 50 years. Able to play both forward and defense, Mark played in 929 NHL games after 426 games in the WHA during a career which began in 1973-74 with the Houston Aeros. Played two years in

Hartford with the WHA Whalers and also three NHL campaigns before general manager Larry Pleau unwisely pulled the trigger on a three-team deal struck on August 19, 1982 involving Philadelphia and Edmonton. In every trade, the key is often which team lands the best player. Hartford did not get equity in the Howe trade. The Whalers dealt Mark to the Flyers with a third-round pick in 1983 (Derrick Smith) for Ken Linseman, Greg Adams and Philadelphia's first-round pick in 1983 (David A. Jensen) and a third-rounder (Leif Karlsson). Hartford then sent Linseman and Don Nachbaur to the Oilers for Risto Siltanen and Brent Loney. With the Flyers, Howe piled up points and ice time, two commodities the Whalers sorely missed from the blue line. It was a void the club never filled until a trade on November 21, 1985 secured Dave Babych from Winnipeg for Ray Neufeld. Like his dad, Mark overcame a serious injury to play 22 years in professional hockey. Against the New York Islanders on December 27, 1980, Howe was crosschecked by John Tonelli during a scramble near the Hartford net. The

Mark Howe has a step here on Bryan Trottier of the New York Islanders. Howe played for the Whalers in both leagues. He was traded to Philadelphia on August 19, 1982, a transaction that also involved Edmonton. In exchange for Howe, the Whalers secured power-play specialist *Risto Siltanen,* third-line winger *Greg Adams* and a number one draft pick. Hartford used the selection to take *David A. Jensen* in 1983 and eventually obtained center *Dean Evason* and goalie *Peter Sidorkiewicz* from *Washington for Jensen.*

Marty Howe may have lacked the offensive talents of his brother and father yet he was a capable defenseman who usually made the safe play. He scored 20 goals over seven campaigns with the Whalers. In the NHL, Howe scored twice, once for Boston in 1982-83 and again with Hartford against New Jersey in 1983-84.

traffic pileup resulted in the cage coming off its moorings and Howe was impaled in the backside by a sharp mental flange and came within inches of permanent paralysis. It took virtually two years for Mark to regain his leg strength and that showed during his years in Philadelphia. He closed out his career with the Red Wings, playing 122 games and 16 Stanley Cup games. Howe's best year was 42-65-107 with the WHA Whalers in 1978-79. His top winter in the NHL was 24-58-92 with the Flyers in 1985-86. He often says that his greatest experience was playing for Team USA in the 1972 Winter Olympiad. At 16, he showcased his skills and helped the Americans capture the silver medal at Sapporo, Japan.

Marty Howe

The oldest son of Gordie Howe, Marty played in 449 WHA games and 197 games in the NHL. A defenseman, Marty played with the Aeros and Whalers in the WHA and with Hartford and Boston in the NHL. Drafted 51st overall by Montreal in 1974, Howe played a steady game at the blue line. Despite being a regular on two AVCO championship clubs in the WHA, early Hartford management opted to ship Marty to the minors for parts of three NHL campaigns, giving him brief trials. Loaned to Boston, the Bruins signed Marty as a free agent on October 1, 1982. He proved to be a sharp acquisition, maturing into a stay-at-home blueliner who was often paired with Raymond Bourque. Returned to the Whalers when Emile Francis was the club general manager. He skated a regular shift in 1983-84 before finishing up in 1984-85.

*The date is March 1, 1980 and **Bobby Hull** addresses the media in front of general manager **Jack Kelley** and **Don Blackburn** following Hartford's acquisition of the Golden Jet from Winnipeg. Hull tallied his final NHL goals as the Whaler, his 609th against Buffalo's Don Edwards and his 610th against Detroit's Rogie Vachon.*

Pat Hughes

A veteran right winger of 10 NHL seasons and three Cup champions in Edmonton, Hughes came to Hartford via waivers on March 10, 1987 in a move for playoff depth. Hughes appeared in two regular season games and three playoff games with the Whalers. Originally picked 52nd overall by Montreal in 1975 where he broke into the league, Hughes went on to play 573 NHL games with additional stops in Pittsburgh, Buffalo and St. Louis.

Bobby Hull

Another Hall of Famer who played in Hartford, the "Golden Jet" played nine games for the Whalers in 1979-80, going 2-5-7 to help the hockey club lock up a playoff berth in its first NHL season. Hull, who starred for 15 years in Chicago, helped expand the game across new markets in North America when he was jumped to the Winnipeg Jets of the upstart WHA in 1972. Played 411 games in the WHA, notching 303 goals and 335 assists. Added 43-37-80 in 60 playoff games. His brief tenure with the Whalers hiked his NHL career totals to 610 goals, 560 assists and 1,170 points in 1,063 games. Had five, 50-goal seasons with the Blackhawks and four with the Jets including 77 in 78 games in 1974-75. Any line with Hull was potent for scoring. Most observers concur a threesome with Hull and Swedish standouts Ulf Nilsson and Anders Hedberg in Winnipeg, a key to the Jets winning AVCO Trophies in 1977 and 1978, ranks among the game's best.

Jody Hull

A first-round pick by Hartford in 1987 (18th overall), Hull played 98 games over two years with the Whalers before getting dealt to the New York Rangers on July 9, 1990 for Carey Wilson and a third-round pick in 1991

***Mark Hunter** played for the Cup champion Calgary Flames in 1989 before joining Hartford in a deal for Carey Wilson on March 5, 1991. In the Cup playoffs against Boston that year, Hunter notched five tallies, the best one-series effort in Stanley Cup play by a Whaler.*

which Hartford used to draft Michael Nylander. A good two-way player, Hull has played in over 400 NHL games. After spending a year with the Ottawa Senators in 1992-93, Hull signed with the Florida Panthers and has found a regular shift. He enjoyed his best pro year in 1995-96, going 20-17-37 in 70 games in helping the Panthers to the Stanley Cup Finals.

Mark Hunter

The seventh overall pick by Montreal in 1981, "Big Game" Hunter played 74 games over two years with the Whalers highlighted by five goals in the 1991 playoffs against Boston. A member of the 1989 Cup champions in Calgary, Hunter played in 628 NHL games with stops in Montreal, St. Louis and Washington. Rang up three years of 30-plus goals including 44 in 1985-86 with the Blues. Joined the Whalers on March 5, 1991 from the Flames in a deal for Carey Wilson. Last year in the NHL was 1992-93 with Washington where he landed following a trade by Hartford. Went to the Caps, along with Yvon Corriveau, on June 15, 1992 for Nick Kypreos. Brother of Dale and Dave Hunter.

Paul Hurley

A member of the original WHA Whalers in 1972-73, Hurley played 251 games with New England before getting dealt to Edmonton for future considerations on January 29, 1976 in a deal involving Kerry Ketter and Steve Carlyle. A steady defenseman noted for his ability to block shots, Hurley played at Boston College en route to a hitch with the 1968 U.S. Olympic Team. Played one game for Boston in the NHL in 1968-69 and eventually left the Bruins when the upstart WHA opened for business. Went on to score 10-76-86 in 275 games, finishing with Calgary Cowboys in 1976-77.

Mike Hyndman

A defenseman in the club's first WHA season, Hyndman played in 59 games for New England during 1972-73 before getting dealt to the Los Angeles Sharks on February 16, 1973 in a trade that brought Mike Byers to the Whalers. Finished up with the Sharks in 1973-74.

Dave Hynes

One of a handful of players to play for both the Whalers and Bruins, this forward played 22 games with the Bruins before jumping to the WHA. Had 5-4-9 in 22 games with New England in 1976-77, his last pro season. Possessed a scoring flair in the American League, averaging 38 goals over a three-year stretch with a career-high 42 at Rochester in 1974-75.

Dave Inkpen

A stay-at-home defenseman, Inkpen played 293 games in the WHA. He wrapped up his career in 1978-79 with the Whalers, coming to New England with Blaine Stoughton once the Indianapolis Racers ceased operations on December 13, 1978. Broke in with Cincinnati in 1975-76 and also played with Edmonton and Quebec. Had seven assists in 41 games with the Whalers to wind up with career totals of 13-76-89 with 273 penalty minutes.

Mark Janssens

A center obtained from Minnesota on September 2, 1992, Janssens anchored a checking line for the Whalers for close to five seasons once arriving from Minnesota in a deal for center James Black. During the 1996-97 season, Janssens reached a couple of career milestones (he crossed the 500-game plateau including 300 games with Hartford) before getting dealt to Anaheim on March 18, 1997 for Bates Battaglia and a fourth-round draft choice in 1998. Originally drafted 72nd overall by the New York Rangers

*Center **Mark Janssens** was noted for his faceoff abilities and physical style during his five-year tenure with the Whalers. Janssens anchored the checking threesome for Hartford until a swap to Anaheim on March 18, 1997 brought prospect Bates Battaglia and a draft choice to the Whalers.*

Doug Jarvis set hockey's standard for consecutive games. The final 139 games of the 964 straight the centerman played were for the Whalers. Jarvis broke Garry Unger's mark of 915 on December 26, 1986 against Montreal, the team the streak started with in 1975-76.

in 1986, Janssens enjoyed his best offensive year in 1992-93, going 12-17-29. A physical player with a willingness to drop the gloves, Janssens nearly had his career end in his first pro season when he suffered a fractured skull and concussion in an altercation in the International League. Has amassed over 1,000 career penalty minutes.

Doug Jarvis

Hockey's "Ironman" spent 13 seasons in the NHL, playing 964 consecutive games from 1975-76 until his retirement two games into the 1987-88 season. A reliable center whose role was often to mark the opposing club's top scorer, Jarvis was noted for his play without the puck, winning the coveted Selke Trophy in 1984 as the league's top defensive forward. In 1987 as a member of the Whalers, he was voted the league's Masterson Trophy as the player who best exemplifies the qualities of perseverance, sportsmanship and dedication to hockey. It was during his days in Hartford that Jarvis, on December 26, 1986, played in his 915th consecutive game to break the standard established by Garry Unger. It only seemed fitting that Jarvis set the record against Montreal, a team where his defensive abilities contributed to four consecutive Stanley Cup winners (1976-1979). Taken 24th overall by Toronto in the 1975 Entry Draft, Jarvis wound up with the Canadiens shortly after the lottery in a deal with Toronto for forward Greg Hubick, a winger who played 77 games over parts of two NHL campaigns. Hubick was out of hockey for six years when Jarvis eclipsed Unger from the record books. In Hartford, Jarvis played a valuable role after coming to the Whalers on December 6, 1985 from Washington for winger Jorgen Pettersson. His special-team contributions with Dave Tippett improved Hartford's overall defensive play, much like the commitment to positional play he brought to the Capitals when he was involved in a significant deal on

September 9, 1982 when Washington added Rod Langway, Brian Engblom, Craig Laughlin and Jarvis for Ryan Walter and Rick Green. Had 139-264-403 career totals and also played in 105 Stanley Cup games. His last NHL game was October 10, 1987.

Grant Jennings

A physical defenseman, Jennings played 163 games with the Whalers where he used his size to clear traffic around the net. He is best remembered for a tremendous check on Boston's Ray Bourque during the 1990 playoffs in Game 1, a hit that limited Bourque's availability for most of the seven-game series. Jennings came to the Whalers from Washington along with Ed Kastelic for Mike Millar and Neil Sheehy on July 6, 1988. Moved on to Pittsburgh in the famous Francis-Cullen deal on March 4, 1991 where he played for two Stanley Cup winners (1991 and 1992). With also stops in Toronto and Buffalo, Jennings is a testament to hard work considering he was never drafted and is closing in on 400 career NHL games.

David A. Jensen

A promising forward who was taken 20th overall on the first round in 1983, Jensen had exceptional speed but lacked the overall game to blossom in the NHL. A member of the 1984 U.S. Olympic Team, Jensen had four helpers in 13 games in 1984-85 before the Whalers dealt him to Washington for center Dean Evason and goaltender Peter Sidorkiewicz on March 12, 1985. Jensen finished with 9-13-22 in 69 career NHL games, the bulk coming in 1986-87 when he was 8-8-16 in 46 games with the Capitals.

Mark Johnson

"Captain America" had his best years in the NHL with the Whalers where he served a hitch as team captain and produced 85-118-203 in 201 games from 1982 to 1985. In hockey annals, Johnson will be forever remembered for his heroics at Lake Placid in 1980 when Team USA won the gold medal. Johnson, a prolific scorer at the University of Wisconsin, joined the Olympic program where he led the Americans in scoring in a series of exhibition games leading up to the Winter Games and then going 5-6-11 in seven games during the Olympiad. The bigger the game, the bigger the effort. Johnson tallied twice in the epic 4-3 win over the Soviet Union, then assisted on the winning marker and added an insurance tally in the 4-2 clincher against Finland. Drafted by Pittsburgh as the 66th overall pick in 1977, Johnson played for his father, the late "Badger" Bob Johnson who turned Wisconsin into a powerful collegiate program and later directed the Penguins to the Stanley Cup in 1991. Johnson played for the Penguins and Minnesota North Stars before coming to Hartford on October 1, 1982 for Jordy Douglas and Kent-Erik Andersson. His stay in Hartford was highlighted by leading the club in scoring in 1983-84 with 35-52-87. Was involved in a key trade that turned the club's fortunes around in the mid-1980s when he and netminder Greg Millen were shipped to St. Louis on February 22, 1985 for goalie Mike Liut. Overall, Johnson had 203 goals and 305 assists and 508 points in 669 NHL games over 11 seasons, winding up his career in New Jersey in 1989-90.

*Mark Johnson wore the "C" in Hartford and led the club in scoring in 1983-84 with 35-52-87 in 79 games. He was dealt to St. Louis on February 22, 1985 along with goalie **Greg Millen**, a transaction that brought netminder **Mike Liut** to Hartford.*

Bernie Johnston

A splendid playmaker in the minor leagues who anchored the Maine Mariners run to the American League's Calder Cup in 1979, Johnston was plucked by Hartford from Philadelphia in the Expansion Draft. "BJ" played in 57 games over two seasons for the Whalers, going 12-24-36. Never blessed with speed, Johnston had the ability to create plays. Among his helpers with the Whalers was one on February 29, 1980 when he and Greg Carroll assisted on Gordie Howe's 800th career NHL goal at 1:27 of the third period in a 3-0 win over St. Louis.

Eddie Johnston

During a 16-year playing career, goaltender Eddie Johnston often remarked that he got "hit in the head" many times by pucks. During a three-year tenure as general manager of the Whalers (1989-1992), "EJ" seemed to get his revenge, hitting every hockey fan in Greater Hartford with a stick to the shins with an incredible string of trades that gutted the franchise of familiar names. In retrospect, if the Whalers had managed to win Game 4 of the 1990

Stanley Cup playoffs (up 2-1 in games and leading 5-2 after 40 minutes against the Boston Bruins), Johnston would have been annointed a saint rather than the villain of the village. The seven-game loss to the Bruins paved the way for major changes. The biggest was the multi-player swap on March 4, 1991 when Hartford sent Ron Francis, Ulf Samuelsson and Grant Jennings to Pittsburgh for John Cullen, Zarley Zalapski and Jeff Parker. The swap backfired on Johnston. Francis and Samuelsson became key players as the Penguins captured consecutive Stanley Cups (1991 and 1992) at the same time Hartford went into a deep freeze. Fired on May 10, 1992, Johnston, amazingly, resurfaced in Pittsburgh. He eventually mended the situation with Francis and Samuelsson and served as a coach until getting the gate during the 1996-97 campaign. Over the years, Johnston has declined comment about his days with the Whalers, only saying that the media "unfairly portrayed him" and "there were other things at work" during his Reign of Error in Hartford. Though Johnston did obtain Pat Verbeek (for Sylvain Turgeon) from New Jersey and Andrew Cassels from Montreal (for a second-round pick), EJ's signature swap was the Francis-Cullen debacle. It's possible the key player in the deal for the Whalers was Zalapski, a mobile and promising defenseman. Johnston had selected Zalapski fourth overall

*Often referred to as the general manager who emptied the Civic Center by trading away popular players, **Eddie Johnston** made a litany of swaps during his three seasons as general manager of the Whalers. EJ fell on his sword a short time later after the big shockwave deal on March 4, 1991 that sent **Ron Francis** and **Ulf Samuelsson** to Pittsburgh for **John Cullen** and **Zarley Zalapski**. Johnston was shown to the exits on May 10, 1992. When he returned to coach against the Whalers, the faithful often serenaded Johnston with the vacuum cleaner chant.*

in the 1986 Entry Draft when he first worked for the Penguins. Maybe if Zalapski had fulfilled the expectations forecasted, the Whalers likely would have found a major answer at the blue line. Looking back, the trade involved popular players and the Whalers never recovered. Some believe the Francis-Cullen trade ranks as one of the worst in NHL history. That's debatable because every team has made clunkers but in Hartford annals, it is certainly the most memorable trade, and most lopsided one, ever made.

Ric Jordan

A defenseman who played in the WHA for the Whalers, Jordan had 1-8-9 in 70 games in New England's early years before getting dealt to Quebec for Guy Dufour on May 15, 1974. A member of consecutive NCAA champions at Boston University (1971 and 1972), Jordan followed coach Jack Kelley to the Whalers. Also played for the Calgary Cowboys to wind up with 11-23-34 in 185 career games.

K

Sami Kapanen

One of the new wave of Europeans to play in the NHL, this native of Finland combined speed and gritty play to bring an exciting flavor to the Whalers. His only problem was staying healthy. In 80 career games with Hartford over two seasons, Kapanen had 18-16-34. Tabbed 87th overall in the 1995 Entry

Draft, Kapanen missed most of training camp that year with a concussion and lower back woes after getting floored by Gerald Diduck during a team scrimmage. Once cleared to play, Kapanen showed promise and notched a two-goal game against Montreal on February 21, 1996 to key a 5-3 victory. Bitten by the injury bug three times in 1996-97, Kapanen missed 36 games yet was a factor in games he played, going 13-12-25 in 45 games. Kapanen, who led the club in plus-minus (a plus 6), had two game-winning markers including the last one against Boston in a 6-3 decision on March 12, 1997.

Al Karlander

Drafted by Detroit in 1967, Karlander played 212 games with the NHL Red Wings before jumping to the WHA, signing with New England for the 1973-74 season. Went 20-41-61 in 77 games with the Whalers and followed up with 7-14-21 in 48 games in 1975-76 before getting dealt to Indianapolis where he completed his 269-game WHA career in 1976-77. A steady, two-way forward, Karlander centered New England's top line down the stretch in 1973-74, a trio that also consisted of John French and Mike Byers. Had a hat trick for Michigan Tech during the 1969 NCAA semifinals against Cornell which featured future Hall of Famer Ken Dryden between the pipes.

Peter Karmanos

An entrepreneur who parlayed an IRS tax return of $10,000 to launch the international computer giant Compuware Corporation in 1973 that now grosses more than $250 million annually, Karmanos ended Hartford's association with the NHL on May 6, 1997 when he officially relocated the Whalers to North Carolina. The decision to exit Connecticut will be debated for many years, especially in the offing when early returns for

*On June 2, 1994, **Peter Karmanos**, flanked by Connecticut governor **Lowell P. Weicker Jr.** and general manager **Jim Rutherford**, speaks to the press from the governor's office regarding his $47.5 million purchase of the NHL Whalers. It seemed the only other time Karmanos smiled was the day he kicked the shins of hockey fans in Connecticut and set sail for North Carolina.*

*There were hopes that the Whalers could remain in Hartford in early 1997 when Connecticut Governor **John Rowland** (left) and NHL commissioner **Gary Bettman** (right) met with team owner **Peter Karmanos** to discuss possible options. The focus of the talks involved construction of a new sports arena. The state proposed construction of a $250 million arena with 20,000 seats and guaranteed revenue streams of $50 million to the Whalers. The plan seemed workable but eventually met deaf ears once Karmanos announced in late March he would relocate the hockey club. Before selecting North Carolina, Karmanos pondered using an airport hangar in Columbus, Ohio as a temporary site.*

season tickets for the upstart Hurricanes are less than 3,000. Most fingers will point at Karmanos as the main culprit but some blame should be directed at NHL commissioner Gary Bettman and Connecticut governor John Rowland. Both paved the way for the hockey club to desert a marketplace which had over 8,000 season tickets in 1996-97. "For the 800th time, my preference is to stay in Hartford," was a common refrain from Karmanos as much as "I can't keep pouring money down the drain" during his stormy, three-year, non-playoff run as CEO of the Whalers. In a nutshell, Karmanos raised ticket prices, overpaid players and complained about a poor lease agreement he inherited when he spent less than $25 million to secure the franchise in 1994. And with Bettman's goal to land a major television contract and the NHL's obvious abandonment of medium-sized markets curiously linked to WHA heritage (Quebec in 1995 and Winnipeg in 1996), the moving vans in Hartford were waiting for a destination. When Karmanos announced on March 26, 1997 that negotiations with Rowland had broken down over a new arena (the state had offered to build a $147 million complex and guarantee revenue streams of $50 million annually to the hockey team), it became a bizarre comedy as Karmanos tried to save face and relocate his club. Minnesota had been the front-runner for the Whalers until the NHL toured the St. Paul Civic Center and deemed the aging structure "unfit for hockey." Karmanos then toured a vacant airplane hangar in Ohio where he claimed that he could install 20,000 portable seats in an effort to bring pro hockey to Columbus. Though a number of communities sought pro hockey, Karmanos had embarrassed the NHL which was simultaneously in the process of examining markets for a planned expansion to 30 teams by the year 2000. Atlanta, Nashville and Houston quickly became off-limits for Karmanos. With few options, there was growing hope that local businessmen in Hartford could put together a package at the 11th hour to buy the Whalers and keep the NHL team. Talks abruptly broke down when the NHL braintrust made it clear that it would not approve such a transaction. Painted into a corner, Karmanos found an atlas and his compass and pointed south. Thus, the Carolina Hurricanes were born. The hockey club is

scheduled to play two years in Greensboro before moving into a new facility in Raleigh that is projected to open in 1999. With an average ticket price of $41 and conservative estimates that the club must average 17,500 to break even in a market that has not supported minor league hockey, it seems difficult to justify why the NHL thumbed its nose at Hartford. Karmanos complained loudly that the Whalers, the only major sports team in Hartford, did not get the daily media attention that the men's and women's college programs at the University of Connecticut did. Moving to a congested market where the Atlantic Coast Conference has four hoop programs (N.C. State, Duke, North Carolina and Wake Forest) and agreeing to be the second tenant in the newly constructed arena, it will be interesting to see the amount of ink and television coverage the upstart Hurricanes generate on a daily basis as well as the day that Karmanos plots to break a 20-year lease he signed with Raleigh officials.

Ed Kastelic

A veteran of seven NHL seasons and 220 games, Kastelic joined the Whalers on July 8, 1988 in a swap with Washington along with Grant Jennings for Neil Sheehy and Mike Millar. Taken 110th overall by the Capitals in the 1982 Entry Draft, Kastelic used his 6-foot-4, 215-pound size to supply a dose of toughness at forward, best exemplified in 1990-91 with 211 penalty minutes in 45 games. Had 10-11-21 with 719 career penalty minutes before winding up his NHL days with Hartford in 1991-92.

*A close-up look at coach **Jack Kelley** who was the first U.S.-born general manager to lead a professional hockey team to a title. In 1972-73, Kelley's Whalers were champs of the upstart WHA.*

Dan Keczmer

A defenseman taken on the 10th round (201st overall) by Minnesota in 1986, Keczmer came to the Whalers on October 2, 1991 in a deal with San Jose for Dean Evason. Played 36 games in Hartford, going 4-5-9 before moving on to Calgary for goalie Jeff Reese on November 11, 1993. Notched two goals in a 7-3 win at Edmonton on February 20, 1993 and then had the game-winner in a 4-2 decision against Buffalo on March 5, 1993. Shuttled to New Jersey by the Flames in a multi-player swap on February 26, 1996 that included Phil Housley. Signed with Dallas as a free agent for 1996-97 and notched one helper in 13 games.

Mike Keeler

A defenseman for the Whalers in the WHA in 1973-74, Keeler appeared in one regular season game and also one playoff game for New England that season.

Jack Kelley

The first U.S.-born native to be a general manager in professional hockey, Kelley holds the distinction of serving in that executive capacity for the Whalers in both the WHA and NHL. After leading Boston University to consecutive NCAA titles (in 1971 and 1972), team founder Howard Baldwin of the upstart Whalers coaxed Kelley to leave the collegiate ranks and take the helm as the franchise's first coach and director of hockey operations. Kelley organized the hockey club from its infant days and gave New England a base of identity in the Boston market. He assembled a strong cadre of players, ranging from NHL veterans Ted Green and Tom Webster to local collegians such as Jake Danby and John Cunniff. When the first season in club history closed on May 6, 1973, the Whalers had secured the first-ever AVCO Trophy, capping a splendid regular season by winning 12 of 15 postseason games. The clincher was a 9-6 rout of Winnipeg at Boston Garden. Larry Pleau, who was the first player signed by the team back on April 19, 1972, notched the hat trick in Game 5. Kelley, who was voted the league's top coach for his efforts, stayed with the organization for much of the club's seven-year history in the WHA as its general manager. When the Whalers moved into the NHL, Kelley lasted two winters. Unlike the WHA where he could stock a team by outbidding an NHL club or plucking a proven player when a rival franchise folded, the task to build a competitive organization was colossal. In Hartford's case, it was compounded by the fact that the Whalers had one of the oldest teams in pro hockey, a roster that included legends such as Gordie Howe and Dave Keon. In addition, a key surrender term in the WHA-NHL merger (limiting the Entry Draft to just six rounds in 1979 and having the established 17 clubs pick before the four WHA teams would) was designed to keep the maverick clubs from being too good at the outset. Despite the obstacles, Kelley did put together a representative squad, one that set a handful of expansion club records and surprised the established NHL by making the playoffs in 1980. What put the Whalers over the top that first year were two deals. Kelley smartly signed free agent defenseman Larry Giroux who solidified the blue line and snared gritty winger Pat Boutette from the Toronto Maple Leafs for forward Bob Stephenson. Both newcomers contributed to a 27-win, 73-point campaign. In retrospect, age and Kelley's shortcomings at the draft table indirectly led to the club's plunge in the standings and a change in leadership in Hartford. Kelley's top draft selections in 1979 and 1980, Ray Allison and Fred Arthur, were busts. So were second-round choices Stuart Smith and Michel Galarneau in the same respective years. Of the 16 players named by the Whalers, Ray Neufeld and Kevin McClelland, both fourth-round selections, went on to have extensive NHL careers with the latter never playing for Hartford. Unlike the deals made in 1979-80, Kelley misfired in two transactions in his second term. He signed free agent Thommy Abrahamsson who had played for the Whalers in the WHA but Abrahamsson's skills had badly eroded some three years later. Kelley, who also signed Gordie Roberts at 17 in the WHA, sensed he had watched the young pro make a litany of miscues over the years and deemed the blueliner with mobility expendable. Kelley's gaff to send Roberts, who went on to play more than 1,000 NHL games, in a 1-for-1 deal with Minnesota for forward Mike Fidler, ranks among the club's worst-ever swaps. Cut adrift on April 1, 1981 when Pleau was installed in the dual capacity of coach and general manager, Kelley has remained in hockey. He served an 11-year hitch as overseer of the Glen Falls Civic Center in New York where the Adirondack Red Wings operate. In 1993, Kelley took a front office post with the Pittsburgh Penguins.

Kevin Kemp

A defenseman drafted by Toronto 138th overall in 1974, Kemp was taken by the Whalers in the 1979 Expansion Draft. He picked up two minor penalties in three games with Hartford in 1980-81, his only time in the NHL.

Dave Keon

Elected to the Hall of Fame in 1986, this classy professional wrapped up his 22-year career in hockey with Hartford in 1981-82. Keon broke in with Toronto in 1960-61 and captured the league's Calder Trophy as the NHL's top rookie. In 1962 and 1963, Keon captured the league's Lady Byng Trophy for sportsmanship. In 1967 when the Maple Leafs won their last Stanley Cup, Keon gained the Conn Smythe Trophy for his playoff performance. With Toronto, Keon played on four Cup winners and remained with the Leafs through 19974-75 as the squad's top centerman until a contract dispute led to his exit. Moved on to the WHA and signed wtih the Minnesota Fighting Saints. He also played with Indianapolis before coming to New England during the 1976-77 season when the Saints disbanded for a second time. Won the WHA's Most Gentlemanly Player Trophy in 1977 and 1978. Played 384

*One of hockey's greatest two-way players, **Dave Keon**. Voted to the hockey Hall of Fame in 1986, Keon captained the Whalers in 1981-82, his last pro season. He was often called "the perfect hockey player."*

games for the Whalers in both leagues. Often called the "Perfect Hockey Player." Captained the Whalers in 1981-82 which was his last NHL season. Announced his retirement on June 30, 1982, ending a three-decade career that summed 1,597 games. Ranks among all-time NHL leaders in games played and career points. Most memorable night with the Whalers came on October 31, 1979 in his first return to Maple Leaf Gardens since Harold Ballard deemed the club's former captain "washed up" five years earlier. In a 4-2 win by Hartford, Keon scored a goal as did Gordie Howe, lamplighters that drew standing ovations from the home crowd.

Three of hockey's greatest played for the Whalers during their first NHL season. The date is February 29, 1980 and Gordie Howe and Dave Keon welcome Bobby Hull to Hartford. Behind Hull wearing the goalie mask is John Garrett.

Forward Derek King joined the Whalers late in the 1996-97 season in a deal with the New York Islanders. He managed just 3-3-6 in a dozen games but his first helper with Hartford was a personal milestone, the 500th point of his career.

Tim Kerr

A four-time, 50-goal scorer, Kerr wrapped up his 13-year, 655-game career with the Whalers in 1992-93, notching just six assists in 22 games before retiring on January 21, 1993. Overlooked in 1979 when the Entry Draft consisted of just six rounds, Kerr signed with the Flyers and tallied the bulk of his 370 goals and 304 assists by parking his 6-foot-3, 230-pound frame in the low slot. Dynamic on the power play, Kerr scored a league-record 34 power play goals in 1985-86. A series of knee and shoulder injuries curtailed Kerr's effectiveness in later years, limiting him to just 32 games with the New York Rangers in 1991-92. Joined the Whalers on July 8, 1992 in a deal for future considerations. Ranks among the NHL's Top 100 career goal scorers and among the Top 10 in goals-per-game average (.565). Won the Masterton Trophy in 1988-89 for perseverance, sportsmanship and dedication to hockey.

Derek King

A veteran of 10 years in the NHL, King came to the Whalers on March 18, 1997 from the New York Islanders for a fifth-round draft pick in the 1997 Entry Draft. Taken 13th overall by the Islanders in 1985, King has played in 650 NHL games and has tallied over 200 goals during his career. King's best production came in 1991-92 when he reach the 40-goal marker, the highwater of a three-year stretch where he tallied 108 goals and averaged 75 points per season. His first point as a Whaler turned out to be a career milestone. In a 4-1 loss at St. Louis on March 20, 1996, King's helper on a goal by Keith Primeau was his 500th NHL point. Signed as a free agent with Toronto for 1997-98.

Larry Kish

In what was his only chance at coaching in the NHL, Larry Kish was behind the Hartford bench for the opening 49 games of 1982-83. Able to churn out consistent and successful clubs at the minor league level, Kish seemed poised to parlay "Coach of the Year" honors in the American League to the next level. "I don't have all the answers but if I have a strength, it's organizing," Kish said on June 2, 1982, the day he was hired. "You start slowly and hopefully, things start to surface." In the coming weeks, another chapter of bad hockey resurfaced with the Whalers and Kish was soon back in the bus leagues. Losing early and often, Hartford got off to a 5-16-3 start and eventually docked home with a 54-loss season, the most ever in franchise annals. Fired following a 4-2 loss in Chicago on January 23, 1983, the Whalers were a woeful 12-32-5 under Kish. His situation was best summed up by defenseman Ed Hospodar who remarked, "it's amazing how fast it can end. Larry spent 18 years trying to get to the top (the NHL) and it was all over in 49 games. I don't think it's ever one man's fault because there are a lot of circumstances involved."

Larry Kish took the coaching reins in 1982-83. In what was the darkest winter in Hartford as the Whalers lost a franchise-record 54 games, Kish got the gate 49 games into the campaign. With just 10 wins in the first 31 games, the Whalers took on water heavily, going 2-15-1 from December 18, 1982 to January 23, 1983. The slump resulted in Kish walking the plank.

Scot Kleinendorst

A defenseman obtained from the New York Rangers on February 27, 1984 for Blaine Stoughton, Kleinendorst played parts of five seasons in Hartford, going 9-30-39 in 186 games. Bothered by leg injuries throughout his career, Kleinendorst had his best year in 1986-87 with the Whalers with 3-9-12 in 66 games. In the playoffs against Quebec, he figured keenly in a 5-4 win in Game 2 on April 9, 1987 with a goal and three helpers. Drafted 98th overall by the Rangers out of Providence College in the 1980 Entry Draft, Kleinendorst played 281 career NHL games. He finished up with Washington when Hartford sent the blueliner to the Capitals on March 6, 1989 for winger Jim Thomson.

Steve Konroyd

A steady defenseman throughout a 15-year career that consisted of 895 games, Konroyd played 92 games over two seasons in Hartford before getting shipped to Detroit on March 22, 1993 for a sixth-round draft choice which was later returned to the Red Wings. Had 5-21-26 with the Whalers after coming over from Chicago on January 24, 1992 in a deal for Rob Brown. Originally tabbed 39th overall by Calgary in the 1980 Entry Draft, Konroyd also played for the New York Islanders and Ottawa Senators. Appeared in 97 Stanley Cup games over 12 consecutive years, a string which ended when Detroit dished him to Ottawa for goalie Daniel Berthiaume on March 21, 1994.

Chris Kotsopoulos played 241 games at the blue line over four years with the Whalers. His best season was 1981-82, going 13-20-33 in 68 games. A contract dispute resulted in a deal to Toronto on October 7, 1985 which brought Stewart Gavin to Hartford.

Chris Kotsopoulos

Hartford obtained "Kotsy" in a three-for-one deal with the Rangers on October 2, 1981 when the Whalers sent Mike Rogers to New York in a trade that also involved Doug Sulliman and Gerry McDonald. Played 241 games with the Whalers, enjoying his best year in 1981-82 with 13-20-33. Possessed a good shot from the point during the power play where he scored 12 of his 29 goals as a Whaler. His tenure in Hartford ended in a contract dispute where Kotsopoulos vowed to go on "coach patrol" in hopes to force management's hand. General manager Emile Francis accommodated the burly defenseman by cutting a deal with Toronto on October 7, 1985 that fetched Stewart Gavin from the Maple Leafs. Kotsopoulos also played for the Red Wings in 1989-90 which was his last year in the NHL, a 10-year stretch of of 479 games with 44-109-153.

*Wayne Gretzky caused plenty of turnovers by blueliners over the years but on this particular play, **Chris Kotsopolous** of the Whalers is able to impede No. 99 from doing some damage in the Hartford zone. "Kotsy" had one of the best shots from the point, low and usually on the net on the power play.*

Robert Kron

A speedy winger from the Czech Republic, Kron came to the Whalers on March 22, 1993 in a deal with Vancouver for Murray Craven and future considerations (Hartford later received Jim Sandlak from the Canucks). The change of address brought out the best in a gritty winger who has turned into a valuable contributor, often scoring big goals. Netted a career-high 24 goals in 1993-94 and followed up with 20 in 1995-96. Tabbed by Vancouver as the 88th overall pick in the 1985 Entry Draft, Kron crossed the 400-game milestone during 1996-97.

Todd Krygier

Few selected in the Supplemental Draft have ever become NHL regulars. Krygier is one who has, a remarkable progression from a Division III college program (Connecticut). Scored 18 goals as a rookie with

Todd Krygier has position here against Boston's Stephan Quintal and does his best to protect the puck. Krygier, selected out of Division III UConn by the Whalers in the 1988 Supplemental Draft, launched what has become a 500-plus game career with Hartford in 1989-90, scoring 18 goals.

the Whalers in 1989-90, the start of what has become a career of better than 500 games. Good speed and anticipation has enabled Krygier to find regular work in stints with Washington and Anaheim. The turning point in his career was learning to play without the puck. Dealt to the Capitals on October 3, 1991 for a fourth-round pick.

Frantisek Kucera

A lanky defenseman from the Czech Republic, Kucera played parts of three seasons with the Whalers and holds the distinction of being involved in two multi-player swaps, one coming to Hartford and a second one leaving the capital city. On March 11, 1984, Kucera came aboard in a deal with Chicago along with Jocelyn Lemieux for Gary Suter, Randy Cunneyworth and a third-round draft pick (later transferred to Vancouver which selected Larry Courville). Kucera played in 48 games during 1994-95, enjoying his best pro season with 3-17-20 during the shortened campaign. Though he began 1995-96 with the Whalers, Kucera was parceled to Vancouver in a three-team transaction that also involved New Jersey (Jim Dowd and draft picks), a deal which brought Jeff Brown to Hartford. Has played in over 350 NHL games after being drafted 77th overall in the 1986 Entry Draft by the Blackhawks.

Robert Kron joined the Whalers in a trade with Vancouver on March 20, 1993 that involved Murray Craven. Kron had a couple of 20-goal campaigns with Hartford including a career-high 24 in 1993-94.

Nick Kypreos

A rugged forward who led the Whalers in penalty minutes (325) in 1992-93, Kypreos notched 17 goals in what was his only full year with Hartford. Played in 10 games the following season before getting dealt to the New York Rangers in a swap that brought James Patrick and Darren Turcotte to the Whalers on November 2, 1993. Wound up playing for the Stanley Cup champions in 1994 when the Rangers ended their 54-year drought. Has also played for Toronto and Washington. A free agent who signed with Philadelphia in 1984, the Capitals claimed Kypreos in the 1989 Waiver Draft which launched what has been a career of close to 500 games. Joined the Whalers on June 15, 1992 in a deal for Mark Hunter.

Frantisek Kucera, a defenseman with a big reach, played 86 games over parts of three winters with Hartford. His most consistent stretch was in the shortened season of 1995 when the Czech Republic native led the blueliners in scoring with 3-17-20 in 48 games.

Andre Lacroix finished as the WHA's all-time scoring leader and wrapped up his extensive hockey career with the Whalers. After scoring 32 goals for New England in 1978-79 to finish among the club leaders, Lacroix played in 29 games for the Whalers in their first NHL season before deciding to retire. For many years, he assisted Chuck Kaiton in the radio booth with expert analysis.

Defenseman Randy Ladouceur came to Hartford on January 14, 1987 in a swap with Detroit for Dave Barr. "Laddy" played 377 games for the Whalers, a reliable member of the blue line corps. He was among the players lost in the 1993 Expansion Draft, taken by the Mighty Ducks of Anaheim.

*Goaltender **Mike Liut** gets help here from defenseman **Randy Ladouceur** to rob Philadelphia's Rick Tocchet from close range. Liut had a 3.37 GAA over 252 appearances during a five-year hitch with the Whalers.*

L

Andre Lacroix

A playmaking center who enjoyed his best days in the WHA, Lacroix played for the Whalers in both leagues, completing his seven-year WHA career with New England in 1978-79 with 32-56-88 in 78 games to wind up as the league's all-time scoring leader with 798 points. Played in 29 games with Hartford in the club's first NHL campaign in 1979-80 to wrap up a 325-game tenure which began with stints in Philadelphia and Chicago. Jumped to the WHA when the new league began in 1972-73 and was the first WHA scoring champion with 124 points. Started with the Philadelphia Blazers, then played for the New York Golden Blades, Jersey Knights, San Diego Mariners and Houston Aeros before coming to the Whalers. Voted the top Canadian Athlete in 1975. Led the WHA in scoring in two of its first three campaigns including a record 106 assists in 1974-75 when amassing 147 points.

Pierre Lacroix

A defenseman who led the Whalers in scoring at the blue line with 5-31-36 in 1982-83, Lacroix's four-year NHL career ended during the summer of 1983 when he sustained a severe chest injury in a car accident. Selected by Quebec as the 104th overall pick in the 1979 Entry Draft, Lacroix was voted CCM's Top Canadian Junior Player of the Year after racking up 137 points for the Trois-Rivieres Draveurs. Came to the Whalers in a deal with the Nordiques on December 3, 1982 for Blake Wesley. Had 24-108-132 in 274 NHL games.

Randy Ladouceur

One of a handful to play and later serve in a coaching capacity for the Whalers, Ladouceur played 930 NHL games with 30-126-156 career totals. Noted for his positional play at the blue line, Ladouceur signed with Detroit as a free agent in 1979. He eventually broke in with the Red Wings in 1982-83 before coming to Hartford on January 12, 1987 for forward Dave Barr. Steady and reliable summed up his playing style which included 377 games with the Whalers. Seemed to take his game up a notch in the playoffs, particularly against the Boston Bruins where he had 2-4-6 in 13 games. In the first two playoff series between the teams, Ladouceur is credited with the game-winning goal in the first ever Stanley Cup

Bruce Landon, who played five years with the Whalers in the WHA, adjusts a contact lens. These days, the former netminder operates the Springfield franchise in the American League.

game between the New England rivals, a 4-3 decision on April 5, 1990. Served as captain of the Whalers in 1991-92. Claimed by Anaheim in the Expansion Draft on June 24, 1993. Captained the Mighty Ducks for two seasons before winding up his 14-year NHL career in 1995-96. Returned to the Whalers in 1996-97 as an assistant coach to work with the defense.

Marc Laforge

The Whalers took Laforge with the 32th overall pick in the 1986 Entry Draft. A rugged player in juniors where he twice put up over 200 penalty minutes with Kingston, Laforge logged nine games in Hartford where he picked up 43 penalty minutes. Moved on to Edmonton in a deal for Cam Brauer on March 6, 1990. Played five games with the Oilers in 1993-94, his last chance in the NHL.

Bruce Landon

The backup goaltender is a difficult role because he usually plays the odd game or must be ready at a moment's notice when the starter gets injured. During his five-year career with the WHA Whalers, Landon filled the No. 2 netminder slot and proved capable. He played in 122 games with a respectable 3.46 GAA. Left the Los Angeles Kings organization to join the Whalers. In 1972-73, Landon played in 26 games as Al Smith's backup during New England's run to the first-ever AVCO Trophy. Played in a career-high 38 games in 1975-76. Settled in western Massachusetts following his playing days and has been involved in minor league hockey for many years, most recently as co-owner of the Springfield Falcons of the American League.

Pierre Larouche

"Lucky" played 83 games over two seasons with the Whalers, producing 43-47-90. Hartford was one of several stops for Larouche who played 14 seasons in the NHL. Among a short list to score 50 goals for two different clubs (53 with Pittsburgh in 1975-76 and 50 with Montreal in 1979-80), Larouche was the eighth overall pick in the 1974 Entry Draft by the Penguins after he scored a then-record 251 points in 67 games for the Sorel Black Hawks. Dealt to Montreal for Pete Mahovlich and Peter Lee on November 29, 1977, Larouche had plenty of explosive nights at the Forum for the Habs where he played for two Stanley Cup champions (1978 and 1979). Fell into

*A number of Whalers wound up playing for the New York Rangers and here, **Blaine Stoughton**, left, and **Pierre Larouche**, right, chuckle following Stoughton's acquisition by the Blueshirts for Scot Kleinendorst. Stoughton joined New York on February 27, 1984, a few months after Larouche, who had 43 goals over two years in 83 games with Hartford, was dropped in training camp by general manager **Emile Francis**. That winter, Larouche struck for 48 goals.*

*Taken first overall in the 1983 Entry Draft by Minnesota, **Brian Lawton** made one of his many stops around the NHL with the Whalers and played 48 games over two seasons. Clever with the puck, Lawton had a dozen goals and 17 assists with Hartford. Today, he is a sports agent.*

disfavor with management and was shipped to Hartford on December 21, 1981 in a flip-flop of draft choices. Supplied instant offense in Hartford before a back injury curtailed his effectiveness and availability. Released by the Whalers once Emile Francis took charge of the personnel decisions, Larouche signed with the New York Rangers and scored 48 goals in 1983-84, enjoying one of his best seasons as he was New York's All-Star Game representative. Wrapped up his 812-game career in 1987-88 with 395-427-822 career points to rank among hockey's all-time Top 100 in goals and points.

Jack Lautier

A journalist who began chronicling hockey when the New England Whalers moved to Hartford in the 1970s, Lautier chased the ups-and-downs of the Whalers for close to two decades. He has contributed to virtually every publication that covers the world's fastest game. His by-line has appeared in a number of Connecticut newspapers over the years, chiefly with *The Bristol Press*. *Forever Whalers* is his fourth book on hockey. The others are *15 Years of Whalers Hockey, Whalers Trivia Compendium* and *Same Game, Different Name, A History of the World Hockey Association.* The latter was co-authored with Frank Polnaszek.

Paul Lawless

Taken 14th overall in 1982, this speedy left winger notched 39 goals over two seasons during Hartford's best years in the 1980s. Iin 1986-87, Lawless had his best year with 22-32-54 including a 2-4-6 performance in an 8-3 rout of Toronto on January 4, 1987 when he became the first Hartford player to score six points in a game. Traded to Philadelphia on January 22, 1988 for Lindsay Carson. Later moved on to Vancouver and Toronto where he finished his career, winding up with 49-77-126 in 239 NHL games.

Brian Lawton

The first American-born player picked as the top choice in the Entry Draft (in 1983 by the Minnesota North Stars), Lawton played nine seasons and 483 games in the NHL, notching 112-154-266 from 1983-84 to 1992-93. Besides a 48-game stint over two seasons with the Whalers, Lawton also played for the New York Rangers, Quebec Nordiques, Boston Bruins and San Jose Sharks. Came to Hartford with Don Maloney and Norm Maciver on December 26, 1988 for Carey Wilson and a fifth-round draft choice (the Rangers picked Lubos Rob in 1990). Claimed by Quebec on waivers on December 1, 1989. Drifted from team to team and never really reached projected expectations that profile first-round draftees usually deliver. Currently a sport agent.

Jamie Leach

Drafted 47th overall by Pittsburgh in 1987, the son of Reggie Leach, a feared sniper during his career with the Flyers, never really carved his own niche in the NHL. Had a handful of trials including a 19-game stop with Hartford in 1992-93 after getting claimed on waivers on November 21, 1992. Played his final two NHL games in 1993-94 with Florida for a total of 81 career games, with 11-9-20.

Jocelyn Lemieux

Tabbed 10th overall by St. Louis in the 1986 Entry Draft, "Jocko" also made stops in Montreal and Chicago before coming to Hartford on March 11, 1994 in a multi-player swap with the Blackhawks that also involved Frantisek Kucera, Gary Suter, Randy Cunneyworth and a draft choice. A punishing force in the corners who was voted the Fan's Favorite player in 1994-95, Lemieux reached the 500-game NHL milestone of his career in Hartford early in the 1995-96 campaign before moving on. He was traded in a three-way deal involving Vancouver and New Jersey on December 19, 1995, an exchange that brought defenseman Jeff Brown to the Whalers. Played 18 games with the Devils before another shuttle sent him to Calgary for the final 20 games of 1995-96.

Mike Lenarduzzi

One of the many "goalies of the future" in club annals, Lenarduzzi caught the organization's attention with a 19-8-3 mark and a 3.27 GAA in Canadian juniors during 1990-91, a season after the Whalers picked him 57th overall in 1990 Entry Draft. Like a litany of others who just passed through, Lenarduzzi made just a cameo in the NHL,

Jocelyn "Jocko" Lemieux was voted the Fan's Favorite Players in 1994-95. His feisty play and punishing checks along the wall needled rivals and drew cheers from the faithful. The winger had 13-8-21 in 86 games.

*Several players scored a goal in their first game as a Whaler over the years but no one did as dramatically as defenseman **Curtis Leschyshyn** did on November 9, 1996. Hours after being acquired from Washington for center **Andrei Nikolishin**, Leschyshyn took a pass from **Geoff Sanderson** and scored from the slot at 4:55 in overtime to nip Buffalo 4-3. Leshchyshyn finished with 4-13-17 in 64 games in his only season with the Whalers.*

playing four games in Hartford over two seasons. His last appearance was in relief of Mario Gosselin in 1993-94, allowing one goal in the 21 minutes of a 6-3 loss at Philadelphia on November 18, 1994.

Louis Levasseur

In his only year with the Whalers, Levasseur went 14-11-2 with a 3.30 GAA and three shutouts, playing his best in the 1978 WHA playoffs (8-4 with a 2.59 GAA) in sparking a run to the AVCO Trophy Finals. Levasseur finished his 85-game WHA career with Quebec in 1978-79. Levasseur came to the Whalers on June 16, 1977 in a

*Often called the "heart and soul" of the Whalers as a player and one of three players in club annals to have his uniform jersey retired, **Rick Ley** played 559 games for the franchise in both the WHA and NHL. A seven-year veteran in the WHA, Ley served as captain of the hockey club in both leagues. On defense, he could play it rough. On June 7, 1989, Ley was introduced as coach of the club and in 1989-90, he directed Hartford to 38 wins en route to the league's seventh best record. Ley lasted only two seasons behind the bench yet had the best winning percentage (.494) among anyone to coach the Whalers in the NHL.*

trade with Edmonton that sent the contract rights to veteran goaltender Dave Dryden to the Oilers. Though he was with the Whalers for just one winter, Levasseur had to appreciate the stability of getting a paycheck considering the rocky tenure he had with the Minnesota Fighting Saints, a team which ceased operations in consecutive years when management abruptly called it quits before completing the 81-game schedule. Between the pipes, Levasseur was a remarkable story in 1976-77. At midseason, he compiled the league's best GAA (2.73) and was voted to the league's All-Star Game. The problem was that he had no team to represent when the Saints disbanded. That dilemma did not stop Levasseur from dipping into his savings for plane fare to Hartford. He was voted MVP in a 4-2 win by the East in a 4-2 win over the West on January 18, 1977. Levasseur resurfaced to again play in Minnesota with the NHL North Stars. In his only start on February 24, 1980, he dropped a 7-5 decision at Detroit.

Curtis Leschyshyn

Taken third overall (behind Mike Modano by Minnesota and Trevor Linden by Vancouver) in the 1988 Entry Draft by the Quebec Nordiques, Leschyshyn remains an effective, "make the safe play" type defenseman. He came to the Whalers on November 8, 1996 from Washington in a one-for-one swap involving center Andrei Nikolishin. In his first game with Hartford, Leschyshyn won the hearts of the Civic Center faithful when he scored a dramatic goal in the closing seconds of overtime to beat Buffalo 4-3 on November 9, 1996. Has played over 500 NHL games and was a member of Colorado's Stanley Cup champions in 1996.

*Defenseman **Rick Ley**, who captained the Whalers in both leagues, keeps his position against Birmingham's Frank Maholvich in front of goaltender **Al Smith** during a WHA game in 1977-78.*

Rick Ley

One of three Whalers to have his jersey number retired (No. 2) and considered the best defenseman to ever play for the franchise, "Pluggy" holds a number of distinctions in club annals including wearing the "C" of the team in both the WHA and NHL and having the best winning percentage behind the Hartford bench in the NHL. Taken by Toronto as the 16th overall pick in 1966, Ley spent four seasons with the Maple Leafs then jumped to the WHA along with teammates Brad Selwood and Jim Dorey to form the cadre at the New England blue line along with Ted Green. Spent seven years in the WHA and was voted to a number of All-Star Teams. A member of the WHA All-Stars who played against the Soviet Union in 1974, Ley got into an altercation with Valeri Kharlamov moments after a game and pummelled the Russian legend. Ley later apologized for his actions. Had 35-210-245 in 478 WHA games and 12-72-84 in 310 NHL games during his career. His best season was 1977-78 with 3-41-44. Also played in 73 AVCO Trophy playoff games and 14 Stanley Cup games. A mainstay in Hartford's first NHL season (1979-80), Ley's playing career prematurely closed because of a knee injury. That injury came at a practice hours before the Whalers were to play their first-ever Stanley Cup game on April 9, 1980 against Montreal. Ley made a gallant recovery and did dress for 16 games in 1980-81 but soon announced his retirement. Hartford honored its long-time leader when the Whalers raised his number to the rafters on December 26, 1982. After a number of successful years coaching in the minors, Ley was hired to pilot the Whalers on June 7, 1989. His best year was 1989-90 when Hartford won 38 games en route to the seventh-best overall mark in the league. Stressed "playing without the puck" which seemed to bring out the best of the Whalers during a tenure marred by tirades directed at certain players through the media. Later coached the Vancouver Canucks before getting the gate.

*Goalie **Mike Liut** had his best year with the Whalers in 1986-87 when he won 31 games in sparking Hartford to the Adams Division title.*

Mike Liut

A big-play goaltender who helped turned the franchise into a contender in the 1980s, Liut's arrival from St. Louis on February 22, 1984 for Greg Millen and Mark Johnson revived the Whalers into what may have been Emile Francis' best trade during his tenure as general manger in Hartford. Liut, originally drafted by the WHA Whalers, was sent to Cincinnati on May 26, 1977 for Bryan Maxwell and Greg Carroll. Though his draft rights were also owned by St. Louis of the NHL, Liut opted for the WHA after his collegiate career at Bowling Green. With the Stingers, Liut played in 81 WHA games and led the

The Nordiques seemed to have too many Stastnys (Peter, Anton and Marian) against the Whalers in their yearly battles. Here, Peter Stastny finishes a breakaway in Game 6 of the Stanley Cup playoffs in 1987 against **Mike Liut** *to oust Hartford in overtime.*

league in shutouts (3) in 1978-79. Reclaimed by the Blues in the 1979 Reclaimation Draft, Liut anchored the club's revival which included a 107-point season in 1980-81 when he was voted a first-team NHL All-Star. Hartford fans witnessed a similar run in 1986-87 when Liut won 31 games as the Whalers copped the Adams Division flag with a franchise-high 93 points. Played in 252 games for the Whalers before moving to Washington on March 6, 1990 for Yvon Corriveau. Holds a number of club marks including the first netminder to post a shutout in Stanley Cup competition (a 1-0 decision over Montreal on April 27, 1986).

Mike Liut *was usually at his best in games against Adams Division clubs like the Boston Bruins and here he plays the angle to stop Bobby Carpenter. Liut posted 31 victories in 1986-87 to anchor Hartford's Adams Division title club.*

George Lyle played for the Whalers in both leagues, a total of 232 games with 92 goals and 185 points. The big left winger broke into the WHA in 1975-76 with 39 goals to capture the league's Rookie of the Year honors. Though claimed by Detroit when the leagues merged in 1979, Lyle later returned to the Whalers to play 30 games over two seasons, going 6-18-24 before suffering an eye injury during a practice which ended his career.

Chuck Luksa

A defenseman who played eight games in Hartford's first NHL season, Luksa was originally drafted by Montreal in 1974 (172nd overall) and spent four years in the American League at Nova Scotia. Unable to crack the parent club, Luksa signed on with the Cincinnati Stingers for 1978-79 and had 8-12-20 in his only WHA campaign. Signed by Hartford on August 3, 1979, Luksa had one assist for the Whalers in their first-ever NHL win, a 6-3 decision over Los Angeles on October 19, 1979.

Dave Lumley

A four-year standout at the University of New Hampshire, Lumley enjoyed his best years with the Edmonton Oilers. Originally Montreal's 12th round pick (199th overall) in 1974, Lumley appeared in three games with the Habs in 1978-79 before getting dealt with Dan Newman on June 13, 1979 to the Oilers where he eventually played for two Stanley Cup winners. Developed into a solid, two-way forward for the Oilers and wound up with the Whalers in the 1984 Waiver Draft. Played 48 games in Hartford during 1984-85 before returning to the Oilers later in the campaign. Lumley concluded his NHL career in Edmonton in 1986-87, a run of 437 games with totals of 98-160-258.

Gilles Lupien

One of the tallest in club annals to patrol the blue line for the Whalers, the 6-foot-6 Lupien had a remarkable debut with Hartford. In his first game on February 22, 1981, Lupien potted a pair of goals to help the Whalers nip the New York Rangers 6-5. The goals came in a historic occasion for hockey since it marked the first time in NHL annals that two American-born coaches (Larry Pleau of the

Whalers and Craig Patrick of the Rangers) ever matched lines in a game. Lupien wound up playing 21 games over two seasons in Hartford to conclude his 226-game career. Originally selected by Montreal (33rd overall in 1974), Lupien was a member of two Stanley Cup winners with the Canadiens. Dealt on September 26, 1980 to Pittsburgh for a future draft choice, Lupien came to the Whalers on February 20, 1981 from the Penguins for a sixth-round 1981 draft choice (Paul Edwards). Had 5-25-30 with 416 penalty minutes during his career. Currenlty a sports agent.

George Lyle

A winger with a good touch around the net who capped an impressive collegiate career at Michigan Tech with 47 goals in 1975-76, Lyle broke into the WHA with much fanfare with the Whalers in 1976-77 as he went 39-33-72 in 75 games to win Rookie of the Year honors. Followed up with a 30-goal campaign and notched 17 goals in his final WHA season. Claimed by the Red Wings in the Expansion Draft on June 19, 1979 (he was originally drafted by Detroit as the 123rd overall pick in 1973), a knee injury limited Lyle's effectiveness in the Motor City. He eventually returned to Hartford via waivers on November 11, 1981. Played 30 games over two seasons with the Whalers in the NHL but his career suddenly ended when he took a high stick in the face during a practice session and sustained an eye injury.

*An honest player who rattled the foes with his "avalanche checks" along the wall, **Paul MacDermid** often chipped in with a key goal or marked a rival scorer during his 373-game tenure with the Whalers. Here, MacDermid reacts after Hartford slipped one past Boston netminder Andy Moog.*

M

Paul MacDermid

In a career that fell 10 games shy of 700, MacDermid was an honest hockey player who was good along the wall, able to chip in with a memorable goal and often punished rivals with his patented "avalanche checks." MacDermid broke in with the Whalers in 1981-82 and soon became a dependable, two-way forward, logging 373 games in Hartford before going to Winnipeg on December 31, 1989 for Randy Cunneyworth. A third round (61st overall) selection in the 1981 Entry Draft, MacDermid tallied his first NHL goal on April 3, 1982 in a 3-3 tie against the New York Rangers. Assisting on the tally was Davey Keon, a helper which was the Hall of Famer's final career point. Also played for Washington and Quebec. Best year was 20 goals in 1987-88 with the Whalers. Notched a pair of Stanley Cup game-winning goals for Hartford against Quebec, a 4-1 win in Game 2 on April 10, 1986 and a 3-2 overtime decision in Game 1 on April 8, 1987.

Gary MacGregor

A swift centerman who played 251 games over five years, MacGregor spent 30 games with the WHA Whalers in 1976-77 after arriving on November 12, 1976 from Indianapolis for Rosaire Paiement. Became expendable when Dave Keon, John McKenzie and the Carlson brothers arrived with the folding of the Minnesota Fighting Saints and was shipped to the Calgary Cowboys on January 19, 1977. Also played for Chicago, Denver-Ottawa and Edmonton in the WHA. Died from leukemia complications in 1995.

Randy MacGregor

A forward who put up some impressive numbers in the minor leagues, MacGregor played two games in the NHL with the Whalers in 1981-82. Had a goal and an assist in a 5-2 win over Philadelphia on March 20, 1982.

Norm Maciver

A veteran of over 400 NHL games, Maciver patrolled the blue line for 37 games in Hartford in 1988-89 after coming over from the New York Rangers with Don Maloney and Brian Lawton for Carey Wilson on December 26, 1988. Dealt to Edmonton on October 8, 1989 for Jim Ennis, Maciver has also played for Ottawa, Pittsburgh, Winnipeg and Phoenix where he has proven to be a valuable pointman on the power play. Best year was 1992-93 when he went 17-46-63 in 80 games for the expansion Senators. Never drafted but signed with Rangers after solid four-year collegiate career at Minnesota-Duluth.

Rick MacLeish

A standout on two Stanley Cup winners in Philadelphia (1974 and 1975), MacLeish was near the end of his 846-game NHL career when he came to Hartford on July 3, 1981 with Don Gillen and Blake Wesley for Ray Allison and Fred Arthur, Hartford's top selections in the 1979 and 1980 Entry Drafts, and a flip-flop of 1982 first-round picks. MacLeish went 6-16-22 in 34 games before departing for Pittsburgh for Russ Anderson and an eighth-round pick (Hartford picked Chris Duperron in 1983) on December 29, 1981. Finished up with Detroit in 1983-84 to wind up with 349-410-759 career totals. Originally drafted fourth overall by Boston in the 1970 Entry Draft, MacLeish was sent to the Flyers on February 1, 1971 with Danny Shock for Mike Walton. Potted 50 goals in 1972-73 for Philadelphia where he often finished passes from Bobby Clarke.

Marek Malik

A 6-foot-5 defenseman from the Czech Republic with a good reach, the Whalers drafted Malik on the third round (72nd overall) in 1993. Malik had brief trials with Hartford in 1994-95 and 1995-96 before getting into the mix at the blue line during 1996-97. Made his first NHL goal count as the game-winner in a 6-2 decision over New Jersey on October 19, 1996.

Merlin Malinowski

The "Magician" came to the Whalers on October 15, 1982 from the New Jersey Devils along with the rights to collegian Scott Fusco for Rick Meagher and Garry Howatt. Had 5-23-28 in 75 games at center, the final hitch of a 282-game NHL career where he tallied 54-111-165. Best year was 25-37-62 in 69 games with the Colorado Rockies in 1980-81. Originally drafted 27th overall by the Rockies in 1978.

Greg Malone

A veteran center who played 704 NHL games over 11 seasons, Malone checked into Hartford on September 30, 1983 in a deal with Pittsburgh for a third-round pick in 1985. Supplied consistent two-way play in the pivot for close to three seasons before getting dealt to Quebec on January 17, 1986 for Wayne Babych. Taken 19th overall by the Penguins in the 1976 Entry Draft, Malone notched a career-high 35 goals in 1978-79. Had four game-winners to lead the Whalers in 1984-85 with the highlight being a five-point game on November 22, 1984 in a 9-3 rout of his former club. Completed his tenure with the Nordiques in 1986-87 with career totals of 190-310-501.

Don Maloney

A good cornerman during his career with the New York Rangers, Maloney played just 21 games of his 765-game tenure with Hartford in 1988-89 once coming over from the Blueshirts in a multi-player swap with Norm Maciver and Brian Lawton for Carey Wilson on December 26, 1988. A shoulder injury limited his contributions with the Whalers to 3-11-14 in his only season. Signed on with the New York Islanders for 1990-91, his 13th and final NHL season to wind up with 214-350-564. Also played in 94

*One of the taller rear guards in team annals, 6-foot-5 **Marek Malik** had brief trials with the Whalers over three seasons. The Czech Republic native played 47 of his 55 games for Hartford in 1996-97. His only goal proved to be the gamer in a 6-2 win over New Jersey on October 19, 1996.*

Stanley Cup games. Eventually served a term as general manager of the Islanders (August 17, 1992 to December 14, 1995). Originally drafted 26th overall by the Rangers in 1978, Maloney is currently in management with the Rangers. He will oversee operations of the Hartford Wolf Pack, New York's farm club in the American League. With the departure of the NHL Whalers, the Rangers and Madison Square Garden officials cut a deal with the Connecticut Development Authority to relocate their affiliate from Binghamton, N.Y. to Hartford Civic Center.

Kent Manderville

A big forward who had 137 NHL games experience over five seasons, Manderville won a job in training camp for 1996-97 with his spirited play after signing a free agent deal with Hartford. With the Whalers, Manderville began the season in the minors and soon regained the playing style that resulted in another chance. He emerged as one of the better comeback stories on the Hartford squad during 1996-97. His first three-goal game was also the last by a Whaler, a hat trick on March 12, 1997 in a 6-3 wipeout of Boston. Drafted 24th overall (in 1989) by the Calgary Flames, Manderville moved to Toronto in a blockbuster

Kent Manderville is credited with the last hat trick in Whalers history. The three-goal game came on March 12, 1997 in a 6-3 pasting of the Boston Bruins. The centerman wound up with 6-5-11 in 44 games with Hartford.

Defenseman Bryan Marchment played it tough with a snarl on the ice but he wound up having a short stay in Hartford. The Whalers lost Marchment to the Edmonton Oilers as compensation for signing free agent winger Steven Rice.

swap on January 2, 1992 that involved 10 players including Doug Gilmour. Spent 1992 with the Canadian Olympic Team and emerged as a valuable checker for the Maple Leafs who reached the Western Conference Finals in 1993. Moved on to Edmonton on December 4, 1995 and sustained a wrist injury that set back his career.

Bryan Marchment

Taken 16th overall in 1987 by Winnipeg, this hard-hitting defenseman has been a regular in the NHL since 1991-92 once dealt to Chicago for Troy Murray and Warren Rychel on July 22, 1991. Came to the Whalers from the Blackhawks on November 2, 1993 with Steve Larmer for Eric Weinrich and Patrick Poulin. Played with passion and a mean streak at the blue line but his stay in Hartford was limited to just 42 games when he was lost as compensation to Edmonton when the Whalers signed free agent forward Steven Rice.

Paul Marshall

A forward who was drafted 31st overall by Pittsburgh in 1979, Marshall completed his 95-game NHL tenure in 1982-83 with Hartford, going 1-2-3 in 13 games after coming to the Whalers on October 5, 1982 for a 10th-round pick (Greg Rolston) in a deal with Toronto. The Maple Leafs dealt Dave Burrows and Paul Gardner to the

Penguins on November 11, 1980 to obtain Marshall and Kim Davis. Marshall had 9-12-21 in 46 games during his rookie year in 1979-80 before a shoulder injury ended his season and diminished his effectiveness.

Tom Martin

A physical presence at left wing, Martin was the 74th overall pick in the 1982 Entry Draft by Winnipeg. Martin played for three different clubs in the NHL, a total of 92 games over six years, the bulk coming in three different tours with Hartford. Signed by the Whalers as a free agent on July 29, 1987, Martin was taken in the 1988 Waiver Draft by Minnesota. A short time later, Hartford reclaimed the winger (via waivers on December 1, 1988) and he put together the best stretch of hockey of his career, going 7-6-13 in 38 games with 113 penalty minutes. The highlight was a game-winning goal in a 5-4 over the New York Rangers on December 8, 1988.

Steve Martins

Plucked by Hartford in the 1994 Supplemental Draft, this pesky centerman out of Harvard played in 23 of his 25 career games with the Whalers in 1995-96. The highlight was a goal in his first game, a 7-4 win over the New York Islanders on December 6, 1995. One of the smaller players at 5-foot-9, 170 pounds, Martins played with more heart, gusto and grit than many who possess a more physical presence.

Paul Maurice

The 10th and final coach in the NHL annals of the Hartford Whalers, Maurice replaced Paul Holmgren who was canned just 11 games into the 1995-1996 season. The Whalers beat San Jose 7-3 in Maurice's debut on November 7, 1995. They also won their finale for him on April 13, 1997, a 2-1 decision over Tampa Bay. Overall, Maurice finished with a two-year mark of 61-72-19 in

*The 10th and final NHL coach of the Whalers, **Paul Maurice**, at 27, was the youngest to ever call line changes for the hockey club. Maurice was a winner in his debut, a 7-3 win over San Jose on November 7, 1995 when he replaced **Paul Holmgren** as bench boss. He also won his last game, a 2-1 decision over Tampa Bay on April 13, 1997. Overall, Maurice's totals were 61-72-19.*

Hartford and no trips to the playoffs. At 28, he became the youngest to serve as bench boss of the Whalers. Unproven at the pro level, Maurice served an extensive apprenticeship with the junior hockey programs of the Compuware Corporation. A promising two-way forward with the Windsor Spitfires who had dreams of playing in the NHL, Maurice suffered an eye injury during an Ontario Hockey League game in 1988 and was forced to retire. Given an opportunity to stay with the club, Maurice eventually moved into the coaching ranks for Compuware and capped his tenure with a trip to the Memorial Cup Finals with the Detroit Junior Whalers in 1995. In Hartford, Maurice's highlight was the first half of 1996-97 when he had the Whalers atop the Northeast Division and was named an assistant coach for the East at the All-Star Game. The lowlight had to be Hartford's dismal road record of 9-24-8 in 1996-97, a grade that looks even sorrier when a 4-0-4 unbeaten streak from November 12, 1996 to December 5, 1996 on foreign ice (the longest ever in club history) is subtracted.

Bryan Maxwell

A rugged defenseman who played for teams in both leagues, the Whalers acquired Maxwell with Greg Carroll from Cincinnati on May 26, 1977 for the draft rights to goalie Mike Liut. Released by New England on February 9, 1978, Maxwell signed with the NHL Minnesota North Stars. He also played for St. Louis and Winnipeg in the NHL. Maxwell began his career with the Cleveland Crusaders in 1975-76, opting for the WHA rather than go with the NHL North Stars which had tabbed the 6-foot-3, 210-pounder on the fourth round in the 1975 Entry Draft.

Jim Mayer

A journeyman forward who played 74 games in the WHA, Mayer played a career-high 51 games with the Whalers in 1977-78. Among his 11 goals with New England was a game-winner on January 28, 1978 in a 3-0 win at Houston. Moved on to Edmonton for 1978-79. Originally taken 239th overall by the New York Rangers in 1974, New York claimed the winger from the Oilers where he logged a four-game stint with the Blueshirts in 1979-80.

Steve Martins played 25 games for the Whalers over two seasons. The little centerman scored his first NHL goal in his first game, one of several by Hartford in a 7-4 romp of the New York Islanders on December 6, 1995.

Jason McBain

A fourth-round pick in 1992 (81st overall) by Hartford, McBain has filled in at the blue line for the Whalers since 1995-96. Rates among the better mobile defensemen in the organization and is effective manning the point during power-play situations. His future hinges on his development since there exists a logjam on defense, a position stocked with several experienced veterans.

Rob McClanahan

A member of the 1980 United States Olympic gold medal team, McClanahan joined the Whalers in the 1981 Waiver Draft when plucked from Buffalo. In 1978, the Sabres made the speedy winger from the University of Minnesota the 49th overall choice in the Entry Draft. McClanahan had three helpers in 17 games with Hartford before getting dealt to the New York Rangers on February 2, 1982 for a 10th-round pick (Reine Karlsson). Reunited in New York with U.S. coach Herb Brooks. McClanahan prospered with the Rangers where he wrapped up his 224-game career with 38-63-101 career totals including 22 goals in 1982-83.

Brad McCrimmon

Taken 15th overall by Boston in 1979, McCrimmon played 17 years in the NHL, a span of over 1,200 games which wound up in 1996-97 with the Phoenix Coyotes. Played 156 games with the Whalers over a three-year hitch after coming to Hartford on July 1, 1993 from Detroit for a draft pick. Also played for Philadelphia, Calgary and Detroit. A member of Caglary's Stanley Cup team in 1989, McCrimmon played for the Western League's Brandon Wheat Kings in 1978-79, a club that lost only five games that winter and produced several first-round draft selections in 1979 (Laurie Boschman by Toronto, Brian Propp by Philadelphia and Ray Allison by Hartford).

Gerry McDonald

One of three players obtained from the New York Rangers for Mike Rogers on October 2, 1981, McDonald played eight games at the blue line for Hartford during 1981-82 in his only stop in the NHL. Worked with Chuck Kaiton in the radio broadcast booth during 1996-97.

Mike McDougal

A gritty winger who was taken 76th overall by the New York Rangers in 1976, McDougal came to Hartford in the 1981 Waiver Draft. Played 58 games with the Whalers over two seasons, producing all of his career points (8-10-18) during a 55-game stint with the Whalers in 1982-83.

Mike McEwen

A 12-year NHL veteran who played in New York with the Rangers and Islanders, Colorado Rockies, Los Angeles Kings, Washington Capitals and Detroit Red Wings, "Cue-Ball" finished up his 716-game tenure with 67 games over three years with the Whalers. Tabbed 42nd overall by the Rangers in 1976, McEwen put up his best numbers in 1978-79 (20-38-58) before heading to the Rockies in a 5-for-1 deal on November 2, 1979 for Barry Beck. Returned

Pierre McGuire coached Hartford for 67 games during the 1993-94 campaign. Though a 13-11-1 mark in McGuire's first 25 games suggested the Whalers had found the right mentor, Hartford limped home with just 10 wins in its final 42 games, greasing the skids for McGuire's departure.

to New York in a deal for Chico Resch on March 10, 1981 and added power-play juice to help the Islanders to the 1983 Stanley Cup. Joined Hartford on March 11, 1986 from the New York Rangers for winger Bobby Crawford.

Bob McGill

A defenseman who played 705 NHL games, McGill checked out of the NHL with the Whalers after a 30-game stop in 1993-94. Drafted 26th overall by Toronto in 1980, McGill also played for Chicago, San Jose, Detroit and the New York Islanders, winding up with 17-55-72 and 1,764 penalty minutes. Had the ability to skate as well as drop the gloves, a major reason why he lasted 13 seasons despite limited offensive numbers. Went scoreless in 49 Stanley Cup games but did amass 88 penalty minutes.

Pierre McGuire

During a 57-game tenure behind the Hartford bench in 1993-94, McGuire was the master second-guesser especially at questioning players' hearts and then wondering why the team quit on him. Elevated from assistant to head coach on November 16, 1993 when Paul Holmgren decided to concentrate solely on the general managerial duties, the Whalers, a dismal 4-11-2 out of the gate, regrouped during McGuire's first 25 games. Hartford went 13-11-1 to move within five games of breakeven (17-23-3) at the midseason break before a disastrous 0-4 road trip in mid-January opened a 10-26-6 slate for the second term. McGuire's fate, in retrospect, had been sealed on March 24, 1994 when a number of players and assistant coaches were arrested in an after-hours scuffle with patrons and police in a Buffalo nightclub. The off-ice altercation convinced team owner Richard Gordon that it was time to sell his stake in the hockey club. McGuire got the boot on May 19, 1994, a few weeks before Peter Karmanos purchased the franchise. McGuire, however,

Tempers flare here against Quebec as New England's **John McKenzie** *(left with stick raised),* **Mike Antonovich** *and* **Dave Keon** *come to the aid of* **Thommy Abrahamsson**.

Johnny "Pie" McKenzie, the first Whaler to ever have his jersey number retired, finished his pro hockey career with New England of the WHA, logging 189 games over three seasons. McKenzie was the last to ever wear No. 19, a jersey number officially retired by the hockey club on February 27, 1980. Gordie Howe and Rick Ley are the only other Whalers to be so honored.

A fan favorite during his years with the Whalers, Jim McKenzie has carved out an extensive NHL career since breaking into the league in 1990. McKenzie was dealt to Florida on December 16, 1993, a swap that secured blueliner Alexander Godynyuk from the Panthers.

quickly got his revenge on the Whalers. When general manager Jim Rutherford signed free agent Steven Rice of Edmonton, it was McGuire who turned over confidential scouting information to the Oilers who used the data to sway arbitrator George Nicolau who awarded defenseman Bryan Marchment to the Alberta club.

Jack McIlhargey

A gritty defenseman who played eight years in the NHL, McIlhargey broke in with the Philadelphia Flyers in 1974-75 and moved on to Vancouver on January 20, 1977 with

Larry Goodenough for Bob Dailey. "Black Jack" came to the Whalers in a deal with defenseman Norm Barnes on November 21, 1980 for a second-round draft choice in 1982. He played 98 games in two years at the blue line for Hartford to complete a 393-game stay that ended in 1981-82. Went 11-36-47 with 1,102 penalty minutes. Usually played it tough and rough.

Ross McKay

A netminder signed by the Whalers as a free agent, McKay wore the highest numeral by a Hartford goalie (No. 49) in his only NHL game, a 35-minute relief effort during a 5-4 loss against Buffalo on February 23, 1991.

Jim McKenzie

Another big winger who was willing to drop the gloves, Hartford secured McKenzie with the 73rd overall pick in the 1989 Entry Draft. Scored the occasional goal while racking up 468 penalty minutes during a 203-game stay with the Whalers where he became a fan favorite. Dealt to Florida on December 16, 1993 for Alexander Godynyuk. The Panthers then shipped McKenzie to Dallas for a draft choice. The Stars moved McKenzie to Pittsburgh where he logged 50 games over two seasons before moving on to the New York Islanders. Obtained by Winnipeg in the 1995 Waiver Draft, McKenzie spent 1996-97 with the Phoenix Coyotes where he notched his first career hat trick.

John McKenzie

"Pie" played 691 NHL games and 477 in the WHA, starting in 1959-60 with Chicago and winding up in 1978-79 with the Whalers. An imp with a tenacious streak, McKenzie eased the physical play to a degree once moving to the WHA. After helping Boston to Stanley Cups in 1970 and 1972, McKenzie was among several Bruins who exited the Hub for the upstart league. He signed with Philadelphia where he served as player-coach of the Blazers. McKenzie also played in Vancouver, Minnesota and Cincinnati in the WHA, going 163-250-413. Also played with Detroit and the New York Rangers in the NHL. Played the game with a robust passion and had his jersey (Number 19) retired by the Whalers because of his on-ice feistiness. Helped form one of New England's best trios with Dave Keon and Mike Antonovich which was dubbed "Capital City Lightning Company" for its effectiveness and energy.

Bob McManama

A center who played 99 NHL games over three seasons with the Pittsburgh Penguins, McManama signed on with the WHA Whalers on December 9, 1975 and contributed steady play for the final 37 games of 1975-76. Finished with a strong playoff, going 4-3-7 in a dozen AVCO Trophy games.

Rick Meagher

A center who played 691 games over a dozen seasons in the NHL, Meagher proved that there is a spot in the league for small forwards who can play without the puck. Earned the league's Selke Trophy in 1990 as the top defensive forward. Signed by Montreal as a free agent out of Boston University on June 27, 1977, the speedy center came to Hartford on June 5, 1980 in a flip-flop of third and fifth round draft picks in 1981. Emerged as a dependable checker and scorer, netting a career-high 24 goals and 43 points in 1981-82. Dealt to New Jersey on October 15, 1982 with Garry Howatt for Merlin Malinowski and Scott Fusco. It was with the Devils and later the Blues where Meagher solified his specialist role.

Glenn Merkosky

A speedy center from Edmonton, Merkosky signed with Hartford as a free agent on August 10, 1980 and reached the NHL during 1981-82, going scoreless in seven games. Released following the season, Merkosky hooked on with New Jersey where he went 4-10-14 in 34 games in 1982-83. Also played with Detroit where he finished his career in 1989-90 for a career total of 66 games.

Gerry Methe

Taken on the fourth round by Pittsburgh in 1971, this defenseman logged his only pro time with the WHA Whalers, a five-game trial in 1974-75. Was a member of the Hershey Bears who copped the American League's Calder Cup in 1974.

Mike Millar

Taken 110th overall by the Whalers in 1984, Millar had strong goal-scoring years in Canadian juniors (50 and 66 goals) and with the Canadian Olympic Team (50) but never could crack the lineup as a regular in Hartford, Washington, Boston or Toronto. Played 78 NHL games over five seasons with 1990-91 being his last year. Millar's best numbers came in 1987-88 when he went 7-7-14 in 28 games with the Whalers. Dealt by Hartford to Washington with Neil Sheehy on July 6, 1988 for Grant Jennings and Ed Kastelic.

Greg Millen

Drafted 102nd overall by Pittsburgh in 1977, Millen established himself as a goalie of promise with the Penguins in his first three seasons. The Whalers, desperately seeking help between the pipes, signed Millen on June 15, 1981 in what became a major case around the NHL involving free agency. When Hartford and Pittsburgh failed to agree on fair compensaton, the case went before an arbitrator. Judge Joseph Kane awarded Pat Boutette and Kevin McClelland to the Penguins rather than John Garrett and Jordy Douglas, the two players Hartford had offered as payment. The ruling proved costly as the Whalers lost their top left winger and a gritty center who eventually

*Speedy center **Rick Meagher**, featured here defending against Edmonton's Paul Coffey, played 96 games over three seasons in Hartford before exiting in a swap with New Jersey.*

*The puck rolls to the right of goaltender **Greg Millen** in action here against New Jersey at the Hartford Civic Center. Millen appeared in 219 games between the pipes for the Whalers. Among his shutouts as a Whaler was an 11-0 blanking of Edmonton on February 12, 1984.*

played on four Stanley Cup championship teams over a 12-year NHL career. The ruling also sent a message around the league for teams to shy away from signing free agents. Millen, meanwhile, toiled through the dark years of Hartford's early NHL history, a major reason for a 46-98-27 record with the Whalers. After leading the league in appearances (60) and minutes played (3,583) in 1983-84, Millen had another iron streak in 1984-85, fashioning a 16-22-6 record in 44 games before being dealt to St. Louis along with Mark Johnson in a big swap that brought Mike Liut and Jorgen Pettersson to Hartford. While Liut keyed a revival of the Whalers, Millen put together some impressive years with the Blues including a league-high six shutouts in 1988-89. Also played for Quebec, Chicago and Detroit before calling it quits in 1991-92, a 14-year career where he made 604 appearances with a 215-284-89 record, 17 shutouts and a 3.87 GAA.

Warren Miller

A blue-collar forward who played for the Whalers in both leagues, "Grumpy" had his best year offensively when he scored 22 goals for Hartford in 1980-81. Drafted 241st overall by the Rangers in 1974, Miller was a four-year standout at the University of Minnesota and opted to sign with the WHA Calgary Cowboys. Miller also played for Edmonton and Quebec before coming to the Whalers in 1978-79 where he finished a 238-game tenure in the WHA with 65-83-148. In the 1979 Reclaimation Draft, the Rangers plucked Miller from Hartford. Miller went 7-6-13 in 55 games with the Blueshirts before coming back to the Whalers for a modest sum of $100 under terms of the NHL-WHA merger. Finished his 262-game career in Hartford, wrapping it up in 1982-83.

Chris Murray

A rugged winger, the Whalers obtained Murray in a three-way deal involving Montreal and Phoenix on March 18, 1997 at the trading deadline. Drafted 54th overall by

*Forward **Warren Miller**, better known for his play without the puck, notched a career-high 22 goals for Hartford in 1980-81.*

the Canadiens in 1994, Murray had three seasons of 200-plus penalty minutes with Kamloops of the Western League and continued his feisty play in the American League. Played over 100 NHL games with the Canadiens who shipped the winger and Murray Baron to the Coyotes for Dave Manson. Murray joined Hartford moments later for Gerald Diduck. Had his only points as a Whaler, notching a goal and an assist, in a 4-1 win over the Habs on April 5, 1997, the last road triumph in Hartford team annals.

Dana Murzyn

Drafted fifth overall in 1985 by the Whalers, Murzyn had the size and discipline to play a positional game and launched in Hartford in 1985-86 what has become an 800-game NHL career. Has also played for Calgary and Vancouver. Though his best offensive years were with the Whalers, Murzyn was dealt to the Flames on January 3, 1988 with Shane Churla for Carey Wilson, Neil Sheehy and the draft rights to Lane McDonald. Played on Calgary's Stanley Cup team in 1989. Dealt to the Canucks on March 5, 1991 for Ron Stern, Kevan Guy and future considerations. Made it back to the Cup Finals with Vancouver in 1994.

Jason Muzzatti

A Waiver Draft claim by the Whalers on October 6, 1995, Muzzatti had what is best described as a streaky tenure over two seasons with Hartford. When goalie Sean Burke was sidelined with assorted injuries, Muzzatti often carried the load. He was either brillant or fought the puck. A standout at Michigan State where he won more games (83) than any Spartan goalie in school annals, Muzzatti was drafted by the Calgary Flames on the first round (21st overall) in 1988. He stopped 40 shots in 1-0 win over New Jersey on April 4, 1996 for his first NHL shutout. Had a 4-8-3 mark with a 2.90 GAA in his rookie season (1995-96) when he also compiled the most penalty minutes (33) among the league's netminding fraternity. In 1996-97, Muzzatti had a 9-5-3 mark in his first 17 decisions but failed to earn a point in his last 10 decisions (0-8-2).

*Drafted fifth overall in 1985, **Dana Murzyn** launched his NHL career with the Whalers. Though he had his best offensive years with Hartford, Murzyn earned a Stanley Cup ring with Calgary in 1989 and made it back to the Cup Finals with Vancouver in 1994.*

Jason Muzzatti served in a backup role between the pipes in Hartford for two seasons, going a combined 13-21-8 in 53 appearances. He notched his first NHL shutout on April 4, 1996, a 1-0 win over New Jersey.

*The date is March 11, 1976 and **Harry Neale**, most recently the bench boss of the Minnesota Fighting Saints, speaks at a press conference in Hartford which announced his appointment as coach of the Whalers. Looking on are managing general partner **Howard L. Baldwin** and **Don Blackburn**. Neale had a strong run with the Whalers, reaching the semifinals in 1976 and the league finals in 1978.*

Don Nachbaur

A center taken 60th overall by Hartford in the 1979 Entry Draft, Nachbaur played 233 games over eight NHL campaigns. With the Whalers in 1980-81, he registered career highs in goals (16) and points (83). A regular the following season, Nachbaur moved on to Edmonton with Ken Linseman in a swap that brought Risto Siltanen and Brent Loney to Hartford (in conjunction with a trade involving Mark Howe to Philadelphia). Nachbaur played four games with the Oilers and soon joined Los Angeles when the Kings claimed him in the 1983 Waiver Draft. He eventually signed with Philadelphia where he made fill-in appearances for the Flyers over five winters before calling it quits in 1989-90.

Harry Neale

One of several coaches who launched a pro career with the birth of the WHA, the Whalers secured this quotable bench boss once the Minnesota Fighting Saints disbanded on March 10, 1976. Neale stayed at the helm of the Whalers for better than two seasons, exiting for the NHL's Vancouver Canucks following New England's drive to the 1978 AVCO Trophy Finals. Neale, who also coached the Detroit Red Wings in the NHL, may have posted his greatest victory with the Whalers in an exhibition game on December 27, 1976, a 5-2 decision when New England ambushed the Soviet Nationals at the Hartford Civic Center. That triumph came on the heels of a dynamic postseason in 1976 when a kid netminder named Cap Raeder anchored New England to within one game of the Finals. "Just how crazy this game can be is summed up in the playoffs that year," recalls Neale, a current hockey broadcaster. "In the series against Cleveland, we lost both our goalies, Christer Abrahamsson and Bruce Landon, to injuries. When I was told the club was going to bring up Cap Raeder from the Cape Cod Cubs, my reaction was 'Cap who from where?' What a performance that young man put forth. If you could have had the choice of any

*Right winger **Ray Neufeld** goes to the net here against Buffalo's Tom Barrasso. "Neufy" had the ability to play a power game and created room with his size and strength yet lacked the "hands" scorers usually have. His best year with the Whalers came in 1983-84 with 27-42-69 in 80 games. It was during that season Neufeld scored a milestone goal in club NHL history, the 1,000th by a Whaler.*

goalie in hockey at the time, be it Ken Dryden, Gerry Cheevers or Bernie Parent, none would have played any better than Raeder did. We really had a beaten-up team that really played right up to its potential."

Ray Neufeld

A big winger who was taken by the Whalers on the fourth round (81st overall) in 1979, Neufeld was the first black player to play for the franchise, appearing in 331 of his 595-game NHL career with Hartford. He rang up goal seasons of 26, 27 and 27 before moving to Winnipeg on

November 21, 1985 in a deal that brought big-play defenseman Dave Babych to the Whalers. Neufeld added goal seasons of 20, 18 and 18 for the Jets before he was dispatched to Boston for Moe Lemay on December 30, 1988. Had 157-200-357 over his career with 816 penalty minutes. Neufeld had great size and strength at 6-foot-3 and 210 pounds and likely would have scored more goals if he had been blessed with better hands.

John Newberry

A forward who made his NHL debut in the 1983 Stanley Cup playoffs with Montreal, Newberry signed with the Whalers as a free agent in 1985. He went scoreless in a three-game trial during the 1985-86 season.

*Colorful and quotable, **Harry Neale** coached the Whalers to the AVCO Finals in 1978 before moving on to Vancouver of the NHL. "Everyone in the National League underrated the WHA," Neale says. "It was a better league than people thought or gave credit to. My team in Hartford with the Howes, Keon, McKenzie, Antonovich, Pleau, Ley, Selwood, Smith and Webster was a far superior team than the NHL team I was going to. In my heart, I knew I had a better chance at winning with the Whalers than I did with the Canucks."*

Barry Nieckar

A rough-playing forward out of the East Coast League, Nieckar signed a free agent contact with Hartford on September 25, 1992. Though he sustained a serious skate cut on his hand during workouts, Nieckar recovered to play two games for the Whalers during 1992-93 before getting his release. Nieckar racked up a league-high 491 penalty minutes with Saint John of the American League in 1994-95, the type of enthusiasm which resulted in a second look in the NHL for three games with Calgary that winter.

Andrei Nikolishin

A high-skilled player who honed his playmaking ability with the Moscow Dynamo and captained the Soviet team at 21, Nikolishin loomed as a player of promise when the Whalers drafted him 47th overall in the 1992 Entry Draft. A standout for the Confederation of Independent States (CIS) in the 1994 Winter Olympics, Nikolishin crossed the Atlantic and in 1994-95 when "Niko" made his debut with Hartford, he became the second Russian to ever play for the Whalers. A regular for two seasons highlighted by a career best 14-37-51 in 61 games in 1995-96, general manager Jim Rutherford deemed Nikolishin expendable. The Whalers sent the centerman to Washington on November 9, 1996 for defenseman Curtis Leschyshyn.

*Once asked where he grew up in Russia, **Andrei Nikolishin** said "white bear" referring to Arctic Circle area north of the Ural Mountains known for cold temperatures and polar bears. "Niko" learned to skate in rinks of the city of Vorkuta. Drafted by the Whalers in 1992, Nikolishin earned a gold medal for the Soviets in the 1994 Winter Olympiad.*

Lee Norwood

A motorcycle-driving defenseman, Norwood rode into Hartford in a deal with New Jersey on October 3, 1991 for a fifth-round pick in 1993 (John Guirestante). He played in six games for the Whalers before exiting for St. Louis on November 13, 1991 for a fifth-round pick in 1993 (Hartford took Nolan Pratt). Drafted 62nd overall by the Nordiques in 1979, Norwood played 503 NHL games once breaking in with Quebec in 1980-81. He also made stops in Washington and Detroit before finishing up with Calgary in 1993-94.

Michael Nylander

Drafted 59th overall in 1991, Nylander came over from Sweden and looked like a playmaker for seasons to come, especially after debuting with 11-22-33 in 59 games in 1992-93. Though he had 44 points in 58 games the following winter, Nylander began feuding with coach Pierre McGuire. The disagreement resulted in Hartford including the clever centerman in a multi-player deal with Calgary on March 10, 1994 where Zarley Zalapski and James Patrick also went to the Flames in exchange for Gary Suter, Paul Ranheim and Ted Drury. Nylander notched six game-winners while rolling up 17-38-55 in 73 games with Calgary in 1995-96, his best pro season to date.

*Sweden's **Michael Nylander** looked to be a playmaker of promise once joining the Hartford lineup in 1992-93. Within two years, he was dished to Calgary along with **James Patrick** and **Zarley Zalapski** for **Gary Suter, Paul Ranheim** and **Ted Drury** on March 10, 1994.*

O

Fred O'Donnell

After three seasons in Boston playing for the AHL Braves and NHL Bruins, O'Donnell jumped to the WHA and signed with New England where he played 155 games over two seasons. Originally drafted by Minnesota, the North Stars sent the centerman to Boston where he went 11-15-26 over 115 games before being deemed expendable. The Bruins, on February 7, 1974, packaged O'Donnell and Chris Oddleifson to Vancouver for Bobby Schmaultz but O'Donnell balked at reporting. O'Donnell wound up with the Whalers where he produced 32 goals during his tenure highlighted by a hat trick in a 6-2 win over Winnipeg on March 11, 1975. He finished in 1975-76 with 11 goals before the Whalers shuttled O'Donnell off to Minnesota with Bob McManama for the rights to Wayne Connelly on June 29, 1976.

Jeff O'Neill

A promising center, O'Neill got a first-hand look at the city and fans of Hartford on June 28, 1994 when the Whalers hosted the NHL Entry Draft and made the Ontario League standout the fifth overall selection on the first round. Though he racked up points in bushels in the OHL at Guelph and went 8-19-27 in 65 games during his rookie NHL campaign in 1995-96, O'Neill has yet to have a break-through year for the Whalers. In 1996-97, O'Neill had 14-16-30 in 72 games for Hartford with the highlight being his first NHL hat trick in a 6-3 loss at Anaheim on January 31, 1997.

Ted Ouimet

A goaltender who played one game in each pro league, Ouimet was third on the depth chart behind Jacques Plante and Glenn Hall in St. Louis. He made his only NHL start during the 1968-69 campaign. Ouimet resurfaced in the WHA with New England where he made his only appearance in 1974-75, allowing three goals in relief during a lopsided loss to Quebec.

Fred O'Donnell joined the Whalers after exiting the NHL and produced 32 goals during his 155-game tenure over two seasons in the WHA. His best night came in a 6-2 win over Winnipeg on March 11, 1975, a hat trick.

*Hartford's top pick in 1994 was **Jeff O'Neill** of the Guelph Storm. When the centerman was taken fifth overall on June 28, O'Neill immediately heard the cheers from the Civic Center faithful since the Whalers hosted the Entry Draft that year.*

*Don Blackburn chases down the puck to the right Winnipeg goalie Joe Daley as **Fred O'Donnell** skates to the net during this WHA game at the Civic Center. Blackburn played for the Whalers and eventually coached the hockey club. After interim chances behind the bench in the WHA, "Blackie" was named as Hartford's first coach in the NHL.*

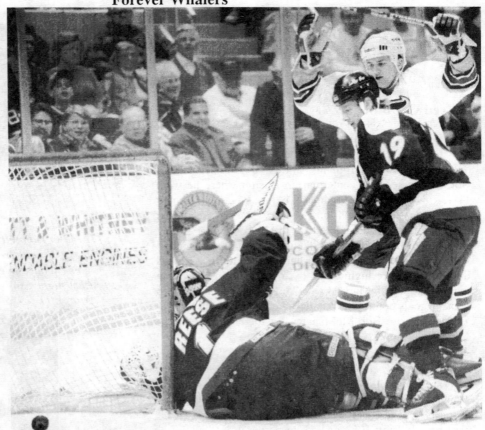

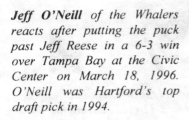

Jeff O'Neill of the Whalers reacts after putting the puck past Jeff Reese in a 6-3 win over Tampa Bay at the Civic Center on March 18, 1996. O'Neill was Hartford's top draft pick in 1994.

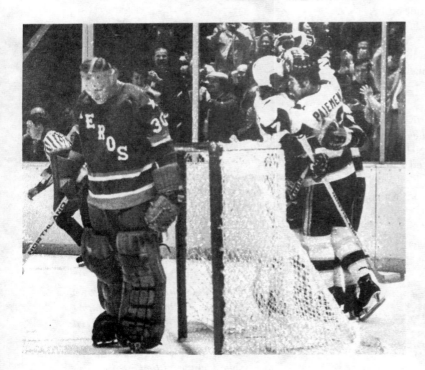

*Houston goalie Wayne Rutledge looks down at the ice after the Whalers celebrate a goal. That's **Gordie Roberts** and **Rosaire Paiement** celebrating after a goal by **Danny Bolduc**. Paiement had a big year with New England in 1975-76, going 28-43-71 in 80 games, the only Whaler to play in every game that season.*

P

Rosaire Paiement

A gritty forward, Paiement logged 190 games for Philadelphia and Vancouver in the NHL before jumping to the WHA. Spent two seasons with the Chicago Cougars and eventually landed in New England on July 1, 1975 from Denver in a deal for a first-round pick. Played 93 games with the Whalers, going 33-45-78 before being dealt to Indianapolis on November 12, 1976 for center Gary MacGregor. Notched a pair of game-winning playoff goals in 1976 when New England reached the AVCO Trophy semifinals.

Jeff Parker

One of the extras in the Francis-Cullen swap with Pittsburgh on March 4, 1991, Parker played only four games for Hartford before his career ended following a wicked crash into the boards during a 4-3 overtime loss at New Jersey on March 27, 1991. It was one of several concussions that Parker, a forward, sustained from his physical play, resulting in his exit from hockey. Drafted 111th overall by Buffalo in 1982, Parker played 137 games over four seasons with the Sabres before moving to Winnipeg on June 16, 1990 in a multi-player swap involving Phil Housley and Dale Hawerchuk.

Mark Paterson

Every team has a vacancy for a hard-hitting defenseman and in 1982, the Whalers figured they had landed one with the 35th overall pick when they secured Paterson from Ottawa of the Ontario League. In Canadian juniors, Paterson was paired with Brad Shaw, who later played for the Whalers. The tandem was one of the best in the OHL. In brief trials over four years, Paterson had 3-3-6 in 29 games but lacked the quickness to excel at the NHL level. On October 7, 1986, the Whalers sent Paterson to Calgary for winger Yves Courteau.

James Patrick

The ninth overall pick in the 1981 Entry Draft by the New York Rangers, Patrick was the subject of two multi-player trades during the 1993-94 season. In between the deals, Patrick logged 47 games at the blue line for the Whalers. A mobile defenseman who was effective at the point on the power play, Patrick notched seven consecutive years of 10 or more goals with the Rangers before falling out of the regular mix. Deemed expendable when Mike Keenan took the helm, Patrick was dealt to Hartford along with Darren Turcotte on November 2, 1993 for Steve Larmer, Nick Kypreos, Barry Ritcher and a 1994 sixth-round draft pick (Yuri Litvinov). He stabilized the Hartford backline until a deal to Calgary on March 10, 1984 with Zarley Zalapski and Michael Nylander brought Gary Suter, Paul Ranheim and Ted Drury to the Whalers. Likely would have reached the 900-game milestone with the Flames during 1996-97 but sustained a knee injury. Has tallied 120 goals and 428 assists in 875 career games.

*James Patrick went 8-20-28 in a 47-game hitch with the Whalers in 1993-94 after coming over from the New York Rangers. Before the season reached a conclusion, the veteran defenseman was on the move again, this time in a six-player swap with Calgary for **Gary Suter, Paul Ranheim** and **Ted Drury**.*

Jim Pavese

Obtained from Detroit on March 7, 1989 for Torrie Robertston, Pavese logged just five games with the Whalers, the final lap of his 328-game career over eight seasons. Able to play defense or forward, Pavese also played for the St. Louis Blues and New York Rangers. He was originally drafted 54th overall by the Blues in 1980.

Allen Pedersen

Taken 105th overall by Boston in 1983, Pedersen played eight seasons in the NHL with stops in Minnesota and eventually Hartford where he wrapped up his career in 1993-94, playing 66 games over two years. Pedersen never scored a point in 64 career playoff games and had only 5-36-41 over 428 regular season games. Had a good reach at 6-foot-4, 205 pounds. A reliable blueliner who made the safe play. Had 1-4-5 with the Whalers with his only goal helping Hartford knot the Rangers 4-4 on December 19, 1992.

Barry Pederson

One of several players who had a short stay (five games in 1991-92) in Hartford, Pederson had his best NHL seasons in Boston but was never the same player after undergoing surgery to remove fibrous tumor tissue from his right bicep. Pederson played 701 NHL games over a dozen seasons, going 238-416-654. He also had 22-30-52 in 34 playoff games. Taken 18th overall in 1980 by the Bruins, Pederson was traded to Vancouver on June 6, 1986 in what turned out to be a swap that tilted heavily in favor of Boston. In exchange for Pederson, the Canucks shipped Cam Neely and a first-round pick to the Bruins. In passing seasons, Neely emerged as a power forward, scoring 50 or more goals three times. Boston, meanwhile, also plucked defenseman Glen Wesley (third overall) in 1987 Entry Draft with the prime selection it obtained from the Canucks. Vancouver eventually put Pederson on the move again, dealing him to Pittsburgh on January 8, 1990 in a multi-player move.

Andre Peloffy

A point-per-game scorer in the American League with plenty of foot speed, the WHA Whalers signed Peloffy out of Springfield on August 28, 1977. He appeared in 10 games for New England, notching a pair of goals in 1977-78, his only stop in the big leagues.

Brent Peterson

A 10-year NHL veteran, Peterson filled the role of a checker on every team he played for, starting with Detroit, then Buffalo, Vancouver and eventually Hartford where he completed a two-year hitch, logging 118 games before retiring after the 1988-89 season. Taken 12th overall in 1978 by the Red Wings, Peterson was dealt to Buffalo on December 2, 1981 with Mike Foligno and Dale McCourt for Danny Gare, Jim Schoenfeld and Derek Smith. Moved on to Vancouver in the 1985 Waiver Draft and was plucked from the Canucks by the Whalers in the 1987 Waiver Draft. An honest player whose faceoff and positional play were keys to Hartford having the best penalty-killing unit in the NHL during 1987-88.

Robert Petrovicky

Hartford's top pick in 1992 (9th overall) played 77 games over three years but never found a fit with the Whalers. Drafted by Brian Burke, Petrovicky, a centerman with good speed, came out of Slovakia with strong credentials. Like many young players, Petrovicky was pushed into the lineup ahead of schedule. Both the organization and player paid a price. Petrovicky showed promise by going 30-52-82 in 74 games at Springfield of the American League in 1994-95, but was unable to crack the Hartford lineup. GM Jim Rutherford dealt Petrovicky to Dallas for Dan Kesa on November 29, 1995. He hooked on with St. Louis for 1996-97 and wound up skating on a line with Brett Hull and Pierre Turgeon, going 7-12-19 in 44 games.

Jorgen Pettersson

A Swedish forward who had some prolific goal-scoring years with St. Louis, the Whalers received Pettersson as future considerations to complete a Februry 22, 1984 deal with the Blues involving Mike Liut, Greg Millen and Mark Johnson. In 23 games with Hartford, Pettersson failed to deliver offense and was soon dispatched to Washington on December 6, 1985 for center Doug Jarvis. Pettersson wrapped up his career with the Caps, a 435-game tenure with 174 goals and 192 assists.

Michel Picard

A low-round pick (178th overall) by the Whalers in 1989, Picard began to open eyes by leading the American League in goals (56) in 1990-91 but fell short in trials in Hartford. Played 30 games with the Whalers over two seasons before going to San Jose on October 9, 1992 for future considerations (Hartford reacquired Yvon Corriveau from the Sharks on January 21, 1993). Picard has also played for Ottawa, a total of 96 NHL games with 15-19-34. Spent 1996-97 in the International League.

Checker Brent Peterson was secured in the 1987 Waiver Draft from Vancouver and completed a 10-year career in the NHL with the Whalers, retiring after the 1988-89 season.

Randy Pierce

Taken 47th overall in 1977 by the Colorado Rockies, Pierce, a forward, completed his 277-game NHL career by playing 34 games over two seasons in Hartford, wrapping it up in 1984-85. Also played for the New Jersey Devils when the Rockies relocated operations. Signed with the Whalers as a free agent on October 6, 1983.

Frank Pietrangelo

Pittsburgh took Pietrangelo as the 63rd choice in 1983 and the netminder completed his 141-game NHL career (46-59-6, 4.12 GAA) with the Whalers in 1993-94. Though his career numbers in Hartford suggest little impact (12-27-3), Pietrangelo looked like the real deal once coming over in a trading-deadline swap on March 10, 1992 from the Penguins for two draft picks. Finished 3-1-1 down the stretch and then went 3-4 with a 2.68 GAA in a memorable seven-game series with Montreal. Stopped 53 shots in Game 7, a 3-2 loss on May 1, 1992 in double overtime, a game that ended at 25:26 on a goal by Russ Courtnall. Not only was it the longest playoff game in club annals but also the last ever played by the Whalers.

Larry Pleau

The first player ever signed by the WHA Whalers, Pleau played 468 games over seven seasons and another 66 playoff games before retiring prior to the franchise joining the NHL in June, 1979. A native of Lynn, Mass., Pleau played in the Montreal chain before jumping to the WHA where he proved to be a capable, two-way performer in the pivot. In New England's AVCO Trophy championship season, Pleau scored a hat trick in the clincher, a 9-6 win over Winnipeg on May 6, 1973. Stayed with the organization for 17 years, serving stints as coach and general manager before getting dismissed as bench boss on May 16, 1989. His firing was debated at the time since

*The Whalers used their first-round draft choice to select Czech centerman **Robert Petrovicky** as the ninth overall pick in 1992.*

*The first player ever signed by the Whalers was **Larry Pleau** who stayed 17 years with the organization, in positions from coach to general manager to television announcer.*

Pleau proved more capable in his second try as coach, a direct contrast from his early stewardship which were easily the darkest hours in Hartford. Traded Mike Rogers and Mark Howe in deals during his term as general manager, swaps that badly tilted in favor of other clubs. Pleau, however, drafted better than most. In his two years as general manager, he selected Ron Francis, Paul MacDermid and Randy Gilhen in 1981 and followed up in 1982 by netting Kevin Dineen, Ulf Samuelsson and Ray Ferraro. Once leaving Hartford, Pleau held several slots in management with the New York Rangers. He was director of player recruitment until June 9, 1997 when he left the Blueshirts to take over the general manager duties of the St. Louis Blues.

*Brett Hull of St. Louis eludes **Robert Kron** to reach the slot but is denied here by goalie **Frank Pietrangelo**. The latter had a sharp debut with the Whalers in 1992, going 3-1-1 down the stretch followed by a remarkable seven-game playoff series against Montreal. In 54 games, Pietrangelo's mark was 12-27-3 with a 4.10 GAA.*

*Left wing **Patrick Poulin** was Hartford's top pick in 1991 and tallied 20 goals as a rookie in 1992-93. He was dealt away on November 2, 1993 to Chicago with **Eric Weinrich** for **Bryan Marchment** and **Steve Larmer**.*

*Taken third overall in the 1991 Entry Draft by Detroit, **Keith Primeau** joined the Whalers in a major deal on October 9, 1996 which involved **Brendan Shanahan** and **Paul Coffey**. The towering 6-foot-5 center supplied some power and pop in the pivot for the Whalers, going 26-25-51 in 75 games.*

Ron Plumb

A defenseman who played for the Whalers in both leagues, Plumb came to New England on February 11, 1978 from Cincinnati for Greg Carroll. Played 131 regular season games and 23 playoff games before winding up his career at the conclusion of the 1981-82 season. One of 21 players to play seven seasons in the WHA, Plumg had his best years most notably with the Stingers where he achieved All-Star status. Possessed a solid shot from the point where he was most effective on the power play. Originally drafted by the Boston Bruins (ninth overall in 1970), Plumb jumped to the WHA where he played for Philadelphia, Vancouver and San Diego before moving on to Cincinnati. Had 65-264-329 in 549 WHA games. Second in career WHA games played to Andre Lacroix (551).

Marc Potvin

A forward who liked to drop the gloves, Potvin played in 121 NHL games with the bulk coming in 1993-94 when he racked up 246 penalty minutes in 51 games as Hartford's on-ice enforcer. Originally drafted 169th overall by the Detroit Red Wings in 1986, Potvin moved on to Los Angeles on January 29, 1993 in a multi-player deal involving Paul Coffey. Became a Whaler on November 3, 1993 in exchange for Doug Houda. Though willing to fight, Potvin did sustain a series of concussions during his tenure with the Whalers which hastened his exit from the NHL. Signed on with Boston where he played 33 games over two seasons, amassing only 16 penalty minutes.

Patrick Poulin

Hartford's top pick in 1991 (ninth overall), Poulin led the Quebec Major League in scoring (52-86-138) in 1991-92. He seemed like a player of promise by going 20-31-51 as a rookie with the Whalers in 1992-93, but like many younger players over the years, Poulin was deemed expendable. Hartford sent Poulin in a trade package with Eric Weinrich shipped to Chicago on November 2, 1993 which brought Steve Larmer and Bryan Marchment to Hartford. The Blackhawks eventually dealt Poulin to Tampa Bay on March 20, 1996 for Enrico Ciccone. Poulin had 12-14-26 in 73 games with the Lightning in 1996-97.

Nolan Pratt

Hartford selected this 6-foot-2, 195-pound prospect with the 115th overall pick in the 1993 Entry Draft and the defensive-minded defenseman made his NHL debut on Opening Night in 1996-97, a 2-0 win over the Phoenix Coyotes on October 5, 1996. Unherald coming into training camp after a 2-6-8 season at Springfield in the American League, Pratt soon impressed management with his steady work in the defensive zone. Because of a veteran group on defense, Pratt managed to play in nine games. He also recorded his first NHL point in a 4-4 tie at Boston on October 31, 1996.

Keith Primeau

A 6-foot-5, 210-pounder center who was taken third overall by Detroit in 1990 (behind Owen Nolan and Petr Nedved), the Whalers obtained Primeau from Detroit on October 9, 1996 along with Paul Coffey and a first-round draft pick in 1997 for Brendan Shanahan and Brian Glynn.

Primeau, just 25, yearned the chance to blossom into a No. 1 center, a path that was blocked by Steve Yzerman and Sergei Fedorov who were the best pivots on the Detroit depth chart and keys to their Stanley Cup victory in 1997. Primeau's arrival improved Hartford's size and versatility. In addition, he welcomed a trade and wanted to come to Hartford, a direct contrast to Shanahan who deemed a $4 million contract was not enough to sway his feeling or commitment to the Whalers. Primeau reached the 400-game milestone during 1996-97 while going 26-25-51 in 75 games.

Chris Pronger

Projected as a big-play defenseman by most observers, the Whalers engineered a draft-day deal with San Jose on June 26, 1993 to select Pronger as the second-overall pick, a 6-foot-5, 220-pound blueliner. After two seasons of potential and some off-the-ice shenanigans, Hartford dealt Pronger to St. Louis in a 1-for-1 trade that secured two-time 50-goal scorer Brendan Shanahan. It was a swap of two players with hefty contracts, a deal that virtually rented Hartford a goal scorer for one season. As for the draft day swap, the Whalers sent Sergie Makarov to the Sharks along with their first and third round picks as well as a second round pick (previously obtained from Toronto for John Cullen). The Sharks took Victor Kozlov (sixth overall), Vlastimil Kroupa (45th) and Ville Peltonen (58th). Went 11-24-35 in 79 games with the Blues in 1996-97.

Brian Propp

In a 15-year career which spanned 1,016 games, Propp completed his NHL tenure with the Whalers in 1993-94, reaching a couple of milestones -- 1,000 games played and 1,000 points. Taken 14th overall in 1979 by Philadelphia, Propp, who racked up 511 points in the Western Hockey League as a big-play maker for the Brandon Wheat Kings, brought a solid two-way game to the Flyers. Cracked the 40-goal plateau four times and played in five All-Star Games. Also played for Minnesota and Boston. Reached the 1,000-game milestone against Philadelphia on March 19, 1994 in a 5-3 win at the Spectrum. Wound up with 425 goals and 579 assists. Added 64 goals and 84 helpers in 160 Stanley Cup games en route to becoming this trivia stumper: "Who has played in the most Stanley Cup games without ever playing for a team which won the Cup?"

Chris Pronger played 124 games over two years in Hartford before getting traded to St. Louis on July 27, 1995 for high-scoring veteran Brendan Shanahan. Pronger, a defenseman, was the second overall pick in the 1993 Entry Draft.

Left wing Brian Propp reached the 1,000-game and 1,000-point milestones of this NHL career with the Whalers in 1993-94. It was his last season in the league and maybe the most satisfying considering, at age 34, he came to training camp as a free agent and wound up earning a spot. Propp went 12-17-29 in 65 games.

*Wayne Gretzky keeps his eye on the puck while defenseman **Joel Quenneville** does his best to keep the "Great One" from finding a teammate during action at the Hartford Civic Center.*

Q

Joel Quenneville

One of a select few to have a surname begin with the letter "Q" in NHL history, Quenneville played 13 seasons, a total of 803 games with 457 as a member of the Whalers. Taken 21st overall in 1978 by Toronto, "Quennie" broke in with the Maple Leafs in 1978-79 and was soon dealt to Colorado on December 29, 1979 with Lanny McConald for Pat Hickey and Wilf Paiement. With the Rockies, Quenneville emerged into a consistent blueliner and also spent a season in New Jersey when the franchise relocated from Denver. In a span of three weeks, Quenneville was involved in two trades. On June 20, 1983, he went to Calgary with Steve Tambellini for Phil Russell and Mel Bridgman. Then on July 5, 1983, Quenneville came to Hartford with Richie Dunn, another veteran defenseman, for Mickey Volcan in what was the first trade made by newly-named general manager Emile Francis. It proved to be a smart deal for the Whalers as Quenneville became an on-ice leader noted for blocking shots and positional play. Had 25 goals and 70 assists with Hartford with his most memorable goal coming short-handed at 19:40 of the third period to stun Calgary 6-5 on October 11, 1986. In postseason play, Quenneville had seven helpers in 26 games with the biggest coming on Sylvain Turgeon's game-winner at 2:36 of overtime to beat Quebec 3-2 on April 9, 1986 in Hartford's first-ever Stanley Cup victory. Wound up his career with Washington in a swap for future considerations on October 3, 1990. Moved into the coaching ranks and after serving as an assistant in Colorado's Stanley Cup run in 1996, Quenneville was named bench boss of the St. Louis midway in 1996-97 when Mike Keenan was dismissed.

*Noted for his shot-blocking and positional play, **Joel Quenneville** played 457 games for the Whalers. "Quennie's" most memorable goal in Hartford came while killing a penalty with 20 seconds left in the 1986 lidlifter to beat Calgary 6-5.*

R

Cap Raeder

A goalie who achieved his greatest moments in hockey with the WHA Whalers, "Capper" played 44 games in a career remembered for his postseason heroics during a remarkable run in the 1976 playoffs. When injuries struck down Christer Abrahamsson and Bruce Landon, Raeder stepped in to key a thrilling seven-game series victory over Indianapolis and then carried New England to a seventh-game against two-time defending AVCO Cup champion Houston. The most notable performance by Raeder, however, never counted in the standings. It was another night when the Civic Center faithful cheered loudly, a thrilling 5-2 upset of the Soviet Nationals on December 27, 1976, a game where Raeder and the Whalers outdueled the great Russian netminder Valdislav Tretiak.

***Cap Raeder**, who once outdueled Russian great Vadislav Tretiak in a thrilling 5-2 upset of the Red Army by the Whalers, had a heroic postseason in 1976 when injuries crippled New England's netminding corps. "Capper" carried the Whalers to a seventh game against two-time AVCO Trophy champion Houston before the club expired 2-0 in Game 7.*

*Goalie **Daryl Reaugh** made the save here on Quebec's Tony McKegney and moving in for the rebound is the great Guy Lafleur. The night is November 28, 1990, a frustrating one for the Civic Center faithful as a goal by Joe Sakic gives the visiting Nords a 4-3 win. Reaugh played 20 games for Hartford. His best performance came in a 26-save masterpiece for a 1-0 win over Philadelphia on December 22, 1990. **Rob Brown** had the only strike, with 14:34 left in the third period.*

__Paul Ranheim__ played a total of 202 games for the Whalers over parts of four seasons. The speedy winger usually played his best games against the New York Rangers. His best game as a Whaler was four-point night on October 16, 1995 to key a 7-5 win over the Blueshirts at Madison Square Garden.

Paul Ranheim

The 38th overall pick in 1984 by Calgary, Ranheim completed his fourth year with the Whalers in 1996-97, a hard-working forechecker who often had his best games against the New York Rangers. A regular with the Flames for five seasons, Hartford obtained the right winger on March 10, 1994 in a deal with Calgary that also involved Gary Suter and Ted Drury for Zarley Zalapski, James Patrick and Michael Nylander. Though Ranheim never reached the 20-goal plateau for the Whalers as he did three times for the Flames, he proved to be an effective role player, often shadowing the opposition's top scorer. Blessed with outstanding speed, Ranheim once tallied 68 goals in the International League (1988-89) when he was named the circuit's top rookie. A veteran of over 500 NHL games, Ranheim tallied his 100th career goal with the Whalers in a 4-3 loss to Philadelphia on April 28, 1995.

Daryl Reaugh

A goalie who played 27 games over three NHL campaigns, Reaugh served as Hartford's backup netminder in 1990-91 when he played a career-high 20 games highlighted by a 1-0 shutout of the Flyers at Philadelphia on December 22, 1990 when he stopped 26 shots. Won four of his first five decisions with Hartford and finished with a 7-7-1 ledger with a 3.15 GAA. Taken 42nd overall in the 1984 Entry Draft by Edmonton, "Razor" was a

member of the Oilers Stanley Cup team in 1987-88 when he played six games as Grant Fuhr's caddie but lacked the required quota of appearances to get his name etched on the Cup. Career stats are 8-9-1 with a 3.47 GAA. Hooked on with Hartford as a free agent during training camp in 1989. Remained with the organization through 1991-92. Served a hitch as the club's TV analyst during 1995-96.

Mark Reeds

A checker who finished his eight-year NHL career with the Whalers, Reeds played 45 games over two seasons with the Whalers and had eight assists. Drafted 88th overall in 1979 by St. Louis, Reeds twice had double-digit campaigns and wound up with Hartford on October 5,

Jeff Reese nearly stood on his head as he won his first four decisions for the Whalers once arriving from Calgary on November 19, 1993. Over three seasons, he went 9-17-4 before getting shipped to Tampa Bay for a low-round pick.

1987 for a 1989 third-round draft pick (later upgraded to a second-rounder in a subsequent deal which brought Charlie Bourgeois to the Whalers; the Blues took Rick Corriveau with the 31st overall pick). Reeds helped the Whalers to the top penalty-killing unit in the NHL during 1987-88. Had 45-114-159 in 365 NHL games. Added 8-9-17 in 53 Stanley Cup games including an overtime game-winner on April 7, 1984 to beat Detroit 4-3 in double overtime at 37:07.

Jeff Reese

A career reserve netminder, Reese had the right temperment to keep that role with several clubs over nine NHL campaigns. Drafted 67th overall by Toronto in 1984, Reese played 172 games with stops in Calgary, Hartford, Tampa Bay and New Jersey. Joined the Whalers on November 19, 1993 from the Flames for Dan Keczmer and future considerations. Won his first four decisions as a Whaler and wound up 9-17-4 over three seasons. Dealt to the Lightning on December 1, 1995 for a 10th round pick (Hartford took Askhat Rakhmatullin 231st overall) in 1996. Played three games with the Devils in 1996-97 before joining Detroit of the International League where he helped the Vipers to the Turner Cup.

Mark Renaud

Hartford tabbed this blueliner as the 102nd overall pick in the 1979 Entry Draft and it led to a five-year career in the NHL which wrapped up in 1983-84 in Buffalo. Renaud possessed good mobility for a defenseman and was a regular for the Whalers in 1982-83 when he went 3-28-31 in 77 games. Played in 152 NHL games overall, the final 10 with the Sabres who plucked him from Hartford on October 3, 1983 during the Waiver Draft.

Steven Rice

Taken 20th overall by the New York Rangers in 1989, Rice has shown flashes of being that big-play winger teams covet since becoming a regular in the NHL. Dealt to Edmonton on October 4, 1991 in a deal with Bernie Nicholls and Louie DeBrusk which brought Mark Messier to the Blueshirts, Rice put together a promising season in 1993-94, going 17-15-32 in 63 games. Hartford, in one of general manager Jim Rutherford's first moves, signed Rice as a free agent on August 18, 1994 with compensation owed to the Oilers. Arbitrator George Nicolau awarded defenseman Bryan Marchment as payment to Edmonton on August 30, 1994, a swap which removed Hartford's

Steven Rice got the Whalers off to a good start here on February 10, 1995 in Tampa Bay as he scores against goalie Daren Puppa. At the finish, the Lightning won 4-3 on a goal by Petr Klima.

*Cigar-chomping **Jimmy Roberts** took the helm of the Whalers in 1991-92, his only year in Hartford. His last game was a 3-2 double overtime loss in Game 7 to Montreal on May 1, 1992. It was the longest Stanley Cup game ever played by the Whalers and also proved to be the 49th and final one in franchise annals. Hartford's postseason mark was a subpar 18-31 in the NHL underscored by the fact that it managed to win just one series while losing eight in the first round. That was a dramatic shift from the WHA where the Whalers made the playoffs each year, won eight of 14 series and posted a 59-64 ledger.*

best agitator and body-checker from the blue line. Rice had 21-22-43 in his first 109 games with the Whalers before going 21-14-35 in 1996-97.

Todd Richards

The only Whaler to play in more playoff games (11) than regular season games (8) in Hartford history, Richards was obtained from Montreal on October 11, 1990 for future considerations and had brief trials over two seasons. Drafted 33rd overall by Montreal in 1985 out of Armstrong (Minnesota) High School, Richards had a solid four-year career at the U of Minnesota capped by a berth on the NCAA all-tournament team in 1989 when the Gophers reached the championship game.

Steve Richardson

A center who played 72 games in the WHA, Richardson finished his career with the Whalers, going scoreless in six games during the 1975-76 season. In 1974-75, Richardson divided time with the Indianapolis Racers and Michigan-Baltimore club, going 9-22-31 in 66 games. He is credited with scoring the first goal in the history of the WHA Racers, at 1:18 of the second period on October 17, 1974 in a 4-2 loss to Michigan. Richardson wound up with the Stags on November 27, 1974 along with Steve Andrascik in a deal for Jacques Locas and Brian McDonald.

Terry Richardson

Though he never played a minute for the Whalers, this goaltender was involved in Hartford's first-ever trade once the franchise joined the NHL. Taken 11th overall in the 1973 Entry Draft by the Detroit Red Wings, Richardson

went on to play in 20 games. During the Expansion Draft in 1979, the Whalers claimed Ralph Klassen from the Colorado Rockies and sent the centerman to the New York Islanders on June 14, 1979 for Richardson. In 1979-80, Richardson spent the year in the minors at Springfield of the American League, going 15-22-7 with a 3.65 GAA in 46 appearances.

Doug Roberts

One of the first U.S.-born players to make a mark in the NHL, this Detroit native played 419 games over 10 NHL seasons before jumping to the WHA. Roberts had stops in Detroit, Oakland, California and Boston before joining New England in 1975. He logged two winters with the Whalers. A reliable player at either forward or defense, Roberts had 43-103-146 in the NHL and 7-31-38 in 140 games in the WHA, Doug helped in the hockey introduction of his younger brother, Gordie Roberts, who signed with the Whalers at 17. Doug's son, David Roberts, has also launched a pro career with stops in St. Louis, Edmonton and Vancouver.

Gordie Roberts

The first-U.S. born native to play 1,000 games in the NHL, Roberts began the journey with the Whalers in 1979-80, four years after he signed with New England in the WHA as a teen-ager out of Victoria of the Western League at 17. Roberts played 418 games with the Whalers

Gordie Roberts blossomed into a dependable blue liner for other teams in the NHL. Though he joined the WHA with the Whalers as a teen-ager in 1975 by signing a big-dollar contract, he has often regretted the decision. "There was a bidding war of sorts back then," Roberts says. "I had good times in the WHA but as a young player, I was thrown into a situation where there were a lot of older players. The league tried to make every young guy the second coming of Bobby Orr or Phil Esposito. It wasn't like going to a young organization where you could learn a system and develop."

and was often the target of catcalls from the Civic Center faithful in the early years of his apprenticeship at the blue line. Jack Kelley, who signed Roberts, also dealt the defenseman to Minnesota on December 16, 1980 for Mike Fidler. The newcomer wound up playing just 40 games for the Whalers. Roberts, meanwhile, went on to patrol the blue line for the North Stars, St. Louis Blues and Boston Bruins before retiring in 1993-94, a span of 1,097 NHL games with 61-359-420. His first NHL goal was also the first in franchise annals by Hartford, a tally on October 11, 1979 at 14:15 of the third period in a 4-1 loss at Minnesota. Born in Detroit, Roberts began his career with the Whalers who also signed his older brother (Doug) to assist in the transition from the amateur to pro ranks.

Jimmy Roberts

In his only year behind the bench in Hartford, this veteran of five Stanley Cup winners appeared to have the touch in 1991-92 when the Whalers opened the year at 5-1-1 and eventually played their longest Stanley Cup game in history, a 3-2 loss in double overtime at Montreal on May 2, 1992. That epic Game 7 defeat would be the last playoff match involving Hartford in the NHL and also the finale for Roberts once managerial changes beckoned. Eddie Johnston was dismissed as general manager and new operations director Brian Burke, though he

interviewed Roberts for the position, opted to hire Paul Holmgren as bench boss. Roberts, the seventh coach of the Whalers in the NHL, garnered 65 points in 80 games as the hockey club went 26-41-13. When he was elevated by the Whalers and hired on June 7, 1991, Roberts noted, "I accept this challenge because of a burn that has kindled in me. We have got to get something going in this area. There's enough long faces. We want to give people something to yell about." Three who may have yelled the most were farmhands who helped Roberts succeed in the minors, notably winger Yvon Corriveau, defenseman Marc Bergevin and goaltender Kay Whitmore. All had strong years in Hartford under Roberts, much like they did in the American League where Roberts led the Springfield Indians to back-to-back Calder Cups (1990 and 1991).

Torrie Robertson

Hartford's all-time penalty leader (1,368 minutes in 299 games) was a constant fan favorite as well as a nemesis for opposing clubs during his career which ran 10 NHL seasons. Drafted 55th overall by Washington in 1980, Robertson came to the Whalers on October 3, 1983 in a trade for winger Greg Adams and became a willing combatant, racking up 337 penalty minutes in 1984-85 and a club-record 358 penalty minutes in 1985-86. The

*Center **Mike Rogers** was the first Whaler to finish among the NHL's Top 10 scorers in consecutive seasons, doing that in 1979-80 and 1980-81 by linking 105-point campaigns. His departure to the New York Rangers in a 3-for-1 swap days before the 1981-82 season opener simply fleeced the Whalers of a proven scorer. In exchange for Rogers, general manager **Larry Pleau** added **Doug Sulliman**, **Chris Kotsopoulos** and **Gerry McDonald**.*

Torrie Robertson played in 299 games for the Whalers and usually kept the penalty-clock timer on his toes as he set a franchise record for penalties and minutes (1,368).

following year, Robertson played in just 20 games once he suffered a broken left leg on November 29, 1986 in a scuffle with Montreal's Shayne Corson. Though he came back in 1986-87 to rack up 293 penalty minutes to lead Hartford, the leg injury limited Robertson's overall effectiveness and eventually led to a trade to Detroit. Had 34 goals as a Whaler highlighted by a hat trick in a 6-4 win against Boston on March 4, 1984. Moved on to the Red Wings on March 7, 1989 for Jim Pavese where he wrapped up it up in 1989-90, finishing with career totals of 442 NHL games played, 49-99-148 and 1,751 penalty minutes.

Mike Rogers

"Mighty Mike" played for the Whalers in both leagues, a speedy and creative centerman who led Hartford in scoring during the club's first two NHL seasons when he linked back-to-back campaigns of 105 points to finish among the league's Top 10 scorers. Drafted by Vancouver (77th overall) in 1974, Rogers opted to sign with the WHA's Edmonton Oilers and was voted the league's Most Sportsmanlike Player in 1974-75. Joined the Whalers on January 19, 1976 in a deal for Wayne Carleton and logged the bulk of his 396 WHA games as a steady playmaker,

Mike Rogers has a step here on Cincinnati's Ron Plumb in this WHA game played at the Hartford Civic Center. Rogers had the ability to create points and score in the pivot. The only other centerman in club annals to do it as consistently was Ron Francis.

often pivoting a line bookended by George Lyle and Tom Webster. When the leagues merged in 1979, Vancouver declined to reclaim Rogers, a decision which turned into a bonanza for the Whalers. In Hartford's run to the playoffs in its maiden NHL voyage, Rogers anchored the top line with Pat Boutette and Blaine Stoughton on the flanks. The "Bash-Dash-Stash" trio was among the best ever in club annals. Moved on to the New York Rangers on October 2, 1981 in a questionable deal engineered by general manager Larry Pleau, a 3-for-1 swap which fetched forward Doug Sulliman and defensemen Chris Kotsopoulos and Gerry McDonald to Hartford and fleeced Hartford of an established playmaker to go with promising center Ron Francis. With the Blueshirts, Rogers rang up another 100-point season before finishing up his 484-game career in the NHL with Edmonton in 1985-86. Had 202-317-519 in the NHL to go with 145-222-367 in the WHA.

Tom Rowe

A gritty forward who logged seven NHL seasons, Rowe held a league record for goals (31) by a U.S.born player which has been eclipsed over the years (it is currently held by Jimmy Carson and Kevin Stevens who have both notched 55 goals in a year). Drafted 37th overall by

Right wing Tom Rowe logged 115 games over three seasons with the Whalers. In postseason play, Rowe was the first Hartford player to score two goals in a Stanley Cup game.

Ron Ryan coached the Whalers for close to two seasons, winning the Eastern Division title in 1973-74 and piloting the club to first place the following season before an illness, a few weeks before the 1975 playoffs opened, resulted in Jack Kelley to returning to the bench.

Jim Rutherford became the third former goaltender to hold the post of general manager of the Whalers, taking the baton on June 28, 1994. Rutherford spent 14 years in the NHL as a player once getting tabbed in 1969 as Detroit's top draft selection. His best trade in three years at the Hartford helm was acquiring Kevin Dineen for a conditional pick from Philadelphia.

Washington in 1976, Rowe came to the Whalers on January 17, 1980 in a deal for Alan Hangsleben. Played 115 games over three seasons with Hartford but his physical style led to a number of injuries. He was the first Whaler to score two goals in a Stanley Cup game, netting them in a 4-3 overtime loss to Montreal on April 11, 1980 in what happened to be the final NHL game played by Hall of Famers Gordie Howe and Bobby Hull. Released midway in 1981-82, Rowe rejoined the Capitals and then concluded his playing career in Detroit in 1982-83 for a total of 357 games with 85-100-185. Worked as a scout and later as an executive for the Whalers. Most recently, Rowe was among the organizers who landed an expansion American League club for Lowell, Massachusetts, a club which was to begin play in 1997-98.

Pierre Roy

A defenseman with a good set of fists, Roy played one game for the WHA Whalers. Overall, Roy spent 316 games in the WHA, mostly with the Nordiques. He also played for the Cincinnati Stingers.

Jim Rutherford

Another former goaltender to hold the general manager's position with the Whalers, Rutherford was the last to hold that title in Hartford, taking the helm on June 28, 1994 and remaining in that post until the hockey club relocated to North Carolina for the 1997-98 season. With playing experience from stints with four organizations in the NHL and the guiding force for Compuware's amateur and junior hockey operations, Rutherford seemed ready and able to turn around the sagging fortunes of the Whalers. He made trades to acquire Kevin Dineen, Glen Wesley, Brendan Shanahan, Keith Primeau, Steve Chiasson and Jeff Brown. He signed free agents like Steven Rice and Kent Manderville. He also drafted promising players like forwards Jeff O'Neill and Sami Kapanen and goalie J.S. Giguere. The reshaping of the hockey club may have improved the talent base but not enough to earn a playoff berth during his three-year tenure. In fairness, the Whalers came close, particularly in their final run in Hartford, but sputtered down the stretch when majority owner Peter Karmanos announced he would move the franchise. A congenial person who also owned 10 percent of the hockey club, Rutherford can only be judged on the club's lack of success. His worst deal, sending three first-round draft picks to Boston for Wesley, did not pan out. Nor did elevating youthful Paul Maurice to coach the club. The acquisition of Shanahan from St. Louis for Chris Pronger was a steal at the time before becoming a public relations nightmare throughout the months leading up to the start of the 1996-97 season and subsequent weeks when Paul Coffey came aboard and viewed Hartford as oblivion. Rutherford's best deal was bringing Dineen back to Hartford.

Ron Ryan

During the club's WHA years, Ryan coached the Whalers for two seasons (1973-74 and 1974-75) as New England secured first place in the Eastern Division. A collegiate standout at Colby, Ryan joined the WHA when he was hired by Jack Kelley and went behind the bench in the franchise's second season when Kelley concentrated on general manager duties. In his only postseason chance, the Whalers, crippled by injuries, were upset by Chicago in seven games in 1974. The following season, Ryan collapsed from exhaustion at an airport in Toronto just prior to the 1975 AVCO Trophy playoffs, forcing Kelley to don the coaching hat again. It was during 1974-75 that Ryan coached the Whalers to their first game at the newly-constructed Hartford Civic Center, a 4-3 win in overtime against San Diego on January 11, 1975. In subsequent seasons, Ryan held various management positions with the Whalers until the team was absorbed into the NHL in 1979. He moved into cable television in the early 1980s with the launching of Prism New England, leading to an executive post with the Philadelphia Flyers.

S

Ulf Samuelsson

One of Hartford's most popular players and known to give rivals a "taste of the Swedish elbow" over the years, the Whalers tabbed "Ulfie" on the fourth round (67th overall) in 1982 and he quickly developed into a reliable force at the blue line. One of three Whalers to amass over 1,000 career penalty minutes, Samuelsson played 463 games with Hartford before moving on with Ron Francis and Grant Jennings to Pittsburgh on March 4, 1991 in a swap for John Cullen and Zarley Zalapski. An outstanding shotblocker and a disruptive needler on the ice, Samuelsson often tangled with Cam Neely of the Boston Bruins. Joined by Kevin Dineen to represent the Whalers at Rendez-Vous `87 in a memorable two-game series where the NHL played the Soviet Union at Quebec City. Helped the Penguins to consecutive Stanley Cups (1991 and 1992) before moving to the New York Rangers with Luc Robitaille on August 31, 1995 for Petr Nedved and Sergei Zubov.

Geoff Sanderson

The 36th overall pick in 1990 by Hartford, Sanderson emerged as one of the league's rising standouts at left wing by scoring 46 and 41 goals in back-to-back seaons. Blessed with speed, Sanderson had his best all-around season in 1995-96 as he scored 33 goals and became involved with play along the wall. "Sandy" represented Hartford in the 1997 All-Star Game at San Jose and had a goal in an 11-7 win by the East. He led the Whalers in scoring in 1996-97 with 36-31-67. One of nine players in Hartford annals to score a goal in his first NHL game (a 5-5 tie against Buffalo on March 30, 1991), Sanderson is one of 11 players to log over 400 games as a Whaler. He notched his first of six career hat trick on November 28, 1992 in a 4-3 against Boston.

*Though **Ulf Samuelsson** has left the fighting to others in recent years, that was not the case early in his career as he faced all challenges including this one when Mark Messier was playing for the Edmonton Oilers in the mid-1980s.*

Geoff Sanderson was among the finds in the Entry Draft by the Whalers, a second-round pick in 1990 who rang up a career-high 46 goals at age 20 in 1992-93. "Sandy" led Hartford in scoring in its final season, going 36-31-67 in 1996-97.

Jim Sandlak

Taken fourth overall in the 1985 Entry Draft by Vancouver, Sandlak came to Hartford as the player to be named later in the March 22, 1993 swap where the Whalers sent Murray Craven to the Canucks for Robert Kron and "future considerations." A litany of leg and arm injuries limited Sandlak to just 40 games over two years with the Whalers. "House" notched six goals with Hartford including the game-winner in a 4-1 decision over Washington on December 18, 1993. He also played in the 500th game of his NHL career as a member of the Whalers. Wound up back with Vancouver in 1995-96. Possessed the size (6-foot-4) and beef (219 pounds) to punish rivals in the corners and though he did have a 20-goal season in 1988-89, Sandlak never developed into the power forward which scouts projected him to be.

Dick Sarrazin

An original Whaler, Sarrazin jumped from the Philadelphia Flyers to the WHA. He played in 35 games with New England during 1972-73 before getting dealt to the Chicago Cougars for cash on January 8, 1973. Sarrazin played in 100 NHL games for the Flyers, going 20-35-55 with the bulk of his stats coming 1968-69 when the right wing had 16-30-46 in 54 games.

Jean Savard

Claimed by the Whalers from Chicago in the 1979 Expansion Draft, Savard wrapped up his 43-game NHL career with Hartford, failing to score a point in a game against Detroit, his only appearance as a Whaler. In previous trials with the Blackhawks who made the speedy pivot the 19th overall pick in the 1977 Entry Draft, Savard had 7-12-19 in 42 games, the bulk coming in 1977-78 (7-11-18 in 31 games).

Bob Schmertz

When the Whalers came into the sports world in 1971 as a charter franchise in the upstart World Hockey Association, original founders Howard Baldwin, John Coburn, Godfrey Wood and William Barnes recognized the necessity to find additional funding to make their

*Taken 67th overall by the Whalers in the 1982 Entry Draft, **Ulf Samuelsson** quickly emerged as a game-ready combatant. Samuelsson played with flair, gusto and never shied away from giving a rival winger a taste of his "Swedish elbow." He logged 463 games in Hartford before moving on to Pittsburgh.*

dream become a reality. Baldwin approached the owner of the Boston Celtics of the National Basketball Association and soon had the support of Bob Schmertz's bottomless well. Schmertz, who had never seen a hockey game, invested $350,000 to meet the team's initial financial obligations and a million-dollar line of credit to guarantee player contracts. Simply put, the Whalers had the economic power to be a beacon franchise in the newly-formed WHA. Schmertz, who had made his fortune in building planned retirement communities in New Jersey, was involved with the hockey club for three seasons before a massive occulsion in 1975 took his life. His love of sports, especially the satisfaction of winning, could never be underestimated. Neither will his financial contributions to the Whalers modest beginnings. "Our biggest break," recalls Baldwin, "was getting a guy like Bob Schmertz."

Maynard Schurman

A tall forward who was acquired from Philadelphia in the 1979 Expansion Draft, Schurman logged his only NHL time with Hartford in 1979-80, going scoreless in seven games.

Brit Selby

In one of two player moves during the 1972-73 season to complete the roster that produced the first AVCO Trophy, Selby came to the New England Whalers on November 1, 1972 from the Philadelphia Blazers. In 65 games, Selby, who won the NHL's Calder Trophy in 1966-67, still had the smarts and skills to deliver a career-best 13-29-42 on New England's second line. He added 3-4-7 in 13 playoff games before moving on to Toronto for Bob Charlebois on June 6, 1973. Played 347 NHL games with stops in Toronto, Philadelphia and St. Louis.

*One of the original Whalers, **Brad Selwood** played 431 games for New England in the WHA. He was a reliable force at the blue line. Selwood's best season was 1972-73 with 13-21-34 in 75 games.*

***Jim "House" Sandlak** managed to play only 40 games over two years with the Whalers. The big forward, taken fourth overall by Vancouver in 1985, was hampered by injuries. He notched six goals for Hartford including the game-winner in a 4-1 decision over Washington on December 18, 1993.*

Brad Selwood

A seven-year veteran in the WHA with the Whalers, Selwood came to New England with Rick Ley and Jim Dorey from the Toronto organization for greener pastures and virtually solidified the blue line with Ted Green. Selwood, drafted 10th overall by the Maple Leafs in 1968, used his hard-hitting style and positional play to log 431 games with the Whalers and 65 postseason games. Though injuries limited his play for two seasons, Selwood returned for a spirited year in 1976-77 before eventually moving on. Released on March 13, 1979, he signed with the Montreal Canadiens. Dealt to Los Angeles on September 14, 1979 for a flip-flop of fourth-round picks, Selwood wrapped up his 163-game NHL career in 1979-80 with the Kings.

Brendan Shanahan

"B.S." came to the Whalers on July 25, 1995 in one of the better swaps that actually first tilted in favor of the Whalers, but, within a year, Shanahan turned into a poster boy for the modern, disgruntled athlete. He went public and demanded a trade instead of accepting his fate, going forward and leading a team as most who wear the "C" do. Hartford sent Chris Pronger to St. Louis for Shanahan, a two-time 50-goal scorer who rated among the league's premier offensive threats. Originally drafted second overall by New Jersey in 1987, Shanahan utilized his 6-foot-3, 215-pound frame to play it physical as well as finish plays at the net. He left the Devils as a free agent to sign with the Blues, a move that cost St. Louis defenseman Scott Stevens who was awarded as compensation to New Jersey on September 3, 1991. In St. Louis, Shanahan was a popular player and was a first-team NHL selection in 1993-94 after going 52-50-102 in 81 games. Though many were impressed with his play, his departure with the Blues became obvious when new coach Mike Keenan began to feud with the moody winger and decided to deal him for Pronger, a young defenseman who could eventually blossom into a franchise-type cornerstone. Despite missing 10 games with a hand injury, Shanahan captained the Whalers in 1995-96 and led the Whalers in scoring with 44 goals and 78 points. Still, he seemed to float during games. His lack of focus caused more than one in management to accuse Shanahan of "dogging it" yet it was difficult to justify since the big winger had a rash of two-goal games and eventually notched a first hat trick in a 5-4 loss at Pittsburgh on February 23, 1996. When rumors that the Whalers might be exiting Hartford arose and grew louder over the summer months, Shanahan went public and requested a trade. He knocked the city and the club's loyal faithful. Throughout training camp, the Whalers were linked to several clubs in trade talks involving Shanahan. A likely destination was Detroit, a team which could afford Shanahan's salary as well as compensate the Whalers with a workable package of players. During the exhibition games and the first two home games of the 1996-97 season, Shanahan was booed unmercifully by the Civic Center crowd. One patron, dressed in a diaper, carried a pacifier the size of a baseball bat and taunted the forward, much to the delight of Hartford fans. On October 9, 1996, the Whalers sent Shanahan and rear guard Brian

The seats may have been empty behind **Brendan Shanahan** at a practice session at the Hartford Civic Center for an afternoon skate but the stands were full for his final two games as a Whaler. The Coliseum faithful gave B.S. an earful, extremely upset that Shanahan went public to demand a trade to a winning organization and hinted that the community would lose its hockey team. Outraged by the talk, team owner **Peter Karmanos**, who outraged civic and business leaders in the coming months once his ego proved bigger than common sense, sent "Shanny" to Detroit in a multi-player swap on October 9, 1996. In the end, Shanahan proved to be a great pundit. In addition to playing for a Stanley Cup champion, Shanahan watched from a distance as Karmanos pulled up stakes in Connecticut.

Glynn to the Red Wings for center Keith Primeau, defenseman Paul Coffey and a first-round pick in the 1997 Entry Draft. While Shanahan went on to score 47 goals for Detroit and helped spark a ride to the squad's first Stanley Cup in 42 years, Primeau soon emerged into a fan favorite in Hartford. Coffey, meanwhile, wanted no part of the Whalers. His tenure in Hartford lasted 20 games and ended in a swap to Philadelphia. Unlike Shanahan and Primeau, Coffey seemed to be nearing the end of a remarkable career that would leave the rear guard as the game's all-time scoring leader among the blue line fraternity.

Daniel Shank

An aggressive right wing who played a total of 77 NHL games, Shank wrapped up his stay in 1991-92 by playing 13 regular season and five playoff games with the Whalers. The Montreal native signed with Detroit as a free agent on May 26, 1989 and came to Hartford on December 18, 1991 in a swap for Chris Tancil. Since leaving Hartford, Shank has found a home in the International League. He reached All-Star status most notably in 1992-93 with San Diego, going 39-53-92 with a league-high 495 penalty minutes.

Brad Shaw

A mobile defenseman who played 361 games over parts of 10 seasons, Shaw broke into the NHL ranks with Hartford to launch a three-year stint as a regular with the Whalers which began in 1989-90 when he was voted to the league's All-Rookie team. Shaw capped his freshman season by leading Hartford in scoring in the playoffs. Drafted 86th overall by Detroit in 1982, Shaw refused to sign with the Red Wings and wound up coming to Hartford on May 29, 1984 for an eighth-round pick (Urban Nordin). Shaw played 210 games with the Whalers, was sold to New Jersey in a cash transaction on June 13, 1992 and wound up becoming a member of the Ottawa Senators days later when the NHL held its Expansion Draft. Shaw played three winters in Ottawa, notching a career-best 7-34-41 in 81 games in 1992-93.

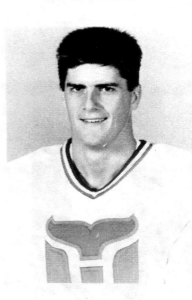

Brad Shaw was a three-year regular at the blue line for the Whalers, playing 11-84-95 in 210 games. Good at moving the puck, Shaw quarterbacked Hartford's power-play unit and capped a berth on the NHL All-Rookie Team in 1990 by leading the squad in postseason scoring in the series against Boston with 2-5-7.

Tim Sheehy had a much longer career than his brother Neil had with the Whalers. Tim played for the Whalers in both leagues, going 92-92-184 in 244 games. One of the keys to New England's title quest in 1973 was a line centered by Larry Pleau with John French and Sheehy on the flanks. Sheehy had 33 goals that winter and "All I had to do," Sheehy recalls, "was shoot the puck. I had two guys giving me great passes." In 1997, the one-time standout for the 1972 United States Olympic squad was enshrined in the U.S. Hockey Hall of Fame.

*Defenseman **Neil Sheehy**, who could play it tough, was the only player in Whalers history to wear the number "0" and did so for 26 games in 1987-88 once arriving in a trade from Calgary. Neil's older brother Tim was a standout for the Whalers in the WHA.*

Neil Sheehy

The brother of Tim Sheehy, Neil played 26 games at the blue line in Hartford once coming over from Calgary in a deal with Carey Wilson on January 3, 1988 for Shane Churla and Dana Murzyn. Had 116 penalty minutes to supply some grit on defense but comments made on television about "I still root for the Flames" made general manager Emile Francis reconsider why he brought Sheehy on board. Francis quickly dealt Sheehy, the only Whaler to wear 0 (zero) as a number, to Washington on July 6, 1988 for defenseman Grant Jennings and forward Ed Kastelic.

Tim Sheehy

Part of a brother tandem who played for the Whalers (Neil played for Hartford in 1987-88), Tim Sheehy had two stints with the Whalers in the WHA and one in Hartford's first NHL season for a total of 244 games. A member of the U.S. Olympic Team in 1972 which won the silver medal in the Olympiad at Japan, Sheehy scored 33 goals during New England's AVCO Trophy season in 1972-73 including a pair of game-winners in the playoffs. Followed up with 29 goals in 1973-74 and had 20 in 52 games in 1974-75 before a deal was made with Edmonton on February 15, 1975 which brought Ron Climie to the Whalers. Moved on to Birmingham before getting dealt to the NHL's Detroit Red Wings for Vaclav Nedomansky on November 18, 1977. Returned to play a dozen games in the NHL with the Whalers in 1979-80. Named to the U.S. Hockey Hall of Fame in 1997.

Gord Sherven

A defensive forward who had stints with Edmonton, Minnesota and with the Canadian Olympic Team, the Whalers secured Sherven from the Oilers in the 1986 Waiver Draft. He logged eight games in 1986-87, was then loaned to the Canadian Olympic squad and never resurfaced in the NHL. Overall, Sherven played in 99 NHL games over five seasons, going 13-22-35. Played collegiately at North Dakota en route to being drafted by Edmonton (197th overall in 1981).

Paul Shmyr

A veteran of 344 NHL games and one of three players to play in over 500 WHA games, this steady blueliner concluded his 14-year career in 1981-82 with the Whalers. Hartford purchased Shmyr from the Minnesota North Stars on October 1, 1981. He played in 66 games, scoring his last NHL goal against the New York Rangers, the same organizataion which signed him in 1967 as a free agent. The Blueshirts dealt Shmyr to Chicago for Camille Henry on August 17, 1967 and he made his NHL debut during the 1968-69 winter. On October 18, 1971, he was shuttled along with goalie Gilles Meloche to California for goalie Gerry Desjardins. Eventually jumped to the WHA where he played for Cleveland, San Diego and Edmonton en route to the third most career games (511) in league history (behind Andre Lacroix's 551 and Ron Plumb's 549). When the leagues merged in 1979, Shymr was reclaimed by Minnesota. He was a key contributor for the North Stars, most notably in 1981 when Minnesota reached the Stanley Cup Finals.

*A native of Poland, **Peter Sidorkiewicz** came to the Whalers along with center **Dean Evason** on March 12, 1985 from Washington for **David A. Jensen**. A member of the NHL All-Rookie team in 1988-89, "Sid" averaged 20-plus wins in his first three seasons with Hartford. He was claimed by Ottawa in the 1992 Expansion Draft.*

Peter Sidorkiewicz

One of the few Polish-born players to reach the NHL, Sidorkiewicz logged the bulk of his 245-game career with the Whalers over a four-year period which was highlighted by being named to the NHL's All-Rookie squad in 1988-89 by virture of a 22-18-4 mark with a 3.03 GAA. Originally tabbed 91st overall by Washington in the 1981 Entry Draft, "El Sid" came to Hartford along with center Dean Evason on March 12, 1985 for David A. Jensen. Made expendable by Hartford for the 1992 Expansion Draft, Sidorkiewicz was picked by the Ottawa Senators where he was voted to the midseason All-Star Game as the new club's first-ever representative. He moved on to New Jersey in a deal on June 20, 1993 and concluded his NHL career with the Devils in 1993-94. His career numbers (79-128-27 with a 3.60 GAA) are somewhat misleading because of Sidorkiewicz's tenure with the sad-sack Senators. In its inaugural campaign, Ottawa won just 10 games a reason for Sid's 8-46-3 ledger that winter.

Risto Siltanen

A speedy but diminutive defenseman who had a hard shot from the point, Siltanen quarterbacked the power-play unit over three years where he netted 29 of his 40 goals as a Whaler. Risto came to Hartford in a three-team deal on August 19, 1982 from Edmonton along with Brent Loney for Ken Linseman and Don Nachbaur in a transaction that also had the Whalers sending Mark Howe to Philadelphia. The first native of Finland to play for the Whalers was picked in 1978 as the 173rd overall pick by

***Risto Siltanen,** the first native of Finland to play for the Whalers, possessed a hard shot from the blue line for the Whalers and was at his best during the power play.*

St. Louis but Siltanen opted to sign with the WHA, playing 20 games with Edmonton in 1978-79. Though reclaimed by the Blues when the leagues merged in 1979, Siltanen wound up staying with the Oilers in a deal for Joe Micheletti on August 19, 1979. He exited Hartford in a memorable swap on March 3, 1986, a trade that brought John Anderson to the Whalers from Quebec, a move that solidified a playoff run including an upset of the first-place Nordiques in the opening round of the Stanley Cup tournament. Siltanen completed his eighth and final NHL campaign with the Nords in 1986-87 to wind up with 90-265-355 in 562 NHL games.

Al Sims

A 10-year veteran who joined the list of ex-Whalers to coach a team in the NHL (San Jose went 27-47-11 in 1996-97), Sims spent two years with Hartford, supplying steady play at the blue line while scoring 26 goals. Drafted in 1973 by Boston (47th overall), Sims played 310 games with the Bruins who deemed him expendable in the 1979 Expansion Draft which led to his arrival in Hartford. Sims missed only four games in his stay with the Whalers and wound up his 475-game career in 1982-83 with the Los Angeles Kings. Also played in 41 Stanley Cup playoff games, the most agonizing being Game 7 with Boston in 1979, a semifinal where the Bruins led 3-2 in the third period, only to see the Habs rally to tie on a bench-minor for too many men on the ice and win it in overtime on a goal by Yvon Lambert at 9:33. "I had always heard of players living all summer with one game," Sims recalls about that 5-4 loss. "I never believed it could happen. It did."

Benjamin Sisti

A general partner with Richard Gordon in ownership of the Whalers for less than two years, Sisti's association with hockey was cut short when his financial company fell under probe by federal investigators for fraudulent practices and collapsed. Authorities charged that Sisti and his partners bilked roughly 7,000 clients of $350 million over the years to bankroll West Hartford-based Colonial Realty with steady streams of monies each week for investment in Connecticut real estate holdings. Sisti was indicted as were Jonathan Googel, Frank Shuch and William Canderlori. The case came to trial and over 20 involved in the scandal have been hit with jail terms. Sisti, whose posh estate in Farmington was once valued at $32 million and contained gold fixtures in the many restrooms in the compound, was convicted on several counts. He has been incarcerated for a number of years.

Dale Smedsmo

One of the WHA's most-feared fighters, Smedsmo played 15 games with the Whalers during the 1976-77 season, netting two goals and 54 penalty minutes. New England purchased this bruising forward from the Stingers on November 29, 1976 and eventually sold him back to the Stingers on February 1, 1977. Smedsmo, who played four games for the Toronto Maple Leafs in 1972-73, completed his WHA tenure with Indianapolis in 1977-78.

*Claimed by the Whalers from the Boston Bruins in the 1979 Expansion Draft, **Al Sims** played 156 games for Hartford with his best winter being 16-36-52 in 1980-81. In 1996-97, Sims, after a long tenure coaching in the minors, was hired as bench boss of the San Jose Sharks.*

Al Smith

A competent goalie who played 290 games for the Whalers in both leagues, "Smitty" proved to be among the better netminders in WHA annals, leading the Whalers to the AVCO Trophy in 1973 and to the Finals in 1978. Played in the NHL with Toronto and Detroit before jumping to the WHA. After three seasons with New England, he returned to the NHL with Buffalo but after playing 21 games over two seasons, Smith rejoined the Whalers where he regained his winning form en route to first-team status on the league's All-Star Team. Eventually helped the Whalers launch their maiden NHL season in 1979-80. On September 4, 1980, Smith was sold to the Colorado Rockies. Holds several "firsts" in club annals, ranging from first shutout (both leagues), first win (WHA) and first postseason victory (WHA). Went 141-98-15 with a 3.25 GAA in 260 WHA games and 68-99-36 in 233 NHL games with a 3.46 GAA. Posted 10 shutouts in each league.

Guy Smith

A defenseman who played 40 games in the WHA and was a member of New England's AVCO Trophy champions in 1973, Smith was the nephew of TV's Tonto (Jay Silverheels). Had a strong collegiate career at New Hampshire and was one of several collegians recruited by Jack Kelley in the club's early history to play for the Whalers.

Al Smith was a pillar for the WHA Whalers, guarding the nets for most of the franchise's early years underscored in 1972-73 when the Whalers captured the league's first championship. After holding off Winnipeg 9-6 in the playoff title clincher, "Smitty" deadpanned, "the difference is that we had the better field goal kicker. Winning is the big thing." Smith played in more goals than any goalie in team annals and finished with a 152-108-16 record with a dozen whitewashes. Smith, in fact, holds the club mark for fewest saves by a netminder to post a shutout, turning aside 14 shots in a 3-0 win over the St. Louis Blues on February 29, 1980. The performance also was overshadowed that night as *Gordie Howe* notched his 800th NHL goal in the Hartford victory.

Stuart Smith

A blueliner who played 77 games over four seasons with the Whalers, Smith never reached the expectations of team scouts who made him the club's second overall pick (39th overall) in 1979. Such is the risk with underage prospects such as Smith who actually had an impressive junior career with the Peterborough Petes of the Ontario League, a powerhouse that won the Memorial Cup in 1979. Smith was a defensive-minded defenseman whose positional play did not completely compensate for slow skates. Notched his first goal as a Whaler with 4:52 left in regulation to tie the score at 4-4 against Los Angeles on January 14, 1980, a game which Hartford lost 5-4 on a late tally by Larry Murphy. Notched his final NHL goal against Detroit in 1981-82.

Kevin Smyth

The 79th overall pick in the 1992 Entry Draft, Smith, at 6-foot-2 and 217 pounds, seemed to have the size and hands to contribute on a regular basis. The big forward had trials over three NHL years in Hartford, going 6-8-14 in 58 games. He opted to sign with the International League in 1996-97 and went 14-17-31 in 38 games with the Orlando Solar Bears, a season where the forward suffered a serious facial injury when a puck hit him near the eye.

Scored his first NHL goal in a 4-2 win at Vancouver on February 6, 1994. Older brother of Ryan Smyth who was taken sixth overall by Edmonton in the 1994 Entry Draft, a winger who led the Oilers in goals (39) in 1996-97.

Ed Staniowski

One of several former Blues to resurface in Hartford during Emile Francis' stewardship, Staniowski played 19 games over two winters for the Whalers to wind up his 10-year NHL career. Taken 27th overall by St. Louis in 1975, Staniowski had an incredible junior career in 1973-74 at Regina as he was voted the top junior player in Canada on the strength of a 3.06 GAA in 62 regular season games and a 2.92 GAA in the playoffs in leading the Pats to the Memorial Cup championship. Besides the Blues, Staniowski also played for the Winnipeg Jets, the club Hartford shipped Mike Veisor to in a deal involving backup netminders on November 10, 1983. Though his mark was just 6-9-1 with Hartford, Staniowski did post the first-ever victories in franchise annals at the Spectrum in Philadelphia (6-5 on December 4, 1983) and the Met Center in Minnesota (5-3 on December 23, 1983). Played his final game in relief of Greg Millen on January 22, 1985, allowing one goal in an 8-5 loss to Montreal. Finished with career numbers of 67-104-21 in 219 appearances with a 4.06 GAA.

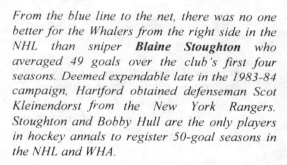

*From the blue line to the net, there was no one better for the Whalers from the right side in the NHL than sniper **Blaine Stoughton** who averaged 49 goals over the club's first four seasons. Deemed expendable late in the 1983-84 campaign, Hartford obtained defenseman Scot Kleinendorst from the New York Rangers. Stoughton and Bobby Hull are the only players in hockey annals to register 50-goal seasons in the NHL and WHA.*

Bob Stephenson

A forward who had 30-30-60 in 117 games over two seasons with the Birmingham Bulls, the Whalers obtained Stephenson in the WHA Dispersal Draft. He played four games with Hartford in 1979-80 before being shuttled to Toronto on December 24, 1979 for Pat Boutette, a deal which may rank as the best swap ever made by general manager Jack Kelley. Boutette racked up big numbers on a line with Mike Rogers and Blaine Stoughton, a reason why the Whalers qualified for the playoffs in their inaugural NHL campaign. The Leafs, meanwhile, thought they had obtained a netminder from Hartford, believing that general manage George "Punch" Imlach had secured Wayne Stephenson (then with Washington) from the Whalers. The Leafs managed to insert the newcomer from the Whalers into 14 games. Stephenson went 2-2-4 in his last chance in the NHL.

John Stevens

A good organizational player who played his best hockey in the American League, Steven logged 44 of 53 NHL games over three seasons in Hartford. Drafted 47th overall by Philadelphia in 1984, Stevens played nine games with the Flyers before signing with the Whalers as a free agent on July 30, 1990. Spent the bulk of his time in Springfield of the AHL during his tenure with the Whalers and was often recalled for fill-in duty. During one of those emergency promotions, Stevens assisted on goal by

Andrew Cassels at 3:15 of overtime to defeat Chicago on March 30, 1994. Rejoined the Flyers on August 6, 1996 and went 2-18-20 in 74 games for the Phantoms in helping Philadelphia's farm affiliate take the AHL's Mid-Atlantic Division crown at 49-21-10.

Jim Storm

A forward who played 74 games over two seasons in Hartford, Storm was a regular in 1993-94 but was released by the organization on August 28, 1995. He hooked on with Dallas where he played 10 games in 1995-96, going 1-2-3, after a stint in the International League with the Detroit Vipers. Originally drafted 75th overall in 1991 by the Whalers, Storm passed up his senior year at Michigan Tech to join the U.S. Olympic Program for the 1994 Winter Games. Following an exhibition game in Hartford on November 9, 1993, Storm decided to turn pro and signed with the Whalers.

Blaine Stoughton

One of the greatest goal-scorers to ever play for the Whalers, Stoughton ranks with Bobby Hull as the only players in hockey annals to score 50 goals in both the NHL and WHA. Originally a first-round pick by Pittsburgh (seventh overall in1973), "Stash" had just 34 goals over three NHL seasons with the Penguins and Toronto Maple Leafs before jumping to the WHA and scoring 52 goals in 1976-77 with the Cincinnati Stingers. Also played with Indianapolis in the WHA before signing with the Whalers as a free agent on December 13, 1978 when the Racers ceased operations. Remained with the club and emerged as one of the game's most prolific snipers during Hartford's first four NHL campaigns, leading the league with 56 lamplighters in 1979-80 and following up with campaigns of 43, 52 and 45. Stoughton's 196 goals over the four-year period was fourth in the league behind Wayne Gretzky (269), Mike Bossy (243) and Marcel Dionne (217). Had the luxury of finishing passes from two outstanding centers with Hartford, Mike Rogers and Ron Francis. Played 393 games with the Whalers before exiting on February 27, 1984 in a deal with the New York Rangers for defenseman Scot Kleinendorst. Had 258-191-449 in 526 NHL games.

Steve Stoyanovich

Taken by the New York Islanders (69th overall) in 1977, Stoyanovich played collegiately at RPI and came to the Whalers on August 23, 1983 for a fifth-round pick. Though the big left winger had a good scoring touch in the minors, in his only chance in the NHL, Stoyanovich had 3-5-8 in 23 games with Hartford.

Garry Swain

A small center in the Pittsburgh organization who played nine games with the Penguins in 1968-69, Swain welcomed a chance to play in the WHA and signed with the Whalers where he became a valuable specialist at killing penalties and pivoting the third line. Played 171 games over three years with New England with his biggest contributions coming in 1976-77 when he had a career-high in goals (10), points (26) and penalty minutes (79). Usually paired with Tommy Earl when New England was shorthanded, it was this duo that put the Whalers atop the pack during 1974-75 in penalty killing.

*One of the best penalty-killers in club annals, **Garry Swain** went 22-33-55 in 171 games over three seasons in the WHA. He scored the winning goal in the first pro game played in Hartford, a short-handed strike at 5:45 of the extra session for a 4-3 win over the San Diego Mariners.*

*Forward **Doug Sulliman** went 57-71-128 in 221 games over three years with the Whalers. His best year in Hartford was 29-40-69 in 77 games in 1981-82 to finish second on the club in scoring that winter.*

Doug Sulliman

The 13th overall pick by the New York Rangers in 1979, Sulliman played three seasons in Hartford once coming to the Whalers on October 2, 1981 along with Chris Kotsopoulos and Gerry McDonald in a trade for center Mike Rogers. Had a career-best 29-40-69 in 1981-82 and followed it up with 22 goals the next season before slipping to 6-13-19 in 1983-84 and getting released. Sulliman revived his career with New Jersey and logged another six years in the NHL, finishing up with Philadelphia in 1989-90, going 160-168-328 in 631 games. Versatile to play either wing and smart enough to play without the puck, Sulliman carved out a two-way career for 11 seasons. Notched his first hat track as Whaler in a 6-5 win at Chicago on January 25, 1982.

Bob Sullivan

A solid two-way player in the minors, Sullivan parlayed a big year in the American League to get the call for his only NHL season, making the Hartford roster in 1982-83. Originally drafted 116th overall by the New York Rangers in 1977, Sullivan signed with the Whalers as a free agent in 1981 and established a point-scoring streak of 28 consecutive games to break the AHL record. Earned league's top rookie honors in 1981-82. Sullivan had the hands to score 18-19-37 points with Hartford but lacked the skating quickness to stay on board for another campaign.

Forward Dave Tippett joined the Whalers after a hitch with he Canadian Olympic Team in 1984 and played 483 games for Hartford. Besides a dependable checker and faceoff man, Tippett holds the club record for most consecutive games played (419).

T

Chris Tancill

Taken by Hartford in the 1991 Supplemental Draft, this speedy winger had a remarkable collegiate career at Wisconsin where he copped MVP honors in 1990 as he led the Badgers to the NCAA championship. Played 19 games with the Whalers, scoring a short-handed tally against Montreal in 1990-91. Shuttled to Detroit on December 18, 1991 for rugged winger Daniel Shank, Tancill emerged as a prolific scorer in the minors, notching a league-high 59 goals to lead the American League in 1991-92 with Adirondack. Has also played for Dallas and San Jose, a total of 132 games over seven NHL seasons entering 1997-98.

Jim Thomson

A punishing forward who played 115 games with six different clubs over seven seasons, Thomson logged five games during 1988-89 in Hartford after coming over from Washington on March 6, 1989 for Scot Kleinendorst. Tabbed 185th overall by the Capitals in 1984, Thomson picked up 416 penalty minutes in his career along with four goals and three helpers. Dealt by the Whalers to New Jersey on October 31, 1989 for Chris Cichocki, Thomson also played for Los Angeles, Ottawa and Anaheim. He finished up with the Mighty Ducks in 1993-94.

Dave Tippett

Maybe the best free agent ever signed by the Whalers in the NHL, Tippett captained Team Canada at the 1994 Winger Games and signed with Hartford immediately after the Olympics. A reliable special-team's player who helped Hartford boast of its top-rated penalty-killing unit in the league in 1987-88, Tippett played 483 games for the Whalers. Though his best offensive year (17-24-41) was registered in 1988-89, the remarkable point about Tippett's stay in Hartford was his durability, dependability and heart. Once turning pro with Hartford and debuting in a 3-3 tie at Boston on March 3, 1984, Tippett logged 419 consecutive games before a thumb injury ended the streak on October 8, 1989. Understood the meaning of commitment to the team as evidence in his style and comment regarding his drop in ice time, "if this team is to improve, I should be playing less. I'm a role player and defensive specialist so I shouldn't get the same ice time as a big scorer or as much as the big name players." Dealt to Washington on September 30, 1990 for a six-round pick in 1992 (Jarrett Reid), Tippett also played in Pittsburgh and Philadelphia to conclude a career of 721 games with 93-169-262. Tippett has moved on to the coaching ranks, serving as bench boss of the Houston club in the International League.

Mike Tomlak

Drafted 208th overall by Toronto in 1983, Tomlak signed with Hartford on November 14, 1988 as a free agent and went on to play 141 games with the Whalers over parts of four seasons. A regular as a checking center for two years (1989-1991), Tomlak used his long reach and positional play to offset his lack of speed. Had two game-winning goals in his rookie year, one at Philadelphia in a 3-2 decision on December 14, 1988 (Hartford's third-ever road victory at the Spectrum) and the clincher in a 4-2 home ice win over Pittsburgh on March 25, 1990. Tomlak's most prolific season was 44-56-100 with Springfield of the American League in 1993-94, the year he also played his final NHL game, a 4-0 loss to New Jersey on March 10, 1994.

Jim Troy

A physical player from Boston who played three seasons in the WHA, Troy earned a look after amassing 201 penalty minutes in 49 games with the Beauce Jaros of the North American Hockey League in 1975-76. He signed with New England where he continued to play it physical, picking up 43 penalty minutes in 14 games. After seven more games in 1976-77 with the Whalers, Troy moved on to Edmonton where he wrapped up his 68-game tenure with two goals and 124 penalty minutes in 1977-78. Went into professional wrestling once his hockey days ended.

Al Tuer

A defenseman who could drop the gloves as fast as anyone, Tuer played 57 games in the NHL for three clubs, winding up with Hartford where he logged six games over two winters, closing out his career in 1989-90. Selected 186th overall by Los Angeles in 1981, Tuer often took a seat in the sin bin around the Western League as he led the circuit in penalty minutes (486) in 1981-82 at Regina. Climbed the minor league ladder and eventually broke in with the Kings in 1985-86. Also played with the Minnesota North Stars where he scored his only NHL goal in a loss to the New Jersey Devils in 1987-88. Signed with the Whalers as a free agent on July 12, 1988 but spent the bulk of his time playing in the American League at Binghamton. Released on July 1, 1990.

Darren Turcotte

A veteran of 524 NHL games entering 1997-98, "Turk" played 66 matches in Hartford over parts of two winters before a swap to Winnipeg on October 5, 1995 brought winger Nelson Emerson to Hartford. Notched 19 goals for the Whalers, often with a flair for dramatic ones such as the overtime strike at 2:59 to dump Boston 3-2 on February 22, 1995 or at 2:04 of the extra session to trim Montreal 4-3 on April 14, 1995. Tabbed by the New York Rangers in the 1986 Entry Draft (114th overall), Turcotte notched four straight years of 25 or more goals for the Blueshirts before coming to the Whalers on November 2, 1993 along with defenseman James Patrick in a multi-player swap that sent Steve Larmer (previously obtained from Chicago), Nick Kypreos and Barry Richter to the Blueshirts. Had 2-8-10 in his first 14 games with Hartford before suffering injuries to both knees which sidelined him for most of 1993-94. Scored a goal in his last game as a

*Center **Darren Turcotte** played 66 games over two seasons in Hartford before getting dealt to Winnipeg on October 6, 1995 for Nelson Emerson. "Turk" had 19 goals with the Whalers with two of his biggest netting wins in overtime - on February 22, 1995 at 2:59 to nip Boston 3-2; at 2:04 on April 14, 1995 to hang a 4-3 loss on Montreal.*

Whaler, a 4-1 loss at Quebec on May 3, 1995, the final regular season game ever played by the Nordiques at Le Colisee. Moved on to San Jose on March 16, 1996 in a deal involving Craig Janney. Reached the 500-game milestone during 1996-97 with the Sharks during a season he went 16-21-37 in 65 games to boost his career totals to 179-204-383.

Sylvain Turgeon

Drafted second overall (behind Brian Lawton) in the 1983 Entry Draft, "Sly" had some impressive goal-scoring seasons for Hartford until a torn stomach muscle in 1986-87 eclipsed his quick first step. Turgeon played 370 games with the Whalers, going 178-150-328 and highlighted by 40 goals as a rookie in 1983-84 and a high of 45 tallies in 1985-86 when representing the Whalers at the All-Star Game on February 4, 1986 (Turgeon was named as Hartford's rep when an injury sidelined Ron Francis) which was played before a sellout crowd of 15,126 at the Civic Center. Often a target of media barbs because of his limited tolerance to play with pain, Turgeon did net one of the biggest postseason goals in franchise annals, an overtime strike at 2:36 at Quebec in Game 1 of the 1986 playoffs to net Hartford's first-ever Stanley Cup victory, a 3-2 decision over the Nordiques which triggered a three-game sweep. Hartford dealt Turgeon to New Jersey on June 17, 1989 in a 1-for-1 transaction with the Devils which brought feisty winger Pat Verbeek to the Whalers. Turgeon potted 30 goals for the Devils but was dealt after one season to Montreal for Claude Lemieux on September 4, 1990. He wrapped up his 669-game NHL career with three campaigns in Ottawa, leading the expansion Senators in goals (25) in their maiden campaign in 1992-93. Finished with career totals of 269-226-495. Older brother of Pierre Turgeon, the first-overall pick in the 1987 Entry Draft by Buffalo.

*Pittsburgh goalie Roberto Romano appears to have the short side covered but is a tad late to stop **Sylvain Turgeon** from scoring.*

*Sylvain Turgeon was Hartford's top pick in 1983, second behind Minnesota's Brian Lawton and one ahead of Pat LaFontaine by the New York Islanders. "Sly" popped in 40 goals as a rookie in 1983-84 and 45 in 1985-86, his best years with the Whalers. His career took a turn following a series of injuries that limited his quick first step. On June 17, 1989, Hartford traded Turgeon for gritty winger **Pat Verbeek** in a 1-for-1 swap with the New Jersey Devils.*

V

Mike Veisor

A goaltender who spent 10 years in the NHL, Veisor played 69 of 139 career games for the Whalers over parts of three seasons. He joined Hartford in a swap with Chicago for a second-round pick (Kevin Griffin) in the 1981 Entry Draft. Though he caddied for Tony Esposito as a member of the Blackhawks for the majority of his stay in the NHL, Veisor enjoyed his best run during 1981-82 with Hartford. After a recall from the minors, "Wormy" won four starts to spark what ranked as the longest unbeaten streak in franchise annals, a stretch from January 20 to February 10, 1982 where the Whalers went 10 games (6-0-4) without a loss. Dealt to Winnipeg on November 10, 1983 for Ed Staniowski, Veisor finished his career with the Jets in 1983-84. He rejoined the Whalers after retiring and took over the club's amateur hockey development program for many years. Originally drafted by the Blackhawks as the 45th overall selection in 1972, Veisor posted a career mark of 41-62-26 with a 4.09 GAA. His son, Mike Jr., was a 12th round pick by St. Louis in 1991. The younger Veisor, also a goaltender, played collegiately at Northeastern.

Mike Vellucci

A defenseman who played for the Belleville Bulls in the Ontario Hockey League, the Whalers took this blueliner with the 131st overall pick in the 1984 Entry Draft. Vellucci missed a valuable year of seasoning when he was forced to rehab a neck injury he sustained in a car wreck. He gallantly returned to the game and eventually made it to the NHL for two games with Hartford. Vellucci collected an assist in his last game, a 4-3 loss to Buffalo on February 27, 1988.

Pat Verbeek

A gritty, feisty and durable player who ranks among the game's all-time Top 75 scorers and Top 20 in penalty minutes, Verbeek came to the Whalers on June 17, 1989 in a deal with the New Jersey Devils for Sylvain Turgeon. It was arguably the best trade that tilted in favor of the

*Goalie **Mike Veisor** views the action from the ice as **Mike Fidler** clears the zone and **Norm Barnes** manages to get his skates tangled up with the stick of Edmonton's Jari Kurri (behind Fidler) during a game at the Civic Center.*

Whalers made during general manager Eddie Johnston's "Reign of Error" yet it should be noted that the deal had been set in motion by outgoing general manager Emile Francis several weeks before it was announced at the Entry Draft in Minnesota. Verbeek played in 433 games for the Whalers and was one of the team's more identifiable players during the darker hours of the early 1990s. He served as team captain from 1992 until he was traded to the New York Rangers on March 23, 1995 in a 4-for-1 deal which brought Glen Featherstone, Michael Stewart and two 1995 draft picks (Hartford used the selections to tab goalie J.S. Giguere as the 13th overall choice and center Steve Wasylko as the 104th overall pick). Verbeek had a pair of 40-goal years with Hartford and led the club in scoring in 1993-94 with 75 points. Originally picked 43rd overall by the Devils in 1982, Verbeek played larger than his 5-foot-9, 190-pound size. He was at his best crashing the net. Often vocal, Verbeek played inspired hockey for the Whalers but suffered one of his worst years in 1991-92 when he was dogged in contract negotiations with management and produced just 22 goals. Off the ice, Verbeek was involved in the infamous after-hours altercation in Buffalo in 1994 when a number of players and coaches were arrested following a game. Once leaving the Whalers, Verbeek had a solid year with the Rangers in 1995-96, notching another 40-goal year, but he opted to exit the Blueshirts and signed as a free agent with the Dallas Stars. He earned over $3 million in 1996-97 when he played in his 1,000th career NHL game in helping the Stars to a 104-point campaign en route to the Central Division crown. In 81 games, Verbeek went 17-36-53.

Mickey Volcan

A mobile defenseman who turned pro once helping the U of North Dakota to the NCAA championship, Volcan signed with the Whalers who picked him as the 50th overall selection in the 1980 Entry Draft. When Volcan, then 18, made his pro debut on October 9, 1980, an 8-6 loss at St. Louis, he was, at the time, the youngest player in

the NHL. During his stay in Hartford, Volcan was empowered to wear the zebra stripes along with New Jersey's Garry Howatt when a blizzard on January 15, 1983 kept the assigned game officials from reaching the Hartford Civic Center rink for the opening face-off. Volcan played 143 games for the Whalers over parts of three winters, going 7-29-36. He was dealt to Calgary on July 5, 1983 for veteran blueliners Joel Quenneville and Richie Dunn in one of the first moves made by general manager Emile Francis. Volcan appeared in 19 games for the Flames in 1983-84, his final chance in the NHL.

Pat Verbeek *played 433 games with the Whalers, a two-way forward who moved on to the Rangers in a 4-for-1 swap on March 23, 1995, a transaction that included a first-round pick which Hartford used in 1995 to tab goalie J.S. Giguere as the 13th overall selection.*

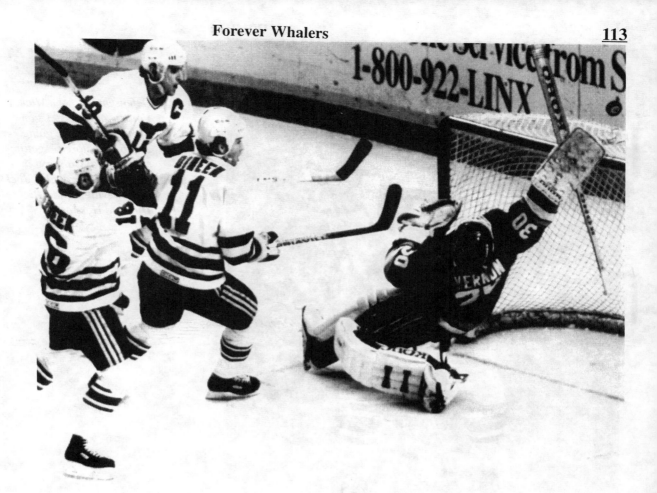

Pat Verbeek, Kevin Dineen and *Ron Francis* showed no fear going to the net and appear to have a clear path to goalmouth only to be foiled on a scoring bid here by Calgary netminder Mike Vernon. The trio had a prolific campaign in 1989-90, combining for 101 of Hartford's 275 goals scored that season.

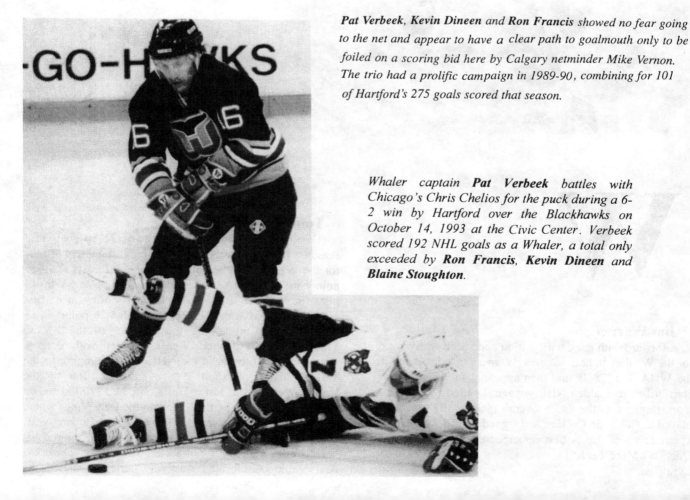

Whaler captain ***Pat Verbeek*** *battles with Chicago's Chris Chelios for the puck during a 6-2 win by Hartford over the Blackhawks on October 14, 1993 at the Civic Center. Verbeek scored 192 NHL goals as a Whaler, a total only exceeded by* ***Ron Francis***, ***Kevin Dineen*** *and* ***Blaine Stoughton***.

*Defenseman **Mickey Volcan** played 143 games with the Whalers, going 7-29-36 before getting shuttled to Calgary in a trade that brought **Joel Quenneville** and **Richie Dunn** to Hartford.*

Jim Warner

A forward with good anticipation and speed who played for the Whalers in both leagues, Warner made his debut in the WHA in 1978-79 and then appeared in 32 games for Hartford in its maiden NHL voyage. Drafted by the New York Rangers as the 245th overall pick in 1974, Warner played at Colorado College and signed with New England as a free agent. He is best remembered for "shadowing" Quebec's Marc Tardif for three games during the 1978-79 campaign.

Tom Webster

The first player in club annals to score 50 goals in a season, "Hawkeye" was one of the biggest gamers to play for the Whalers. He had a remarkable 220 goals and 435 points in 352 WHA games and added 28-26-54 in 43 playoffs. Webster led New England in scoring in its first season (53-50-103 in 1972-73) and added 26 points in 15 playoff games in helping the Whalers to the AVCO Trophy. A right winger, Webster played with Boston, Detroit and California in the NHL before jumping to the WHA Whalers. Drafted 19th overall in 1966 by the Bruins, Webster was lost to Buffalo in the 1970 Expansion Draft and was dealt by the Sabres to the Red Wings where he had a 30-37-67 season in 1970-71. The following year, Webster was traded to the Golden Seals and injured his back. Following surgery, Webster was unable to work a deal with California owner Charles Finley, the

cantankerous mogul who is best remembered for his meddling days with baseball's colorful and powerful Oakland A's of the early 1970s. When the WHA offer came, good fortune smiled on both the Whalers and Webster. "Charlie Finley didn't recognize the World Hockey Association at all and was offering me a contract for the same money I earned the year before I was hurt," Webster recalls. "What he didn't know is that I was negotiating with Jack Kelley. The Whalers offer was nearly three times the amount of coin that Finley had on the table. It was also guaranteed for three years which to me, was money in the bank. I really had no decision to make. In fact, a number of Seals, Gerry Pinder, Gary Jarrett, Paul Shymr, Norm Ferguson and Bobby Sheehan all jumped to the WHA. In all honesty, I was glad to get a chance to resume my career. The Whalers gave me that." After five terrific seasons in the WHA, Webster was off to a great start in 1977-78 on a line with Gordie and Mark Howe (15-5-20 in 20 games) when his lower back gave out. Though his career was virtually finished, Webster did play one game with Detroit in 1979-80 to wind up with 102 NHL games and 33-42-75. He began coaching in the minor leagues a short time later and eventually landed stints in the NHL with the New York Rangers and Los Angeles Kings. He returned to Hartford in 1996-97 as an assistant coach in the club's final run in Connecticut.

Steve Weeks

Taken 176th overall by the New York Rangers in 1978, Weeks was a standout goalie in college, anchoring Northern Michigan's run to the NCAA Finals in 1980. Within two years, he was stopping pucks in the NHL. He debuted with a 23-win season as a rookie with the Blueshirts in 1981-82 and soon settled into a backup role. Dealt to the Whalers on August 31, 1984 for a 1986 third-round pick (Shaun Clouston), Weeks was a reliable netminder, often getting the call moments before a game. He went 40-40-6 over four years with Hartford and might

*The date is March 27, 1975 and **Tom Webster** launches a blast during a 5-3 win over the Cleveland Crusaders. "Hawkeye" scored 220 goals for the Whalers in the WHA including 53 in 1972-73.*

*Goaltender **Steve Weeks** played 94 games for the Whalers over four seasons. Weeks had four shutouts for Hartford including the first one in team annals against the New York Rangers (5-0 on January 26, 1986) and against the Toronto Maple Leafs (3-0 on January 4, 1987).*

Mark Messier, then with the Edmonton Oilers, gets tripped in front of Hartford goalie Steve Weeks by defenseman Ulf Samuelsson. The latter usually found a way to needle a rival player, particularly Boston's Cam Neely.

best be remembered for stepping in when Mike Liut was injured in the 1986 Stanley Cup series against Montreal. Weeks stopped 18 shots in a 2-1 overtime win in Game 4 on April 23, 1986. Hartford shipped Weeks to Vancouver on March 8, 1988 in exchange for Richard Brodeur. He also made stops between the pipes for the New York Islanders, Los Angeles Kings and Ottawa Senators. Weeks had a career mark of 112-119-33 with a 3.73 GAA over 13 seasons. Returned to the Whalers to serve as an assistant coach, working with the goaltenders.

Eric Weinrich

Selected 32nd overall by New Jersey in the 1985 Entry Draft, this mobile defenseman logged 87 games over two seasons with the Whalers of a career which is nearing 500 NHL games. A member of the 1988 U.S. Olympic team, Weinrich soon became a regular for the Devils before moving in a deal on August 28, 1992 which brought goalie Sean Burke to Hartford in exchange for center Bobby Holik. A short time into his second year with the Whalers, Hartford shipped Weinrich and forward Patrick Poulin to Chicago on November 2, 1993 for winger Steve Larmer and defenseman Bryan Marchment. In 1996-97, Weinrich went 7-25-32 in 81 games for the Blackhawks.

Wally Weir

A rugged defenseman who played a combined 470 games in both leagues, Weir became a Whaler in the 1984 Waiver Draft when he was claimed from Quebec. Though Weir added toughness at the blue line, his tenure in

Hartford lasted just 34 games before he was cut loose. He hooked on with Pittsburgh on March 1, 1985 for his final pro season. Weir played the bulk of his games for the Nordiques and finished with 21-45-66 with 625 penalty minutes in 320 NHL games. In 150 WHA games with the Nords, Weir went 5-24-29 with 410 penalty minutes. He was a member of Quebec's AVCO Trophy champs in 1977 when he led all players with 50 penalty minutes in the postseason.

Blake Wesley

A member of the famous Portland (WHL) Winterhawks, a juggernaut Canadian junior team that lost just five games in 1978-79 and included a litany of future pros, Wesley was Philadelphia's second-round pick (22nd overall) in 1979. He played parts of two winters for the Flyers and came to Hartford on July 3, 1981 in a multi-player deal which brought Rick MacLeish and Don Gillen to the Whalers in exchange for Ray Allison, Fred Arthur and a flip-flop of draft slots on the first three rounds in the 1982 Entry Draft. Wesley enjoyed his most productive seasons (9-18-27 in 78 games) in 1981-82 with the Whalers, then slipped on the depth chart and was peddled to Quebec for blueliner Pierre Lacroix on December 3, 1982. He played 119 games for the Nordiques before signing with Toronto where he finished his 298-game career in 1985-86. Blake is the older brother of Glen Wesley, a first-round pick by Boston in 1987 who joined the Whalers in a trade for three draft choices on August 25, 1994.

Tom Webster has position here against Darryl Maggs of Indianapolis during WHA action at the Hartford Civic Center. Webster is wearing the jersey used once the Whalers relocated to Hartford, a crest on the sweater that featured a "W" with a harpoon. The team colors were green, gold and white.

Obtained in a deal with Boston for three first-round draft choices, Glen Wesley had just 16-56-72 in 184 games for the Whalers over three seasons, a dramatic drop in scoring from his days with the Bruins. Though he led Hartford's rearguards in scoring in 1996-97 with 6-26-32, Wesley settled into a more defensive role with the Whalers.

Glen Wesley

Most general managers are reluctant to trade high draft choices yet Hartford's Jim Rutherford agreed to ship three first-rounders to Boston on August 25, 1994 in a swap to land this established defenseman, who at the time, was a free agent and was unable to fetch a big-dollar pact from the Bruins. While the jury will stay out for some time to completely assess this deal (the Bruins took Kyle McLaren in 1995, Jonathan Aiken in 1996 and Sergei Samsonov in 1997), Wesley did supply stability and quality minutes during his three-year tenure with the Whalers yet Hartford failed to make the playoffs each winter. Though Wesley's scoring production sharply dropped from his days with the Bruins, the veteran was the club's most consistent performer at the blue line, especially in the defensive zone. As a player with over 700 NHL games under his belt, Wesley welcomed a trade out of Boston to "finally get out of Ray Bourque's shadow" and establish his own identity. With the Bruins, Wesley, whose older brother Blake also played for the Whalers, felt his game was not appreciated or understood. While he holds a distinction as being one of just three defensemen in club annals to score a hat trick (joining Bourque and Bobby Orr), Wesley also was part of hockey's most lopsided deal in recent memory. On June 6, 1986, the Bruins sent center Barry Pederson to Vancouver in exchange for winger Cam Neely and a first-round pick in 1987. In addition to landing a big-play forward who cracked the 50-goal plateau with regularity before a leg injury prematurely ended his career after the 1995-96 season, the B's used the draft selection to take Wesley in the third overall slot. In the coming years, it remains to be seen if the 3-for-1 package that the Bruins received from Rutherford will result in another coup for Boston.

Whalers

The nickname for a professional hockey team that began play in the WHA in 1972-73 and finished its 25-year history in the NHL in 1996-97 when ownership decided to relocate operations to North Carolina. The moniker was coined by Ginny Kelley, the wife of Jack Kelley who served as general manager of the club in both leagues. Early New England commerce was built around the sea and the whaling industry. The name was a natural fit in Boston and later Connecticut where the team shifted operations to Hartford in 1975. The franchise played 1,975 regular season games and 123 playoffs games in its combined WHA and NHL history. The first game played was October 11, 1972, a 4-3 win over the Philadelphia Blazers. The finale was April 13, 1997, a 2-1 victory over the Tampa Bay Lightning before a sellout crowd of 14,660. The players saluted the faithful by raising their sticks. Over 25 winters, losses often outnumbered wins, but there was always hope for next year. The difference in 1997-98 and beyond, there would be no team called "Whalers" in professional hockey again. An era had ended.

HARTFORD WHALERS

HOME

AWAY

A look at the various logos in club annals, ones used in the World Hockey Association or National Hockey League during the franchise's 25 years in professional hockey. The Whalers began in the WHA and played for virtually two seasons in Boston, using the Boston Garden or Walter Brown Arena for their home games. The team shifted operations to southern New England for the 1974 playoffs and the postseason games against the Chicago Cougars were played at the Big E in West Springfield, Massachusetts. The team remained docked at the Big E for a portion of the 1974-75 campaign before relocating to the Hartford Civic Center. The Whalers played at the HCC until the roof collapse on January 18, 1978 and returned to Springfield to play at the Civic Center until the complex in Hartford was rebuilt. The first game at the "new" Civic Center was played February 6, 1980, a 7-3 crowning of the Los Angeles Kings. The Whalers played the remaining home games of their NHL history at the Civic Center before the lights went out for the final time, a 2-1 win over Tampa Bay on April 13, 1997.

*Goalie **Kay Whitmore** turns away a shot before Adam Graves of the Rangers can do something with the puck in action here on October 12, 1991 at the Hartford Civic Center. The Whalers beat the Blueshirts 5-2 as Whitmore, playing with confidence, keyed a 5-1-1 start out of the gate that season.*

Kay Whitmore

In a draft year that yielded netminders Sean Burke, Troy Gamble and Mike Richter, the Whalers believed they had also tabbed a goalie with a future by taking Whitmore (26th) overall on the second round in 1985. Whitmore had a couple of sharp years at Peterborough of the Ontario League and won his NHL debut on March 29, 1989, stopping 23 shots to beat Boston 4-2. In the playoffs against Montreal, Whitmore started two games, losing twice including 4-3 in overtime in the Game 4 clincher when Russ Courtnall scored at 15:12 of the extra session. Though he logged a handful of starts over the next two years, Whitmore's only real opportunity with the Whalers came in 1991-92 under Jimmy Roberts. It was Roberts who saw what the youngster could do under pressure, especially in the playoffs when Whitmore went 11-4 with a 2.40 GAA in to spark the Springfield Indians to the 1991 Calder Cup and copped MVP postseason honors. Roberts, who was elevated to the NHL following that season, took the reins in Hartford and inserted Whitmore between the pipes, saying it "was his job to lose." After opening the year with a 5-1-1 run that included his first of four career shutouts (a 3-0 verdict over San Jose on October 23, 1991), Whitmore managed just nine wins over his next 34 decisions to finish at 14-21-6, a slump that forced general manager Eddie Johnston to swing a deal to acquire Frank Pietrangelo from Pittsburgh. Though Whitmore played the bulk of his 75 career NHL games that season with the Whalers, a change in command in Hartford was imminent and when Burke was acquired from New Jersey, that made Whitmore expendable. He was dealt away on October 1,

1992 to Vancouver for Corrie D'Alessio and future considerations. Whitmore did revive his career by going 36-22-4 over the next two years with the Canucks. He eventually came to the New York Rangers on March 20, 1996 at the trading deadline for Joey Kocur. Overall, Whitmore's stats were 59-61-16 with a 3.51 GAA over 149 appearances.

Dave "Tiger" Williams

The most penalized player in hockey history wrapped up his career with the Whalers in 1987-88, playing the final 28 games of his career that spanned 14 seasons. Drafted by the Toronto Maple Leafs as the 31st overall pick in 1974, Williams also played for Vancouver, Detroit and Los Angeles before joining Hartford. "Tiger" played a total of 962 games with 241-272-513. With the Whalers, Williams had six goals and 93 penalty minutes to push his time in the sin bin to a staggering 3,966 minutes or roughly 66 games. Williams played with gusto and a passion to finish a check. The only other players in NHL history to reach 3,000 penalty minutes in a career are Dale Hunter and Chris Nilan.

Tommy Williams

One of the first U.S.-born players to make a mark in professional hockey, Williams carved out a 15-year career in the game which included two years with the WHA Whalers. A Minnesota native, Williams came to prominence by helping the Americans capture the gold medal at Squaw Valley in the 1960 Winter Olympiad.

*Forward **Tommy Williams**, a hero on the 1960 U.S. Olympic Team that won the gold medal at Squaw Valley, was among the first members of the Whalers in 1972-73. Williams had 31 goals and 58 assists over 139 games over two seasons. In the franchise's first game on October 12, 1972, Williams scored the first goal in team annals, knocking home a feed from **Tom Earl** and **John Cunniff** at 13:42 of the first period. It helped New England open the campaign with a 4-3 win over the Philadelphia Blazers.*

Shortly after, "Boomer" launched his pro career and in his second NHL year, he netted 23 goals for Boston in 1962-63. Williams spent eight seasons with the Bruins before getting dealt to Minnesota. The centerman also played for the California Golden Seals and wound up back in Boston where he played with the Braves of the American League in 1971-72. When the parent Bruins assigned him back to the minors, Williams elected to jump to the WHA and he proved to be a valuable contributor on the 1972-73 AVCO Trophy champions. In New England's first of 557 regular season games in the team's WHA history, Williams, assisted by Tommy Earl and John Cunniff, is credited with the first goal scored by the Whalers. The tally came against goalie Bernie Parent at 13:42 of the first period on October 12, 1972 in a 3-2 victory at Boston Garden over the Philadelphia Blazers before 14,552 cheering fans. Williams decided to retire after two seasons with the Whalers when he was claimed by Indianapolis in the 1974 WHA Expansion Draft. However, Williams changed his mind and opted to sign with the Washington Capitals of the NHL where he closed out his career in 1974-75, a run of 663 games with 161-269-430.

Carey Wilson

A center who had two tours of duty with Hartford, Wilson played 125 games with the Whalers. It was his 38 points in 36 games after coming over from Calgary on January 3, 1988 (with defenseman Neil Sheehy and the rights to Lane MacDonald for Dana Murzyn and Shane Churla) that fueled a run to the playoffs. After 34 games the following year, the Whalers shuttled Wilson to the New York Rangers on December 26, 1988 to obtain center Brian Lawton, defenseman Norm Maciver and veteran winger Don Maloney, a deal that didn't help either team. Wilson, a creative player inside the blue line, eventually returned to Hartford on July 9, 1990 in a deal for Jody Hull and a draft pick, a choice the Whalers used in 1991 to pick Michael Nylander on the third round. Wilson's stay this time with the Whalers lasted 45 games and ended with a deal to Calgary on March 5, 1991 for winger Mark Hunter. The swap brought Wilson back to the Flames, a club he broke into the NHL with following the 1984

***Dave "Tiger" Williams** holds the hockey record for most career penalty minutes (3,966). A sparkplug for his team and the villian in every ice arena during his 14 seasons, Williams finished his 962-game career with the Whalers in 1987-88, getting six goals and 93 penalty minutes in 26 games.*

*Center **Carey Wilson** came to the Whalers from Calgary in a trade on January 3, 1988 and went 18-20-38 in 36 games to key a drive to the playoffs that winter. Wilson had two cracks with the Whalers, going 37-46-83 in 125 games.*

*Scott **Young** was Hartford's top pick in 1986 out of Boston University. Able to play either wing or defense, Young had 49-89-138 in 197 games before a trade to Pittsburgh on December 2, 1990 brought winger **Rob Brown** to the Whalers.*

Winter Games. Overall, Wilson finished with 169-258-427 in 552 NHL games. Originally drafted 67th overall by Chicago in 1980, Wilson played two years at Dartmouth as well as two years in Europe. His father, Dr. Gerry Wilson, played a key role in hockey's growth in the 1970s. Dr. Wilson served as an executive with the Winnipeg Jets of the WHA and was involved in recruiting Swedish players who had a major impact on the game in North America, most notably Anders Hedberg, Ulf Nilsson and Lars-Erik Sjoberg.

Terry Yake

It took this speedy center several years in the Hartford chain to finally crack the lineup and Yake, after scoring 22 goals in 1992-93 to rank third on the club (behind Geoff Sanderson and Pat Verbeek) in lamplighters, was cut adrift for the 1993 Expansion Draft. Claimed by Anaheim, Yake wound up leading the Mighty Ducks in scoring (21-31-52) in the club's inaugural campaign. Sent to Toronto in a deal for David Sacco on September 9, 1994, Yake eventually wound up with Buffalo and has since played a key role in developing players at Rochester of the American League. In 1996-97, Yake was second in the AHL's scoring race with 101 points. Originally taken 81st overall by the Whalers in the 1987 Entry Draft, Yake had brief trials throughout his days in Hartford. In 1992-93 when the Whalers managed only 14 road wins, Yake had two of the game-winners, the decisive markers in a 4-2 decision at Ottawa on November 19, 1992 and also in a 7-5 decision at San Jose on December 3, 1992. In 205 NHL games, Yake had 48-70-118 including 24-37-61 in 104 for the Whalers.

Ross Yates

One of many who tried to stick in the pivot in Hartford, Yates had several productive years in the minors yet only gained a brief look with the Whalers. Signed by Hartford as a free agent out of Mt. Allison University, Yates led the American League in scoring in 1982-83 (going 41-84-125) and was voted the league's playoff MVP that season. The numbers earned him a chance with the parent club in 1983-84. In seven games, Yates went 1-1-2 with his only goal coming in a 6-5 overtime loss to New Jersey on February 16, 1984.

Scott Young

One of several top draft choices by the Whalers that blossomed once traded to a team blessed with more good fortune and depth, Young has played for Stanley Cup winners in Pittsburgh (1991) and Colorado (1996). Taken 11th overall in 1986 out of Boston University, Young had the ability to play two-way hockey and was versatile to be effective at both forward and defense. A member of the 1988 U.S. Olympic Team which included Brian Leetch and Craig Janney, Young joined the Whalers following the Olympiad and quickly became a regular. Dealt to the Penguins for Rob Brown on December 21, 1990, Young was a contributor in Pittsburgh's first-ever Stanley Cup. Unable to hammer out a contract for the following year, Young went to Europe in 1991-92 and eventually played a second time in the Olympics. He wound up inking a deal with Quebec and in 1992-93, he scored a career-high 30 goals. Young stayed with the franchise which relocated to Denver for 1995-96 and helped the Avalanche to the Cup. A booming shot from the point, Young has played over 600 NHL games. He is a durable and effective player whose ability to play without the puck cannot be overlooked. In 72 games with Colorado in 1996-97, the Clinton, Mass. native went 18-19-37.

Mike Zuke anchored the Hartford special teams for two years once coming over from St. Louis in the 1983 Waiver Draft. The veteran center, along with Mike Crombeen, formed one of the better penalty-killing units in club annals.

Zarley Zalapski, more than John Cullen, may have been the key piece in Hartford's big swap with Pittsburgh on March 4, 1991 involving Ron Francis and Ulf Samuelsson. Zalapski is the only defenseman in club annals to score 20 goals in a season (1991-92) yet played most nights without passion. Within two years, he was gone in a multi-player shuffle with the Flames.

Zarley Zalapski

The possible centerpiece in a deal with Pittsburgh that involved Ron Francis and Ulf Samuelsson on March 4, 1991, Zalapski arrived in Hartford with center John Cullen and easily could have neutralized this one-sided trade if he played with any emotion, pride or passion during his 289-game tenure in Hartford. A tremendous skater with impressive skills, Zalapski put together one of the best offensive years ever by a Hartford rear guard in 1992-93 (a club record 20 goals by a defenseman and also 51 helpers) but it came within limits. Zalapski's on-ice vision was inconsistent and he never seemed to reach the next hurdle to become a player with "impact status." Projected as a take-charge force at the blue line and selected fourth overall (behind Joe Murphy, Jimmy Carson and Neil Brady) in 1986 by Pittsburgh, Zalapski joined the Penguins after playing for Team Canada in the 1988 Winter Olympics. Though he represented the Whalers at the 1993 All-Star Game, Zalapski has yet to fulfill expectations of most NHL observers. On March 10, 1994, Hartford shipped Zalapski to Calgary with James Patrick and Michael Nylander in a deal that brought Paul Ranheim, Ted Drury and Gary Suter to the Whalers (the latter was soon dealt to Chicago for Frantisek Kucera and Jocelyn Lemieux).

Mike Zuke

One of the best face-off specialists to play for the Whalers, Zuke was at his best killing penalties and playing without the puck during his 159-game hitch over three seasons. When Emile Francis took over the general manager position in Hartford, the "Cat" quickly set out to improve the club's inept special steams. In the 1983 Waiver Draft, the Whalers took Zuke, Mike Crombeen and Bobby Crawford from St. Louis. Zuke's arrival revived the penalty-killing unit that soon ranked among the league's best for two seasons (in 1984-85 and again in 1985-86, Hartford was fourth among 21 teams). The Civic Center faithful often howled "Zoooooooooooooook" when coach Jack Evans sent out the "Zuke" patrol, a two-man forward unit anchored by this clever centerman with Crombeen on the flank which often frustrated a rival's power play. Originally a fifth-round pick by St. Louis in 1974 (79th overall), Zuke played at Michigan Tech and sparked the Engineers to the NCAA Finals in 1976. Once turning pro, Zuke opted to sign with the WHA and played for Indianapolis and Edmonton before joining the Blues in 1978-79 to launch a 455-game NHL career. With St. Louis and Hartford, Zuke went 86-196-282.

Moments after the final hour of hockey in Hartford, the last crew to wear the colors of the Whalers raise sticks aloft to salute the Civic Center faithful.

Whalers 2, Lightning 1

April 13, 1997
at the Hartford Civic Center

FIRST PERIOD

1. Hartford, Wesley 6 (Rice, Kron), 2:30.

Penalties: Burke, Hartford (roughing), served by Murray 7:48; Cassels, Hartford (slashing), 19:12.

SECOND PERIOD

No scoring.

Penalties: Burr, Tampa Bay, minor-misconduct (roughing), 0:00; Muzzatti, Hartford, minor-misconduct (roughing), 0:00; Poeschek, Tampa Bay, double-minor (high-sticking), 4;37; Wiemer, Tampa Bay (holding), 11:38.

THIRD PERIOD

2, Hartford, Dineen 19 (Sanderson, Cassels), :24.

3, Tampa Bay, Ciccarelli 35 (Gratton), 2:50.

Penalties: None.

SHOTS ON GOAL

Tampa Bay 13 10 16--39
Hartford   8   9   6--23

POWER-PLAY CHANCES

Tampa Bay 0-for-2, Hartford 0-for-3

GOALIES

Tampa Bay, Tabaracci (23 shots, 21 saves)
Hartford, Burke (39-38)

ATTENDANCE

14,660 (sellout)

REFEREE & LINESMEN

Scott Zelkin, Kevin Collins and Brian Murphy

Seasons in Review, Year-by-Year

1972-73 League Standings

World Hockey Association

Eastern Division	W	L	T	Pts	GF	GA
New England	46	30	2	94	318	263
Cleveland	43	32	3	89	287	239
Philadelphia	38	40	0	76	288	305
Ottawa	35	39	4	74	279	301
Quebec	33	40	5	71	276	313
New York	33	43	2	68	303	334
Western Division	W	L	T	Pts	GF	GA
Winnipeg	43	31	4	90	285	249
Houston	39	35	4	82	284	269
Los Angeles	37	35	6	80	259	250
Alberta	38	37	3	79	269	256
Minnesota	38	37	3	79	250	269
Chicago	26	50	2	54	245	295

League Champion: New England Whalers
League MVP: Bobby Hull, Winnipeg
Top Scorer: A. Lacroix, Philadelphia
(50-74-124 in 78 GP)

Pleau's hat trick powers Whalers to WHA crown

May 6, 1973

BOSTON - People like Larry Pleau are the reason the New England Whalers were founded.

Players like Pleau are reasons why the Whalers won the first-ever WHA championship.

Pleau, a native of Lynn, Mass., left home at 15 to learn to play hockey in Canada. He scored two of the biggest goals of his life Sunday afternoon to scuttle a blossoming comeback by the Winnipeg Jets.

"It seemed like every time they came close," Pleau said after the Whalers held off the Jets 9-6 in the clincher, "we were able to pull away."

The Whalers took the title series in five games. New England built leads of 4-1 and 6-2 but Winnipeg rallied in the third period to close within 6-5 on Bob Woytowich's tally.

Pleau, who scored a short-handed breakaway in the first period, provided relief for the home side by scoring twice within a 1:47 span of the third period to ice it for the Whalers. Both strikes were set up by linemates Tim Sheehy and John French.

Goalie Al Smith stopped 35 saves as he wrapped up his 12th postseason victory. The dozen playoff wins came in 15 games for New England which won a league-best 46 games during the league's inaugural campaign.

"We had a great season," Smith said. "We started from scratch and got all the way here to the victory."

In the jubilance of triumph, team founder Howard L. Baldwin issued a challenge to the National League that his Whalers would play the Stanley Cup winner to determine who has the best team in pro hockey.

"With a team like ours," Smith suggested, "I know we wouldn't be embarrassed."

Whalers Season Statistics

1972-73	Regular Season								Playoffs							
	GP	G	A	Pts	PM	PP	SH	GW	GP	G	A	Pts	PM	PP	SH	GW
Webster	77	53	50	103	83	12	0	6	15	12	14	26	6	3	0	0
Caffery	74	39	61	100	14	7	1	7	8	3	7	10	0	0	0	0
Pleau	76	39	48	87	44	10	0	5	15	12	7	19	15	4	1	2
Sheehy	78	33	38	71	30	6	0	4	15	9	14	23	13	2	0	3
Dorey	74	7	56	63	95	0	1	1	15	3	16	19	41	1	0	0
French	76	24	35	59	59	2	0	3	15	3	11	14	0	0	0	1
Green	78	16	30	46	47	2	0	3	12	1	5	6	25	0	0	0
Selby	72	13	30	43	52	2	0	2	13	3	4	7	13	1	0	1
Ahearn	78	20	22	42	18	1	0	2	14	1	2	3	9	0	0	0
Danby	76	14	23	37	10	1	0	4	8	0	0	0	0	0	0	0
Selwood	75	13	21	34	110	1	0	1	15	3	5	8	22	0	0	0
Williams	69	10	21	31	16	1	0	1	15	6	11	17	2	2	0	1
Ley	77	3	27	30	108	0	0	2	15	3	7	10	24	1	0	0
Earl	77	10	13	23	4	0	4	2	15	2	3	5	10	0	0	2
Hurley	77	3	15	18	60	1	0	0	15	0	7	7	14	0	0	0
Byers	19	6	4	10	4	2	0	0	12	6	5	11	6	0	0	2
Hyndman	59	4	14	18	21	1	0	0	--	--	--	--	--	--	--	--
Sarrazin	35	4	7	11	0	1	0	0	--	--	--	--	--	--	--	--
Cunniff	31	3	5	8	16	0	0	0	13	1	1	2	2	0	0	0
G.Smith	24	3	3	6	6	0	0	1	11	2	0	2	4	0	0	0
Jordan	36	1	5	6	21	0	0	0	4	0	0	0	0	0	0	0
A.Smith	50	0	1	1	37	0	0	0	15	0	1	1	12	0	0	0
Landon	28	0	1	1	8	0	0	0	--	--	--	--	--	--	--	--
Bench	78	--	--	--	18	--	--	--	15	--	--	--	0	--	--	--
Totals	78	318	529	847	852	49	6	46	15	70	120	190	220	14	1	12
Opponents	78	263	418	681	939	45	6	30	15	49	80	129	234	13	0	3

Goaltending	GP	Mts	W	L	T	Sho	GGA	GP	Mts	W	L	Sho	GAA
A.Smith	50	3059	31	19	1	3	3.18	15	909	12	3	0	3.23
Landon	28	1671	15	11	1	1	3.58	--	--	--	--	--	-----
Totals	78	4730	46	30	2	4	3.32	15	909	12	3	0	3.23
Opponents	78	4680	30	46	2	1	4.08	15	906	3	12	0	4.50

Webster becomes first in club annals to reach 50 goals

March 29, 1973

OTTAWA - Tom "Hawkeye" Webster reached the 50-goal plateau for New England but his milestone marker failed to stop a mild slide for the first-place Whalers.

Gavin Kirk's hat trick sparked the Nationals to a 5-2 win over New England which has dropped three straight after clinching the Eastern Division title. Goalie Gilles Gratton had a strong game in goal for Ottawa.

Webster, who leads the Whalers in scoring, became the first in team annals to reach the coveted 50-goal marker. He is the fifth player to accomplish the feat in the WHA's first season.

1973-74 League Standings

1973-74 World Hockey Association

Eastern Division	W	L	T	Pts	GF	GA
New England	43	31	4	90	291	260
Toronto	41	33	4	86	304	272
Cleveland	37	32	9	83	266	264
Chicago	38	35	5	81	271	273
Quebec	38	36	4	80	306	280
New Jersey	32	42	4	68	268	313
Western Division	**W**	**L**	**T**	**Pts**	**GF**	**GA**
Houston	48	25	5	101	318	219
Minnesota	44	32	2	90	332	275
Edmonton	38	37	3	79	268	269
Winnipeg	34	39	5	73	264	296
Vancouver	27	50	1	55	278	345
Los Angeles	25	53	0	50	239	339

League Champion: Houston Aeros
League MVP: Gordie Howe, Houston
Top Scorer: Mike Walton, Minnesota (57-60-117 in 78 GP)

Whalers Season Statistics

	Regular Season								Playoffs							
	GP	G	A	Pts	PM	PP	SH	GW	GP	G	A	Pts	PM	PP	SH	GW
French	77	24	48	72	31	5	1	4	7	4	2	6	2	2	0	2
Webster	64	43	27	70	28	5	0	9	3	5	0	5	7	3	0	0
Pleau	75	26	43	69	35	6	0	3	2	2	0	2	0	0	0	0
Karlander	75	20	41	61	46	4	0	1	7	1	3	4	2	1	0	0
Blackburn	75	20	39	59	18	9	0	2	7	2	4	6	4	1	0	0
Sheehy	77	29	29	58	22	2	0	1	7	4	2	6	4	1	0	0
Williams	70	21	37	58	6	3	0	5	7	4	2	6	4	0	0	0
Harris	75	24	28	52	78	1	0	4	7	0	4	4	11	0	0	0
Byers	78	29	21	50	6	4	0	4	7	2	4	6	12	0	0	0
Dorey	77	6	40	46	134	0	0	0	6	0	6	6	26	0	0	0
Ley	43	6	35	41	148	3	0	1	7	1	5	6	18	0	0	0
Selwood	76	9	28	37	91	0	0	2	7	0	2	2	11	0	0	0
Green	75	7	26	33	42	3	0	2	7	0	4	4	2	0	0	0
Earl	78	10	10	20	29	2	0	0	7	0	2	2	2	0	0	0
Hurley	52	3	11	14	21	0	0	2	--	--	--	--	--	--	--	--
Cunniff	30	4	7	11	6	0	0	2	5	1	1	2	0	0	0	0
Charlebois	74	1	5	6	21	0	2	1	7	0	0	0	4	0	0	0
G.Smith	16	1	5	6	25	0	0	0	--	--	--	--	--	--	--	--
Danby	72	2	2	4	6	0	0	0	7	1	0	1	0	1	1	0
Jordan	34	0	3	3	14	0	0	0	7	0	0	0	6	0	0	0
A.Smith	56	0	3	3	33	0	0	0	7	0	0	0	0	0	0	0
Keeler	1	0	0	0	0	0	0	0	--	--	--	--	--	--	--	--
Landon	24	0	0	0	0	0	0	0	1	0	0	0	0	0	0	0
Berglund	3	0	0	0	0	0	0	0	--	--	--	--	--	--	--	--
Bench	78	--	--	--	18	--	--	--	7	--	--	--	2	--	--	--
Totals	78	291	488	779	875	47	3	43	7	23	42	65	123	7	1	3
Opponents	78	260	414	674	880	36	7	31	7	24	39	63	155	3	0	4

Goaltending	GP	Mts	W	L	T	Sho	GGA	GP	Mts	W	L	Sho	GAA
A.Smith	56	3194	30	21	2	2	3.08	7	399	3	4	0	3.16
Landon	24	1386	11	9	2	0	3.55	1	40	0	0	0	4.50
Berglund	3	180	2	1	0	0	3.33	--	--	--	--	--	-----
Keeler	1	1	0	0	0	0	0.00	--	--	--	--	--	-----
Totals	78	4759	43	31	4	2	3.23	7	439	3	4	0	3.28
Opponents	78	4758	31	43	4	5	3.59	7	440	4	3	0	3.14

Pro hockey coming to city
Civic Center to house WHA Whalers

February 22, 1974

HARTFORD - The Whalers are coming. The Whalers are coming.

Pro hockey's New England Whalers will relocate to Connecticut and base operations at the Hartford Civic Center which is targeted to open Jan. 1, 1975.

Club officials and the Hartford Downtown Council have agreed to a deal where several corporations in banking and insurance have entered a partnership to bring the World Hockey Association team to the capital city.

The Whalers, currently atop the Eastern Division standings, have operated in Boston since Howard L. Baldwin secured a franchise in 1972. The franchise was the best in the WHA's first year, winning the league championship.

The hockey club would become the prime tenant in the $35 million Civic Center which is currently under construction. The building will accommodate 10,507 for hockey.

The Whalers have also announced they will play all postseason games in West Springfield, Mass. at the Eastern States Coliseum this season. The Big E arena also will serve as the club's home for a portion of 1974-75 before Baldwin's sextet docks in Hartford.

1974-75 League Standings

World Hockey Association

Eastern Division	W	L	T	Pts	GF	GA
New England	43	30	5	91	274	279
Cleveland	35	40	3	73	236	258
Chicago	30	47	1	61	261	312
Indianapolis	18	57	3	39	216	338

Canadian Division	W	L	T	Pts	GF	GA
Quebec	46	32	0	92	331	299
Toronto	43	33	2	88	349	302
Winnipeg	38	35	5	81	322	293
Vancouver	37	29	2	76	256	270
Edmonton	36	38	4	76	279	279

Western Division	W	L	T	Pts	GF	GA
Houston	53	25	0	106	369	247
San Diego	43	31	4	90	326	268
Minnesota	42	33	3	87	308	279
Phoenix	39	31	8	86	300	265
Michigan/Baltimore*	21	53	4	46	205	341

*franchise relocated to Baltimore on January 29, 1975

League Champion: Houston Aeros
League MVP: Bobby Hull, Winnipeg
Top Scorer: Andre Lacroix, San Diego

(41-106-147 in 78 GP)

Whalers Season Statistics

			Regular Season							Playoffs						
	GP	G	A	Pts	PM	PP	SH	GW	GP	G	A	Pts	PM	PP	SH	GW
Carleton	73	35	39	74	50	3	0	1	6	2	5	7	11	1	0	0
Webster	66	40	24	64	52	7	0	9	3	0	2	2	11	0	0	0
Pleau	78	30	34	64	50	6	0	6	6	2	3	5	14	0	0	0
French	75	12	41	53	28	2	1	3	4	1	2	3	0	0	0	0
Caffery	67	15	37	52	12	2	0	1	--	--	--	--	--	--	--	--
Blackburn	50	18	32	50	10	7	0	3	5	1	2	3	2	0	0	0
Byers	72	22	26	48	10	1	0	3	6	2	2	4	2	1	0	0
Ley	62	6	36	42	50	1	1	2	6	1	1	2	32	0	0	1
Selwood	77	4	35	39	117	3	0	0	5	1	0	1	11	1	0	0
O'Donnell	76	21	15	36	84	0	0	3	3	0	0	0	15	0	0	0
Sheehy	52	20	13	33	18	3	0	2	--	--	--	--	--	--	--	--
Abrahamsson	76	8	22	30	46	2	0	0	6	0	0	0	24	0	0	0
Hurley	75	3	26	29	36	0	0	1	6	0	1	1	4	0	0	0
Swain	66	7	15	22	18	0	2	2	6	0	3	3	41	0	0	0
Dorey	31	5	17	22	43	1	1	0	--	--	--	--	--	--	--	--
Karlander	48	7	14	21	2	0	0	1	5	0	3	3	0	0	0	0
Green	57	6	14	20	29	1	0	3	3	0	0	0	2	0	0	0
Climie	25	8	4	12	12	0	0	2	6	3	0	3	0	1	0	0
Earl	72	3	8	11	20	0	1	0	6	1	1	2	12	0	1	1
Fotiu	61	2	2	4	144	0	0	0	4	2	0	2	27	0	0	0
Hangsleben	26	0	4	4	8	0	0	0	6	0	3	3	19	0	0	0
Charlebois	8	1	0	1	0	0	0	0	4	1	0	1	0	0	1	0
Methe	5	0	1	1	4	0	0	0	2	0	0	0	0	0	0	0
Berglund	2	0	0	0	0	0	0	0	--	--	--	--	--	--	--	--
Landon	7	0	0	0	0	0	0	0	--	--	--	--	--	--	--	--
Abrahamsson	16	0	0	0	0	0	0	0	--	--	--	--	--	--	--	--
Smith	59	0	0	0	18	0	0	0	6	0	0	0	2	0	0	0
Ouimet	1	0	0	0	0	0	0	0	--	--	--	--	--	--	--	--
Bench	78	--	--	--	8	--	--	--	6	--	--	--	0	--	--	--
Totals	78	291	488	779	875	47	3	43	7	23	42	65	123	7	1	3
Opponents	78	260	414	674	880	36	7	31	7	24	39	63	155	3	0	4

Goaltending	GP	Mts	W	L	T	Sho	GGA	GP	Mts	W	L	Sho	GAA
C.Abrahamsson	16	870	8	6	1	1	3.24	--	--	--	--	--	-----
Landon	7	339	2	3	0	0	3.36	--	--	--	--	--	-----
A.Smith	59	3494	33	21	4	2	3.47	6	366	2	4	0	4.59
Ouimet	1	20	0	0	0	0	9.00	--	--	--	--	--	-----
Totals	78	4758	43	30	5	3	3.46	6	366	2	4	0	4.59
Opponents	78	4757	30	43	5	1	3.43	6	367	4	2	1	2.78

Whalers break ice in Hartford

Swain's overtime goal sinks Mariners

January 11, 1975

HARTFORD - Pro hockey opened with a big splash Saturday night at the Hartford Civic Center.

The new team had to work overtime. Nobody complained about spending more time in the city's new playpen or with the results.

Garry Swain's unassisted short-handed goal at 5:43 of the extra session enabled New England to net a 4-3 victory over the San Diego Mariners.

Swain, a centerman noted for his defensive play, stole the puck from Kevin Morrison near the San Diego blue line. He took a couple of strides toward the cage and just fired at goalie Ernie Wakely.

The red light went on and the sellout crowd of 10,507 roared in approval.

"I took a quick peek and hit the corner," Swain said.

Rick Sentes of the Mariners, assisted by Andre Lacroix, scored the first-ever goal in the newly opened Veterans Memorial Coliseum, beating Al Smith at 5:43 of the opening period.

New England's first goal was scored at 11:21 when Wayne Carleton and Tom Webster collected helpers on a goal by Don Blackburn.

Fred O'Donnell and Carleton also connected for the Whalers. Sentes tallied for San Diego as did Michel Rouleau. The latter's goal, with 5:51 left in the third period, forced overtime.

1975-76 League Standings

World Hockey Association

Eastern Division	W	L	T	Pts	GF	GA
Indianapolis	35	39	6	76	245	247
Cleveland	35	40	5	75	273	279
New England	33	40	7	73	255	290
Cincinnati	35	44	1	71	285	340
Canadian Division	W	L	T	Pts	GF	GA
Winnipeg	52	27	2	106	345	254
Quebec	50	27	4	104	371	316
Calgary	41	35	4	86	307	282
Edmonton	27	49	5	59	268	349
Toronto	24	52	5	53	335	398
Western Division	W	L	T	Pts	GF	GA
Houston	53	27	0	106	341	263
Phoenix	39	35	6	84	302	287
San Diego	36	38	6	78	303	290
Minnesota*	30	25	4	64	211	212
Denver/Ottawa**	14	26	1	29	134	172

*franchise disbanded on March 9, 1976. ** francise relocated to Ottawa on January 2, 1976, disbanded January 16, 1976

League Champion: Winnipeg Jets
League MVP: Marc Tardif, Quebec
Top Scorer: Marc Tardif, Quebec (71-77-148 in 81 GP)

Whalers ambush Soviets
Raeder outduels Tretiak, 5-2

December 27, 1976

HARTFORD - The New England Whalers harpooned the mighty Soviet Nationals, a team considered one of the best in the world that features goalie sensation Valdislav Tretiak.

Rookie goalie Cap Raeder stopped 31 shots to key a stunning 5-2 win by the injury-crippled Whalers who played without blueliners Brad Selwood, Rick Ley and Thommy Abrahamsson.

New England's special teams were also effective, negating seven of eight power plays chances.

The Soviets, who are scheduled to play seven more exhibitions against WHA clubs, should be well over the 14-hour jet lag for the remainder of the series. The long flight may have left coach Boris Kulagin's squad a bit heavy-legged.

Still, it was a remarkable performance by Harry Neale's club which has had difficulty playing break-even hockey since the puck was dropped in early October.

"I wouldn't classify this our usual effort but we didn't play our usual opponent so maybe it brought out the best in us," Neale said.

George Lyle had a pair of goals while Tom Earl, Gary Swain and Gary MacGregor also tallied for New England. Alexander Maltsev and Alexander Yakushev scored for the Red Army.

Whalers Season Statistics

			Regular Season								Playoffs					
	GP	G	A	Pts	PM	PP	SH	GW	GP	G	A	Pts	PM	PP	SH	GW
Webster	55	33	50	83	24	9	0	11	17	10	9	19	6	3	0	0
Pleau	75	29	45	74	21	12	2	4	14	5	7	12	0	1	0	1
Paiement	80	28	43	71	89	4	1	2	17	4	11	15	41	1	0	2
Climie	65	25	20	45	17	6	0	3	--	--	--	--	--	--	--	--
Ley	67	8	30	38	78	2	0	0	17	1	4	5	49	1	0	0
Abrahamsson	63	14	21	35	47	3	2	1	17	2	4	6	15	1	0	0
Backstrom	38	14	19	33	6	0	1	1	17	5	4	9	8	2	0	0
Carleton	35	12	21	33	6	2	0	0	--	--	--	--	--	--	--	--
Rogers	36	18	14	32	10	0	0	2	17	5	8	13	2	1	0	2
Swain	79	10	16	26	46	1	3	0	17	3	2	5	15	0	1	0
Hangsleben	78	2	23	25	62	0	0	0	13	2	3	5	20	1	0	1
O'Donnell	79	11	11	22	81	0	0	3	17	2	5	7	20	0	0	1
G.Roberts	77	3	19	22	102	0	0	1	17	2	9	11	36	1	0	0
Earl	66	8	11	19	26	0	6	2	17	0	5	5	4	0	0	0
Borgeson	31	9	8	17	4	5	0	1	3	1	1	2	0	0	0	1
D.Roberts	76	4	13	17	51	0	0	0	17	1	1	2	8	1	0	0
Arndt	69	8	8	16	10	1	0	1	8	0	0	0	0	0	0	0
Hurley	16	0	14	14	20	0	0	0	--	--	--	--	--	--	--	--
McManama	37	3	10	13	28	1	0	0	12	4	3	7	4	0	0	0
Selwood	40	2	10	12	28	0	0	0	17	2	2	4	27	1	0	0
Bolduc	14	2	5	7	14	0	0	0	16	1	6	7	4	0	0	1
Byers	21	4	3	7	0	0	0	0	--	--	--	--	--	--	--	--
Charlebois	28	3	3	6	0	0	0	1	--	--	--	--	--	--	--	--
Fotiu	49	3	2	5	94	0	0	0	16	3	2	5	57	0	0	1
Blackburn	21	2	3	5	6	0	0	0	--	--	--	--	--	--	--	--
Busnick	11	0	3	3	55	0	0	0	17	0	2	2	14	0	0	0
Gateman	12	0	1	1	6	0	0	0	--	--	--	--	--	--	--	--
Landon	38	0	1	1	0	0	0	0	4	0	0	0	0	0	0	0
Abrahamsson	40	0	1	1	6	0	0	0	1	0	0	0	0	0	0	0
Hoganson	4	0	1	1	0	0	0	0	--	--	--	--	--	--	--	--
Danby	1	0	0	0	0	0	0	0	--	--	--	--	--	--	--	--
Raeder	2	0	0	0	0	0	0	0	14	0	1	1	0	0	0	0
Richardson	6	0	0	0	0	0	0	0	--	--	--	--	--	--	--	--
Troy	14	0	0	0	43	0	0	0	2	0	0	0	29	0	0	0
Caffery	2	0	0	0	2	0	0	0	--	--	--	--	--	--	--	--
Bench	80	--	--	--	32	--	--	--	17	--	--	--	15	--	--	--
Totals	80	255	424	679	1012	46	15	33	17	53	89	142	374	14	1	10
Opponents	80	290	469	759	1072	56	4	40	17	40	67	107	347	15	1	7

Goaltending	GP	Mts	W	L	T	Sho	GGA	GP	Mts	W	L	Sho	GAA
C.Abrahamsson	41	2385	18	18	2	2	3.42	1	0	0	0	0	0.00
Landon	38	2181	14	19	5	0	3.47	4	197	3	0	0	2.13
Hoganson	4	224	1	2	0	0	4.29	--	---	--	--	--	-----
Raeder	2	100	0	1	0	0	4.80	14	819	7	7	2	2.27
Cooley	--	--	--	--	--	--	----	1	0	0	0	0	0.00
Totals	80	4890	33	40	7	2	3.51	17	1016	10	7	2	2.36
Opponents	80	4885	40	33	7	6	3.06	17	1018	7	10	3	3.06

1976-77 League Standings

World Hockey Association

Eastern Division	W	L	T	Pts	GF	GA
Quebec	47	31	3	97	353	295
Cincinnati	39	37	5	83	354	303
Indianapolis	36	37	8	80	267	305
New England	35	40	6	76	275	290
Birmingham	31	46	4	66	289	309
Minnesota*	19	18	5	43	136	129

Western Division	W	L	T	Pts	GF	GA
Houston	50	24	6	106	320	241
Winnipeg	46	32	2	94	366	291
San Diego	40	37	4	84	284	283
Edmonton	34	43	4	72	243	304
Calgary	31	43	7	69	252	296
Phoenix	28	48	4	60	281	383

*franchise disbanded on January 20, 1977

League Champion: Quebec Nordiques
League MVP: Robbie Ftorek, Phoenix
Top Scorer: Real Cloutier, Quebec (66-75-141 in 76 GP)

Goalie without a team shines brighest for WHA

January 18, 1977

HARTFORD - Just another strange chapter in the Wild, er, World Hockey Association.

Goalie Louis Levasseur may be without the team he began the season with but he found a way to get to the Civic Center Tuesday and do what he has been done all season: frustrate those who shoot the puck.

Levasseur made the stops in helping the East to a 4-2 win over the West before 10,337 at the Hartford Civic Center.

The only save Levasseur couldn't make this season involved keeping his Minnesota Fighting Saints afloat.

Either could the front office in St. Paul.

Levasseur had the league's best goals-against average at the break (2.73) but must now move on to another club with the Saints ceasing operations.

The perilous situation forced Levasseur to reach into his own wallet for money for air fare to the game. He refused to buckle and shared MVP honors with Willy Lindstrom of the West in the fifth annual WHA attraction.

Whalers Season Statistics

			Regular Season						Playoffs						
	GP	G	A	Pts	PM	PP	SH	GW	GP	G	A	Pts	PM	PP	SH	GW
Webster	70	36	49	85	43	11	0	5	5	1	1	2	0	0	0	0
Rogers	78	25	57	82	10	6	0	4	5	1	1	2	2	0	0	0
Lyle	75	39	33	72	62	12	0	2	5	1	0	1	4	0	0	1
Backstrom	77	17	31	48	30	0	0	5	3	0	0	0	0	0	0	0
G.Roberts	77	13	33	46	169	0	0	2	5	2	2	4	6	1	0	0
Keon	34	14	25	39	8	1	1	1	5	3	1	4	0	0	0	0
Pleau	78	11	21	32	22	0	1	1	5	1	0	1	0	1	0	0
McKenzie	34	11	19	30	25	0	3	1	5	2	1	3	8	0	1	0
Abrahamsson	64	6	24	30	33	1	0	4	5	0	3	3	0	0	0	0
Earl	54	9	14	23	37	0	1	1	1	0	0	0	0	0	0	0
Ley	55	2	21	23	102	0	0	1	5	0	4	4	4	0	0	0
Hangsleben	74	13	9	22	79	0	0	1	4	0	0	0	0	0	0	0
Arndt	46	8	14	22	11	1	0	0	--	--	--	--	--	--	--	--
Antonovich	26	12	9	21	10	0	0	0	5	2	2	4	4	0	0	0
D.Roberts	64	3	18	21	33	0	0	0	2	0	0	0	2	0	0	0
MacGregor	30	8	8	16	4	3	0	0	--	--	--	--	--	--	--	--
Callighen	33	6	10	16	41	1	0	0	--	--	--	--	--	--	--	--
Selwood	41	4	12	16	71	2	0	0	5	0	0	0	2	0	0	0
S.Carlson	31	4	9	13	40	0	0	0	5	0	0	0	9	0	0	0
J.Carlson	35	7	5	12	81	0	0	1	5	1	1	2	9	1	0	0
Bolduc	33	8	3	11	15	0	0	1	--	--	--	--	--	--	--	--
Busnick	55	1	9	10	141	0	0	0	--	--	--	--	--	--	--	--
Butters	26	1	8	9	65	0	0	0	5	0	1	1	15	0	0	0
Hynes	22	5	4	9	4	1	1		--	--	--	--	--	--	--	--
Swain	26	5	2	7	6	1	0	1	2	0	0	0	0	0	0	0
Paiement	13	5	2	7	12	1	0	1	--	--	--	--	--	--	--	--
Smedsmo	15	2	0	2	54	1	0	0	--	--	--	--	--	--	--	--
Abrahamsson	45	0	1	1	12	0	0	0	2	0	0	0	0	0	0	0
Landon	23	0	0	0	4	0	0	0	3	0	1	1	0	0	0	0
Raeder	26	0	0	0	7	0	0	0	1	0	0	0	0	0	0	0
Troy	7	0	0	0	7	0	0	0	--	--	--	--	--	--	--	--
Climie	3	0	0	0	0	0	0	0	--	--	--	--	--	--	--	--
Hanson	1	0	0	0	9	0	0	0	1	0	0	0	0	0	0	0
Bench	81	--	--	--	12	--	--	--	5	--	--	--	0	--	--	--
Totals	**81**	**275**	**450**	**725**	**1254**	**47**	**8**	**35**	**5**	**14**	**18**	**32**	**72**	**3**	**1**	**1**
Opponents	**81**	**290**	**491**	**781**	**1186**	**65**	**10**	**40**	**5**	**23**	**38**	**61**	**98**	**7**	**1**	**4**

Goaltending	GP	Mts	W	L	T	Sho	GGA	GP	Mts	W	L	Sho	GAA
C.Abrahamsson	45	2484	15	22	4	0	3.84	2	90	0	1	0	3.33
Landon	23	1118	8	8	1	1	3.17	3	152	1	2	0	4.34
Raeder	26	1328	12	10	1	2	3.12	1	60	0	1	0	7.00
Totals	**81**	**4930**	**35**	**40**	**6**	**3**	**3.49**	**5**	**302**	**1**	**4**	**0**	**4.57**
Opponents	**81**	**4929**	**40**	**35**	**6**	**4**	**3.26**	**5**	**302**	**4**	**1**	**1**	**2.79**

1977-78 League Standings

World Hockey Association

Standings	W	L	T	Pts	GF	GA
Winnipeg	50	28	2	102	381	270
New England	44	31	5	93	335	269
Houston	42	34	4	88	296	302
Quebec	40	37	3	83	349	347
Edmonton	38	39	3	79	309	307
Birmingham	36	41	3	75	287	314
Cincinnati	35	42	3	73	298	332
Indianapolis	24	51	5	53	267	353
Soviet All-Stars	3	4	1	7	27	36
Czechoslovakia	1	6	1	3	21	40

League Champion: Winnipeg Jets
League MVP: Marc Tardif, Quebec
Top Scorer: Marc Tardif, Quebec (65-89-154 in 78 GP)

Whalers Season Statistics

	Regular Season								Playoffs							
	GP	G	A	Pts	PM	PP	SH	GW	GP	G	A	Pts	PM	PP	SH	GW
Gordie Howe	76	34	62	96	85	16	0	8	14	5	5	10	15	1	1	2
Mark Howe	70	30	61	91	32	4	0	3	14	8	7	15	18	0	1	0
Rogers	80	28	43	71	46	4	1	2	14	5	6	11	8	2	0	0
Antonovich	75	32	35	67	32	5	0	4	14	10	7	17	4	4	0	1
Keon	78	24	38	62	2	1	5	2	14	5	11	16	4	1	1	1
G.Roberts	78	15	46	61	118	5	0	0	14	0	5	5	29	0	0	0
McKenzie	79	27	29	56	61	9	0	6	14	6	6	12	16	2	1	1
Lyle	68	30	24	54	74	3	0	5	12	2	1	3	13	0	0	0
Ley	73	3	41	44	95	0	0	0	14	1	8	9	4	0	0	0
Pleau	54	16	18	34	4	2	0	1	14	5	4	9	8	0	0	1
Selwood	80	6	25	31	88	2	0	0	14	0	3	3	8	0	0	0
Hangsleben	79	11	18	29	140	0	0	0	14	1	4	5	37	1	0	0
J.Carlson	67	9	20	29	192	0	0	0	9	1	1	2	14	0	0	0
Carroll	48	9	14	23	27	0	2	1	--	--	--	--	--	--	--	--
Webster	20	15	5	20	5	5	0	0	--	--	--	--	--	--	--	--
Mayer	51	11	9	20	21	0	0	1	--	--	--	--	--	--	--	--
Marty Howe	75	10	10	20	66	1	0	3	14	1	1	2	13	0	0	1
Sheehy	25	8	11	20	14	3	0	0	13	1	3	4	9	0	0	0
Butters	45	1	13	14	69	0	0	0	--	--	--	--	--	--	--	--
S.Carlson	38	6	7	13	11	1	1	2	13	2	7	9	2	0	1	0
Bolduc	41	5	5	10	22	0	0	0	14	2	4	6	4	0	0	0
Plumb	27	1	9	10	18	0	0	0	14	1	4	5	16	0	0	1
Smith	55	0	3	3	4	0	0	0	3	0	0	0	0	0	0	0
Maxwell	17	2	1	3	11	0	0	0	--	--	--	--	--	--	--	--
Peloffy	10	2	0	2	2	0	0	0	2	0	0	0	0	0	0	0
Levasseur	27	0	0	0	2	0	0	0	12	0	1	1	2	0	0	0
Bench	80	--	--	--	14	--	--	--	14	--	--	--	0	--	--	--
Totals	80	355	547	882	1255	62	9	44	14	56	89	135	224	11	5	8
Opponents	80	269	432	701	1131	65	4	31	14	47	81	128	252	11	0	6

Goaltending	GP	Mts	W	L	T	Sho	GGA	GP	Mts	W	L	Sho	GAA
Smith	55	3245	30	20	3	2	3.22	3	120	0	2	0	6.99
Levasseur	27	1654	14	11	2	3	3.30	12	718	8	4	0	2.58
Totals	80	4900	44	31	5	5	3.24	14	838	8	6	0	3.22
Opponents	80	4893	31	44	5	2	4.02	14	838	6	8	0	3.93

Forever Whalers

From Houston to Hartford

Whalers offer lands Howe family

May 23, 1977

HARTFORD - At a time when every announcement seemed to be "the biggest in club history," New England Whalers general managing partner Howard L. Baldwin fired a major salvo at the rival NHL Monday.

Baldwin's Whalers of the WHA signed Gordie Howe and his sons, Mark and Marty, for the 1977-78 season. The decision to sign the Howes, who were made free agents by the Houston Aeros, came to pass when the Whalers outbid the Boston Bruins who were hot after Mark Howe and keen on breaking up hockey's first family.

"Security, happiness and the sons playing on the same team," Gordie Howe said. "It was impossible to say no because of the security Howard and the Whalers offered us.

"When all the cards were on the table, it was pretty hard to say no."

Howe, 49, spent 25 years in the NHL before retiring in 1971. He came out of retirement two years later when Bill Dineen coaxed his ex-teammate to join the WHA after drafting the Howe boys who were playing in Canadian juniors.

The Howes led the Aeros to two AVCO Trophies in 1974, 1975 and a trip to the league finals in 1976.

Mark, 22, has blossomed into one of the game's bright stars. Marty, 23, has emerged as a reliable defenseman noted for his positional play.

"Any time you can get two players under 24 years of age that have played four years in a pro league, it's a great day," New England coach Harry Neale said. "Not just getting one, but two in the same day."

Not to mention the arrival of Mr. Hockey.

The signing triggered a flood of calls regarding season tickets. It also increased the blood flow to the neck area of Boston general manager Harry Sinden whose dislike for the WHA dates back to the league's creation in 1972 and havoc it has played in terms of player salaries.

A grand for Gordie

December 7, 1977

BIRMINGHAM, Ala. - When he scored his first goal against Turka Broda on October 16, 1946, few sensed if a rookie from the wheat fields of Saskatchewan had the keen touch around the net or just got a lucky carom.

Over four decades, the puck bounced favorably for Howe who, at 49, still has the knack to the finish.

Mired in a 10-game drought, Howe seemed more frustrated than snake-bitten. Number 9 also seemed more relieved than anyone when goal No. 1,000 found the back of the net Wednesday to open a 6-3 win by New England over Birmingham.

The goal at 1:36 of the first period, assisted by John McKenzie and Mike Antonovich, came on the power play, a quick blast from the slot to beat Bulls goalie John Garrett.

"The goal was identical to my first one," Howe said.

"My dream to play hockey started when I was eight years old and I first laced on a pair of skates," Howe said. "I dreamed at that time, with the help of others, that perhaps I could go on and fare well in this game. I'm a lucky boy. All my dreams have been answered."

Hartford Civic Center roof collapses

January 18, 1978

HARTFORD - The New England Whalers could be on an extended road trip in the offing.

It might be two years.

In another community, it might be forever.

In a league where teams have ceased operations because of poor attendance or nothing in the till to play players, the Whalers must now scramble for a suitable facility following the roof collapse of the Hartford Civic Center.

"Looking back, it was a perfect excuse for anyone who was not totally committed to bail out but no one did," team founder Howard L. Baldwin remembers. "Our partners reassured us they would be completely behind us and the fans just hung in there.

"That dark day gave everyone time to regroup and determine a way to move forward," Baldwin says. "There's no question about the strength of this franchise."

The Whalers announced they would play all games in nearby Springfield, Mass. for the remainder of the season but still base operations in Hartford.

Several structures in southern New England met the same fate as the Civic Center. A blizzard quickly followed by a driving rain storm as temperatures eased resulted in enormous weight for many roofs on buildings to withstand.

1978-79 League Standings

World Hockey Association

Standings	W	L	T	Pts	GF	GA
Edmonton	48	30	2	98	340	266
Quebec	41	34	5	87	288	271
Winnipeg	39	35	6	84	307	306
New England	33	41	6	83	298	287
Cincinnati	33	41	6	72	274	284
Birmingham	32	42	6	70	286	311
Indianapolis*	5	18	2	12	78	130
Soviet All-Stars	4	1	1	9	27	20
Czechoslovakia	1	4	1	3	14	33
Finland	0	1	0	0	4	8

*franchise disbanded on December 15, 1978
League Champion: Winnipeg Jets
League MVP: Dave Dryden, Edmonton
Top Scorer: Real Cloutier, Quebec (75-54-129 in 77 GP)

Whalers Season Statistics

	Regular Season								Playoffs							
	GP	G	A	Pts	PM	PP	SH	GW	GP	G	A	Pts	PM	PP	SH	GW
Mark Howe	77	42	65	107	32	13	6	3	6	4	2	6	6	1	0	1
Lacroix	78	32	56	88	34	7	4	6	10	4	4	8	0	0	0	0
Rogers	80	27	45	72	31	2	0	2	10	2	6	8	2	0	0	0
Keon	79	22	43	65	2	7	0	5	10	3	9	12	2	1	0	1
Roberts	79	11	46	57	113	0	0	4	10	0	4	4	10	0	0	0
Miller	76	26	23	49	44	1	0	3	10	0	8	8	28	0	0	0
Antonovich	69	20	27	47	35	3	0	0	10	5	3	8	14	0	0	0
McKenzie	76	19	28	47	112	3	2	2	10	3	7	10	10	1	0	1
Gordie Howe	58	19	24	43	51	7	0	2	10	3	1	4	4	2	0	0
Lyle	59	17	18	35	54	1	0	2	9	3	5	8	25	0	0	1
Hangsleben	77	10	19	29	148	0	0	2	10	1	2	3	12	0	0	0
Ley	73	7	20	27	135	0	2	0	9	0	4	4	11	0	0	0
Marty Howe	66	9	15	24	31	1	0	1	9	0	1	1	8	0	0	0
Plumb	77	4	16	20	33	0	0	1	9	0	0	0	0	0	0	0
Douglas	51	6	10	16	15	3	0	0	10	0	4	4	23	0	0	0
Selwood	42	4	12	16	47	2	0	0	--	--	--	--	--	--	--	--
Warner	41	6	9	15	20	1	0	1	1	0	0	0	0	0	0	0
Stoughton	36	9	3	12	2	0	0	0	7	4	3	7	4	0	0	1
Pleau	28	6	6	12	6	0	0	0	10	2	1	3	0	1	0	0
J.Carlson	34	2	7	9	61	0	0	0	--	--	--	--	--	--	--	--
Inkpen	41	0	7	7	15	0	0	0	5	0	1	1	4	0	0	0
Smith	40	0	3	3	35	0	0	0	4	0	0	0	0	0	0	0
Brubaker	12	0	0	0	19	0	0	0	3	0	0	0	12	0	0	0
Roy	1	0	0	0	2	0	0	0	--	--	--	--	--	--	--	--
Bench	80	--	--	--	4	--	--	--	10	--	--	--	0	--	--	--
Totals	80	298	502	800	1090	53	15	37	10	39	64	103	175	6	0	5
Opponents	80	287	460	747	1178	63	5	34	10	45	72	117	175	12	0	5

Goaltending	GP	Mts	W	L	T	Sho	GGA	GP	Mts	W	L	Sho	GAA
Smith	40	2396	17	17	5	1	3.31	4	153	1	2	0	4.71
Garrett	41	2496	20	17	4	2	3.58	8	447	4	3	0	4.30
Totals	80	4893	37	34	9	3	3.44	10	600	5	5	0	4.40
Opponents	80	4894	34	47	9	3	4.54	10	598	5	5	0	3.71

Whalers join NHL for 1979-80

March 22, 1979

CHICAGO - Maybe the beer boycott in Canada the last couple of weeks spoke more volumes than the NHL faction that was pushing for a merger with the rival WHA.

On the heels of a "thanks but no thanks" vote in Florida earlier this month, patrons in Quebec City, Edmonton and Winnipeg have stopped buying products from the Molson breweries, the company which bankrolls the Montreal Canadiens.

And while rivals have been unable to solve the Habs on ice during their strangehold on Lord Stanley's silver bowl since 1976, the company brass felt the pressure from stockholders, especially if the current boycott went on for weeks and really dented revenue via beer sales.

Thus, today's 14-3 tabulation assures peach in hockey and the NHL will expand to 21 teams for 1979-80.

"There was a merger and an execution," recalls one NHL owner. "Jacques Courtois of Montreal and Williams Hughes of Vancouver, the two team presidents who had voted against the merger in Key Largo (Florida) were no longer running their teams. The chairman of the breweries represented the Canadiens and the owners of the Vancouver club were at the meeting this time. That's how the merge finally happened."

As much as Toronto, Boston and Los Angeles were against the admission of the Hartford (nee New England) Whalers, Edmonton Oilers, Quebec Nordiques and Winnipeg Jets, the majority of NHL teams figured the time had come to end what has been a costly strain on the treasury fighting this seven-year duel for players.

1979-80 League Standings

PRINCE OF WALES CONFERENCE

Norris Division

Team	GP	W	L	T	GF	GA	PTS
Montreal	80	47	20	13	328	240	107
Los Angeles	80	30	36	14	290	313	74
Pittsburgh	80	30	37	13	251	303	73
Hartford	80	27	34	19	303	312	73
Detroit	80	26	43	11	268	306	63

Adams Division

	GP	W	L	T	GF	GA	PTS
Buffalo	80	47	17	16	318	201	110
Boston	80	46	21	13	310	234	105
Minnesota	80	36	28	16	311	253	88
Toronto	80	35	40	5	304	327	75
Quebec	80	25	44	11	248	313	61

CLARENCE CAMPBELL CONFERENCE

Patrick Division

	GP	W	L	T	GF	GA	PTS
Philadelphia	80	48	12	20	327	254	116
NY Islanders	80	39	28	13	281	247	91
NY Rangers	80	38	32	10	308	284	86
Atlanta	80	35	32	13	282	269	83
Washington	80	27	40	13	261	293	67

Smythe Division

	GP	W	L	T	GF	GA	PTS
Chicago	80	34	27	19	241	250	87
St. Louis	80	34	34	12	266	278	80
Vancouver	80	27	37	16	256	281	70
Edmonton	80	28	39	13	301	322	69
Winnipeg	80	20	49	11	214	314	51
Colorado	80	19	48	13	234	308	51

League Champion: New York Islanders   **Top Scorer:** Marcel Dionne, Los Angeles (53-84-137 in 80 GP)
League MVP: Wayne Gretzky, Edmonton

Whalers Season Statistics

	Regular Season								Playoffs						
	GP	G	A	Pts	PM	PP	SH	GW	GP	G	A	Pts	PM	PP	SH	GW
Rogers	80	44	61	105	10	3	2	0	3	0	3	3	0	0	0	0
Stoughton	80	56	44	100	16	16	2	9	1	0	0	0	0	0	0	0
Mark Howe	74	24	56	80	22	5	2	3	3	1	2	3	2	0	0	0
Keon	76	10	52	62	10	0	0	0	0	3	0	1	1	0	0	0
Douglas	77	33	24	57	39	5	0	5	Injured							
Boutette	47	13	31	44	75	3	0	0	3	1	0	1	6	0	1	0
Gordie Howe	80	15	26	41	42	2	0	0	3	1	1	2	2	0	0	0
Sims	76	10	31	41	30	2	0	1	3	0	0	0	2	0	0	0
Rowe	20	6	4	10	30	1	0	1	3	2	0	2	0	0	0	0
Roberts	80	8	28	36	89	1	0	1	3	1	1	2	2	0	0	0
Carroll	71	13	19	32	24	1	0	1	Injured							
Allison	64	16	12	28	13	0	0	2	2	0	1	1	0	0	0	0
Debol	48	12	14	26	4	0	0	1	3	0	0	0	0	0	0	0
Johnston	32	8	13	21	8	0	0	0	3	0	1	1	0	0	0	0
Ley	65	4	16	20	92	0	0	1	Injured							
Fotiu	74	10	8	18	107	0	0	0	3	0	0	0	6	0	0	0
Hangsleben	37	3	15	18	69	0	0	1	--	--	--	--	--	--	--	--
Lacroix	29	3	14	17	2	1	0	0	--	--	--	--	--	--	--	--
Plumb	26	3	4	7	14	0	0	0	--	--	--	--	--	--	--	--
Hull	9	2	5	7	0	1	0	0	3	0	0	0	0	0	0	0
Giroux	47	2	5	7	44	0	0	0	3	0	0	0	2	0	0	0
Bennett	24	3	3	6	63	2	0	0	--	--	--	--	--	--	--	--
Sheehy	12	2	1	3	0	0	0	0	--	--	--	--	--	--	--	--
Warner	32	0	3	3	10	0	0	0	--	--	--	--	--	--	--	--
Alley	7	1	1	2	0	0	0	0	3	0	1	1	0	0	0	0
Hill	19	1	1	2	4	0	0	0	--	--	--	--	--	--	--	--
Renaud	13	0	2	2	4	0	0	0	--	--	--	--	--	--	--	--
Garrett	52	0	2	2	12	0	0	0	1	0	0	0	0	0	0	0
Neufeld	8	1	0	1	0	0	0	0	2	1	0	1	0	0	0	0
Brubaker	3	0	1	1	2	0	0	0	--	--	--	--	--	--	--	--
Antonovich	5	0	1	1	2	0	0	0	--	--	--	--	--	--	--	--
Marty Howe	6	0	1	1	2	0	0	0	3	1	1	2	0	0	0	0
Luksa	8	0	1	1	4	0	0	0	--	--	--	--	--	--	--	--
A.Smith	30	0	1	1	10	0	0	0	2	0	0	0	0	0	0	0
Stephenson	4	0	1	1	0	0	0	0	--	--	--	--	--	--	--	--
Savard	1	0	0	0	0	0	0	0	--	--	--	--	--	--	--	--
S.Smith	4	0	0	0	0	0	0	0	--	--	--	--	--	--	--	--
Hodgson	6	0	0	0	0	0	0	0	1	0	0	0	0	0	0	0
Schurman	7	0	0	0	0	0	0	0	--	--	--	--	--	--	--	--
Bench	80	--	--	--	16	--	--	3	--	--	--	--	0	--	--	--
Totals	80	303	501	804	875	42	7	27	3	8	12	20	22	0	1	0
Opponents	80	312	512	824	712	64	1	34	3	18	25	43	26	1	0	3

Goaltending	GP	Mts	W	L	T	Sho	GGA	GP	Mts	W	L	Sho	GAA
Smith	30	1754	11	10	8	2	3.66	2	120	0	2	0	5.00
Garrett	52	3046	16	24	11	0	3.98	1	60	0	1	0	8.00
Totals	80	4800	27	34	19	2	3.90	3	180	0	3	0	6.00
Opponents	80	4800	34	27	19	3	3.79	3	180	3	0	0	2.67

Stoughton gets No. 56 in season finale against Detroit

April 6, 1980

HARTFORD - "Stash" struck again.

As so many times this season, Blaine "Stash" Stoughton knows what to do once he's near the net.

"Don't think," he said. "Just shoot."

Stoughton drained No. 56 Sunday as the Whalers beat Detroit 5-3 to finish their first NHL season at 27-34-19.

With the goal, Stoughton reached the 100-point mark for the season.

Gordie calls it quits after 32 seasons

June 4, 1980

HARTFORD - His first retirement lasted two seasons.

That was nine years ago.

In 1973, the WHA gave Gordie Howe a chance to play hockey again. . . first in Houston and eventually in Hartford.

When the Whalers were among the four WHA clubs absorbed by the NHL last summer, Howe was back in the league he starred in for 25 years.

This past season, Gordie was one of four Whalers to play in all 80 regular season games. His 15-26-41 production with 42 penalty minutes was modest compared to many others but most were half his age. Back on March 31, Howe turned 51 years old.

Now it's up to sons Mark and Marty to keep the Howe name in pro hockey.

"No one teaches you how to retire," Gordie told the international gathering of media which covered Howe's press conference as he officially retired. "I found that out the first time.

"I'd hate to go out after 32 seasons and find out in the middle of winter I'd run short," Howe said. "My goal was to provide a proper house and heater for my parents when I started to play hockey. I just wanted to be lucky to have two suits. Truthfully, I still think I'm damned good enough to play."

Howe's last match was Game 3 of the Stanley Cup playoffs against Montreal on April 11, 1980.

Overall, Howe played 2,186 games over 26 NHL seasons and six WHA campaigns. He had 975 regular season goals and 96 more in the postseason for a total of 1,071.

His uniform jersey No. 9 was retired by both the Red Wings and Whalers. Howe was MVP of the NHL six times and achieved that honor once in the WHA. He led the NHL in scoring six times. He played on four Stanley Cup winners and two AVCO Trophy champions. He was enshrined in the Hall of Fame in 1972.

1980-81 League Standings

PRINCE OF WALES CONFERENCE
Norris Division

Team	GP	W	L	T	GF	GA	PTS
Montreal	80	45	22	13	332	232	103
Los Angeles	80	43	24	13	337	290	99
Pittsburgh	80	30	37	13	302	345	73
Hartford	80	21	41	18	292	372	60
Detroit	80	19	43	18	252	339	56

Adams Division

Buffalo	80	39	20	21	327	250	99
Boston	80	37	30	13	316	272	87
Minnesota	80	35	28	17	291	263	87
Quebec	80	30	32	18	314	318	78
Toronto	80	28	37	15	322	367	71

CLARENCE CAMPBELL CONFERENCE
Patrick Division

NY Islanders	80	48	18	14	355	260	110
Philadelphia	80	41	24	15	313	249	97
Calgary	80	39	27	14	329	298	92
NY Rangers	80	30	36	14	312	317	74
Washington	80	26	36	18	286	317	70

Smythe Division

St. Louis	80	45	18	17	352	281	107
Chicago	80	31	33	16	304	315	78
Vancouver	80	28	32	20	289	301	76
Edmonton	80	29	35	16	328	327	74
Colorado	80	22	45	13	258	344	57
Winnipeg	80	9	57	14	246	400	32

League Champion: New York Islanders **Top Scorer:** Wayne Gretzky, Edmonton (55-109-164 in 80 GP)
League MVP: Wayne Gretzky, Edmonton

Whalers Season Statistics

Regular Season

	GP	G	A	Pts	PM	PP	SH	GW
Rogers	80	40	65	105	32	10	4	1
Boutette	80	28	52	80	160	8	2	3
Stoughton	71	43	30	73	56	10	2	6
Mark Howe	63	19	46	65	54	7	2	3
Sims	80	16	36	52	68	5	0	1
Keon	80	13	34	47	26	2	0	1
Miller	77	22	22	44	37	3	3	1
Rowe	74	13	28	41	190	1	0	1
Nachbaur	77	16	17	33	139	2	0	1
Debol	44	14	12	26	0	2	0	0
Douglas	55	13	9	22	29	3	0	1
Fidler	38	9	9	18	4	2	0	0
Meagher	27	7	10	17	19	0	0	0
Abrahamsson	32	6	11	17	16	4	0	1
Neufeld	52	5	10	15	44	0	0	0
Johnston	25	4	11	15	8	0	0	0
Roberts	27	2	11	13	81	1	0	0
Volcan	49	2	11	13	26	0	0	0
Barnes	54	1	10	11	82	0	0	0
Brubaker	43	5	3	8	93	0	0	0
Galarneau	30	2	6	8	9	0	0	1
Smith	38	1	7	8	55	0	0	0
Fotiu	42	4	3	7	79	0	0	0

	GP	G	A	Pts	PM	PP	SH	GW
McIlhargey	48	1	6	7	142	0	0	0
Lupien	20	2	4	6	39	0	0	0
Alley	8	2	2	4	11	0	0	0
Ley	16	0	2	2	20	0	0	0
Renaud	4	1	0	1	0	0	0	0
Allison	6	1	0	1	0	0	0	0
Marty Howe	12	0	1	1	25	0	0	0
Garrett	54	0	1	1	2	0	0	0
Holland	1	0	0	0	0	0	0	0
Arthur	3	0	0	0	0	0	0	0
Kemp	3	0	0	0	0	0	0	0
Veisor	29	0	0	0	0	0	0	0
Bench	80	--	--	--	34	--	--	--
Totals	**80**	**292**	**469**	**761**	**1584**	**60**	**13**	**21**
Opponents	**80**	**372**	**578**	**980**	**1471**	**80**	**15**	**41**

Goaltending	GP	Mts	W	L	T	Sho	GGA
Veisor	29	1580	6	13	6	1	4.48
Garrett	54	3145	15	27	12	0	4.60
Holland	1	60	0	1	0	0	7.00
Totals	**80**	**4785**	**21**	**41**	**18**	**1**	**4.59**
Opponents	**80**	**4800**	**41**	**21**	**18**	**3**	**3.65**

Rogers to Blueshirts in 3-for-1 deal

October 2, 1981

BINGHAMTON, N.Y. - New general manager Larry Pleau pulled the trigger on his first major deal since taking command the Whalers and faithful followers can only wonder if anyone on board can replace the offensive wizardry of center Mike Rogers.

Rogers, who has linked 105-point seasons for the Whalers to rank among the league's Top 10 in scoring, was dealt to the New York Rangers for three potentially talented but so far unproven players.

Forward Doug Sulliman, New York's top draft pick in 1979, comes to the Whalers along with defensemen Chris Kotsopoulos and Gerry McDonald.

"I know Mike Rogers is a popular player with the fans but there is no other way to put it," Pleau said. "My job is to develop a hockey team for three or four years from now. Rogers is one player who had in our organization that had value.

"We may never replace his 105 points but we can do other things to improve our club," Pleau said. "The philosophy is to put a young, competitive team on the ice."

Sulliman is 21. The two blueliners are 22.

"I guess I'm old at 26," Rogers said. "I still can't figure it out."

With the season to open in four days, the Whalers have a large crater at center ice to fill. Veteran Dave Keon, Rick Meagher and Don Nachbaur are the only pivots on the roster.

The trade may open the door for No. 1 draftee Ron Francis, Hartford's top pick in June, to start his career in the offing or force Pleau to swing another deal to get an experienced vet to win face-offs.

1981-82 League Standings

CLARENCE CAMPBELL CONFERENCE
Norris Division

Team	GP	W	L	T	GF	GA	PTS
Minnesota	80	37	23	20	346	288	94
Winnipeg	80	33	33	14	319	332	80
St. Louis	80	32	40	8	315	349	72
Chicago	80	30	38	12	332	363	72
Toronto	80	20	44	16	298	380	56
Detroit	80	21	47	12	270	351	54

Smythe Division

Team	GP	W	L	T	GF	GA	PTS
Edmonton	80	48	17	15	417	295	111
Vancouver	80	30	33	17	290	286	77
Calgary	80	29	34	17	334	345	75
Los Angeles	80	24	41	15	314	369	63
Colorado	80	18	49	13	241	362	49

League Champion: New York Islanders

PRINCE OF WALES CONFERENCE
Adams Division

Team	GP	W	L	T	GF	GA	PTS
Montreal	80	46	17	17	360	223	109
Boston	80	43	27	10	323	285	96
Buffalo	80	39	26	15	307	273	93
Quebec	80	33	31	16	356	345	82
Hartford	80	21	41	18	264	351	60

Patrick Division

Team	GP	W	L	T	GF	GA	PTS
NY Islanders	80	54	16	10	385	250	118
NY Rangers	80	39	27	14	316	306	92
Philadelphia	80	38	31	11	325	313	87
Pittsburgh	80	31	36	13	310	337	75
Washington	80	26	41	13	319	338	65

League MVP: Wayne Gretzky, Edmonton
Top Scorer: Wayne Gretzky, Edmonton (92-120-212 in 80 GP)

Whalers Season Statistics

Regular Season

	GP	G	A	Pts	PM	PP	SH	GW
Stoughton	80	52	39	91	57	13	1	4
Larouche	45	25	25	50	12	11	1	1
Sulliman	77	29	40	69	39	5	0	4
Francis	59	25	43	68	51	12	0	1
Mark Howe	76	8	45	53	18	3	0	1
Howatt	80	18	32	50	242	1	0	2
Meagher	65	24	19	43	51	2	1	2
Kotsopoulos	68	13	20	33	147	5	0	1
Wesley	78	9	18	27	123	3	0	1
Nachbaur	77	5	21	26	117	0	0	0
Miller	74	10	12	22	68	1	1	0
MacLeish	34	6	16	22	16	4	0	0
Keon	78	8	11	19	6	0	1	1
Renaud	48	1	17	18	39	0	0	0
Douglas	30	10	7	17	44	4	0	2
Lyle	14	2	12	14	9	1	0	0
Bourbonnais	24	3	9	12	11	0	0	0
Shmyr	66	1	11	12	134	0	0	0
Neufeld	19	4	3	7	4	1	0	0
Volcan	26	1	5	6	29	1	0	0
McIlhargey	50	1	5	6	60	0	0	0
Barnes	20	1	4	5	19	0	0	0
Gillen	34	1	4	5	22	0	0	0
Millen	55	0	5	5	2	0	0	0
Anderson	25	1	3	4	85	0	0	1

	GP	G	A	Pts	PM	PP	SH	GW
Marty Howe	13	0	4	4	2	0	0	0
Rowe	21	4	0	4	36	1	1	0
McClanahan	17	0	3	3	11	0	0	0
Smith	17	0	3	3	15	0	0	0
MacGregor	2	1	1	2	2	0	0	0
MacDermid	3	1	0	1	2	0	0	0
Lupien	1	0	1	1	2	0	0	0
Fidler	2	0	1	1	0	0	0	0
Fridgen	2	0	1	1	0	0	0	0
Brownschidle	3	0	1	1	2	0	0	0
Garrett	16	0	1	1	2	0	0	0
McDougal	3	0	0	0	0	0	0	0
McDonald	3	0	0	0	0	0	0	0
Merkosky	7	0	0	0	2	0	0	0
Galarneau	10	0	0	0	4	0	0	0
Bench	80	--	--		8	--	--	
Totals	**80**	**264**	**443**	**707**	**1493**	**68**	**6**	**21**
Opponents	**80**	**351**	**577**	**928**	**1658**	**78**	**7**	**41**

Goaltending	GP	Mts	W	L	T	Sho	GGA
Veisor	13	701	5	5	2	0	4.54
Garrett	16	897	5	6	4	0	4.21
Millen	55	3195	11	30	12	0	4.30
Totals	**80**	**4793**	**21**	**41**	**18**	**0**	**4.32**
Opponents	**80**	**4800**	**41**	**21**	**18**	**3**	**3.30**

Whalers trade Mark Howe to Flyers

August 19, 1982

HARTFORD - The Whalers have never been the same team since general manager Larry Pleau traded away high-scoring center Mike Rogers a year ago.

A stunning three-team swap that involved the Whalers, Flyers and Oilers today raises the same type of questions.

Will the Whalers be a stronger team without Mark Howe, their best defenseman, on board for future winters?

Pleau apparently felt so and he agreed to ship Howe to Philadelphia for a couple of draft choices and the rights to Ken Linseman.

Moments later, the Whalers shuttled Linseman and Don Nachbaur, a young forward some in the organization were touting for captain, to Edmonton for defenseman Risto Siltanen and junior player Brent Loney.

"Now we won't always be looking for Mark to pull us out of a hole," Pleau said. "Now it will have to be 20 guys working together. I don't feel Mark reached his potential with us. Whether it was me or the organization, I just don't know."

Howe, who has slowly regained his skating strength after nearly having his career end in a collision midway in the 1980-81 season when the son of hockey icon Gordie Howe was impaled in the lower back by a net flange.

To okay the deal, Howe waived his no-trade clause. He also had mixed emotions leaving Hartford.

"They tell me I'm getting old (28) but I can still out-skate anybody on the team," Mark said. "They obviously felt what they got was more than what I can give them. I hope this trade works out for both of us."

1982-83 League Standings

CLARENCE CAMPBELL CONFERENCE

Norris Division

Team	GP	W	L	T	GF	GA	PTS
Chicago	80	47	23	10	338	268	104
Minnesota	80	40	24	16	321	290	96
Toronto	80	28	40	12	293	330	68
St. Louis	80	25	40	15	285	316	65
Detroit	80	21	44	15	263	344	57

Smythe Division

Edmonton	80	47	21	12	424	315	106
Calgary	80	32	34	14	321	317	78
Vancouver	80	30	35	15	303	309	75
Winnipeg	80	33	39	8	311	333	74
Los Angeles	80	27	41	12	308	365	66

PRINCE OF WALES CONFERENCE

Adams Division

Boston	80	50	20	10	327	228	110
Montreal	80	42	24	14	350	286	98
Buffalo	80	38	29	13	318	285	89
Quebec	80	34	34	12	343	336	80
Hartford	80	19	54	7	261	403	45

Patrick Division

Philadelphia	80	49	23	8	326	240	106
NY Islanders	80	42	26	12	302	226	96
Washington	80	39	25	16	306	283	94
NY Rangers	80	35	35	10	306	287	80
New Jersey	80	17	49	14	230	338	48
Pittsburgh	80	18	53	9	257	394	45

League Champion: New York Islanders

League MVP: Wayne Gretzky, Edmonton

Top Scorer: Wayne Gretzky, Edmonton (71-125-196 in 80 GP)

Whalers Season Statistics

Regular Season

1982-83	GP	G	A	Pts	PM	PP	SH	GW
Francis	79	31	59	80	60	4	2	4
Stoughton	79	45	31	76	27	10	0	8
Johnson	73	31	38	69	28	5	3	5
Neufeld	80	26	31	58	86	4	0	1
Sulliman	77	22	19	41	14	8	0	0
Larouche	38	18	22	40	8	4	0	0
Sullivan	62	18	19	37	18	5	0	0
Lacroix	56	6	25	31	18	1	0	0
Malinowski	75	5	23	28	16	1	0	0
Renaud	77	3	28	31	37	1	0	0
Kotsopoulos	67	6	24	30	125	3	0	0
Siltanen	74	5	25	30	28	3	0	0
Adams	79	10	13	23	216	1	0	1
McDougal	55	8	10	18	43	0	0	0
Volcan	68	4	13	17	73	0	0	0
Lawless	47	6	9	15	4	1	0	0
Miller	56	1	10	11	15	0	0	0
Lyle	16	4	6	10	8	0	1	0
Hospodar	72	1	9	10	199	0	0	0
Galarneau	38	4	5	9	21	0	0	0
Anderson	57	0	6	6	171	0	0	0

	GP	G	A	Pts	PM	PP	SH	GW
Fridgen	11	2	2	4	2	0	0	0
Henderson	15	2	1	3	64	0	0	0
Marshall	13	1	2	3	0	0	0	0
Millen	60	0	2	2	8	0	0	0
Smith	18	1	0	1	25	0	0	0
Gilhen	2	0	1	1	0	0	0	0
Hoffman	2	0	1	1	0	0	0	0
Paterson	2	0	0	0	0	0	0	0
Brownschidle	4	0	0	0	0	0	0	0
MacDermid	7	0	0	0	2	0	0	0
Veisor	23	0	0	0	2	0	0	0
Bench	80	--	--	--	28	--	--	--
Totals	**80**	**261**	**434**	**695**	**1392**	**51**	**6**	**19**
Opponents	**80**	**403**	**567**	**970**	**1279**	**70**	**9**	**54**

Goaltending	GP	Mts	W	L	T	Sho	GGA
Millen	60	3506	14	38	6	1	4.83
Veisor	23	1277	5	16	1	1	5.54
Totals	**80**	**4783**	**19**	**54**	**7**	**2**	**4.59**
Opponents	**80**	**4800**	**54**	**19**	**7**	**1**	**3.26**

Czar Hunt ends with 'Cat'

May 2, 1983

HARTFORD - It was Alternate Governor Donald Conrad of Hartford who coined the term "Czar Hunt" in regards to finding someone with "impeccable credentials" to take the helm of the slumbering Whalers.

By the time Emile Francis was introduced as president and general manager of the Hartford hockey team, the new bossman sees another career opportunity. All the franchise can do is go up. In 1982-83, the Whalers hit rock bottom: 19 wins, 45 points and an alarming 403 goals against.

"What happened here (54 losses) can happen to anyone," Francis said. "I've come here with an open mind."

And the "Cat" also came with his fix-it bag.

The Whalers will be the third club that Francis, a former goaltender, will try to reshape into a winning organization. He accomplished that with the New York Rangers from 1965 to 1975. He basically saved the St. Louis Blues from disaster, arriving in 1976 and reviving the Blues in light of management's tight purse strings.

The Whalers have been basically an also-ran since joining the NHL. The only thing Hartford has played for over the last five months is a prime Entry Draft slot.

If past form continues, the Whalers will likely add a big-play goaltender, a defenseman who can score points and a litany of hard-working forwards that will upgrade the club's special teams.

Unlike St. Louis where the Ralston Purina Company virtually kept the team solvent with a slim budget, the Whalers do have the resources to be a competitive club. The ailing franchise simply has lacked direction in recent winters.

1983-84 League Standings

Whalers Season Statistics

Regular Season

	GP	G	A	Pts	PM	PP	SH	GW
Johnson	79	35	52	87	27	13	1	2
Francis	72	23	60	83	45	5	0	5
Turgeon	76	40	32	72	55	18	0	3
Neufeld	80	27	42	69	97	5	0	5
Crawford	80	36	25	61	32	5	0	3
Malone	78	17	37	54	56	3	0	2
Siltanen	75	15	38	53	34	12	0	0
Stoughton	54	23	14	37	4	7	0	2
Currie	32	12	16	38	4	6	0	2
Zuke	75	6	23	29	36	1	0	0
Dunn	63	5	20	25	30	0	0	0
Dupont	40	7	15	22	12	2	0	0
Robertson	66	7	14	21	198	0	0	2
Sulliman	67	6	12	18	20	0	0	0
Kotsopoulos	72	5	13	18	118	3	0	0
Bourbonnais	35	0	16	16	0	0	0	0
Quenneville	80	5	8	13	95	0	2	0
Brownschidle	13	1	2	3	10	0	0	1
Marty Howe	69	0	11	11	34	0	0	0
Pierce	17	6	3	9	9	0	1	0
Hospodar	59	0	9	9	163	0	0	0
Stoyanovich	23	3	5	8	11	0	0	1
Tippett	17	4	2	6	2	0	1	0
Crombeen	56	1	4	5	25	0	0	0
Fusco	17	0	4	4	17	0	0	0
Lawless	6	0	3	3	7	0	0	0
Millen	60	0	3	3	10	0	0	0
Paterson	9	2	0	2	4	0	0	0
Yates	7	1	1	2	4	0	0	0
MacDermid	3	0	1	1	2	0	0	0
Staniowski	18	0	0	0	2	0	0	0
Hess	3	0	0	0	0	0	0	0
Veisor	4	0	0	0	0	0	0	0
McDonald	5	0	0	0	4	0	0	0
Bailey	12	0	0	0	25	0	0	0
Bench	80	--	--	--	16	--	--	--
Totals	**80**	**288**	**485**	**773**	**1184**	**80**	**5**	**28**
Opponents	**80**	**320**	**538**	**858**	**1260**	**68**	**14**	**42**

Goaltending	GP	Mts	W	L	T	Sho	GGA
Millen	60	3575	21	30	9	2	3.71
Staniowski	18	1040	6	9	1	0	4.27
Veisor	4	240	1	3	0	0	5.00
Totals	**80**	**4855**	**28**	**42**	**10**	**2**	**3.95**
Opponents	**80**	**4852**	**42**	**28**	**10**	**5**	**3.56**

Whalers go on scoring spree

Hartford hammers Edmonton, 11-0

February 12, 1984

HARTFORD - Wayne Gretzky didn't play. Either did Jari Kurri. That didn't stop Ron Francis and Greg Malone of the Whalers from turning in Edmonton-style performances similar to the nightly efforts by the high-octane Oiler duo.

Francis and Malone each rang up five-point games as the Whalers set a host of club records Sunday in waxing Edmonton 11-0.

Greg Millen turned aside 28 shots for Hartford which snapped a 229-game scoring streak for the Oilers dating back nearly three seasons.

The Whalers led 3-0 after the first period. Hartford tacked on five scores in the second period including four during a five-minute major assessed to Kevin McClelland whose high elbow knocked Sylvain Turgeon unconscious at 8:29 of the second period.

Francis had four goals and one helper while Malone added a hat trick in the carnage.

"Everything went our way," Millen said. "We caught them at a good time."

"A total collapse," Oilers coach Glen Sather said. "It was as if the guys were looking for Wayne and Jari."

The date is March 3, 1985 and some hockey fans in Hartford opt for disguise as another season turned into a nightmare for the faithful. Stuck in a winless streak on home ice, the Whalers discovered another way to lose to Vancouver. Leading 6-4 with 1:10 left in the game on what looked like an insurance tally by **Sylvain Turgeon**, *the Canucks rallied in the final 45 seconds on goals by Patrik Sundstrom and Stan Smyl to force overtime. In the extra session, Jean-Marc Lanthier beat goalie* **Mike Liut** *with a wrister at 2:47 for a 7-6 victory as 10,070 sat in disbelief. The futility skein eventually reached 13 games before* **Kevin Dineen** *snapped the dry dock with the game-winner to beat Pittsburgh 4-3 on March 17, 1985.*

1984-85 League Standings

CLARENCE CAMPBELL CONFERENCE
Norris Division

Team	GP	W	L	T	GF	GA	PTS
St. Louis	80	37	31	12	299	288	86
Chicago	80	38	35	7	309	299	83
Detroit	80	27	41	12	313	357	66
Minnesota	80	25	43	12	268	321	62
Toronto	80	20	52	8	253	358	48

Smythe Division

	GP	W	L	T	GF	GA	PTS
Edmonton	80	49	20	11	401	298	109
Winnipeg	80	43	27	10	358	332	96
Calgary	80	41	27	12	363	302	94
Los Angeles	80	34	32	14	339	326	82
Vancouver	80	25	46	9	284	401	59

PRINCE OF WALES CONFERENCE
Adams Division

	GP	W	L	T	GF	GA	PTS
Montreal	80	41	27	12	309	262	94
Quebec	80	41	30	9	323	275	91
Buffalo	80	38	28	14	290	237	90
Boston	80	36	34	10	303	287	82
Hartford	80	30	41	9	268	318	69

Patrick Division

	GP	W	L	T	GF	GA	PTS
Philadelphia	80	53	20	7	348	241	113
Washington	80	46	25	9	322	240	101
NY Islanders	80	40	34	6	345	312	86
NY Rangers	80	26	44	10	295	345	62
New Jersey	80	22	48	10	264	346	54
Pittsburgh	80	24	51	5	276	385	53

League Champion: Edmonton Oilers
League MVP: Wayne Gretzky, Edmonton
Top Scorer: Wayne Gretzky, Edmonton (73-135-208 in 80 GP)

Whalers land Liut in swap with Blues

February 22, 1985

HARTFORD - The goalkeeper who revived the fortunes of the St. Louis Blues will be asked to do it again, this time for the Hartford Whalers.

Mike Liut, a one-time draft pick by the WHA Whalers, will finally wear a Hartford jersey following the swap with St. Louis that involved Whalers captain Mark Johnson and goalie Greg Millen.

With just a handful of games left in the season, Hartford general manager Emile Francis has basically started to plan for next year.

Liut, in Francis' assessment, would be the cornerstone for the Whalers. Such was the case in 1980-81 when the Blues rolled up 45 wins during a 107-point campaign. That year, Liut reached all-league status.

"I know what Mike is capable of," Francis said. "Mike went through a rebuilding stage (in St. Louis) and a big reason for our success when I was running the Blues. You don't often get a chance to trade for a goalie of Liut's caliber. When you can, they don't come cheap."

Under terms of the trade, Hartford will also get a player-to-be named later. Speculation centers on Jorgen Pettersson, a forward, as the final payment.

Millen, who came to the Whalers as a free agent from Pittsburgh in 1981, has been a workhorse for the locals. Johnson, the one-time ace of the gold-medal 1980 U.S. Olympic Team, emerged into a point-per-game player with the Whalers since arriving in a swap with Minnesota in 1982.

Whalers Season Statistics

Regular Season

	GP	G	A	Pts	PM	PP	SH	GW
Francis	80	24	57	81	66	4	0	1
Turgeon	64	31	31	62	67	11	0	3
Neufeld	76	27	35	62	129	12	0	2
Malone	76	22	39	61	67	6	0	4
Johnson	49	19	28	47	21	10	0	2
Siltanen	76	12	33	45	30	8	0	2
Dineen	57	25	16	41	120	8	4	2
Robertson	74	11	30	41	337	1	0	3
Crawford	45	14	14	28	8	2	0	0
Ferraro	44	11	17	28	40	6	0	2
Lumley	48	8	20	28	98	1	0	0
Quenneville	79	6	16	22	96	0	0	2
Tippett	80	7	12	19	12	0	0	0
Zuke	67	4	12	16	12	0	0	1
Boutette	33	6	8	14	51	0	0	2
Fenton	33	7	5	12	10	0	0	2
Cote	67	3	9	12	17	1	0	1
MacDermid	31	4	7	11	29	0	0	0
Crombeen	46	4	7	11	16	0	1	1
Currie	13	3	8	11	2	1	0	0
Fusco	63	3	8	11	40	0	0	0
Kleinendorst	35	1	8	9	69	0	0	0
Kotsopoulos	33	5	3	8	53	1	0	0
Samuelsson	41	2	6	8	83	0	0	0

	GP	G	A	Pts	PM	PP	SH	GW
Evason	2	0	0	0	0	0	0	0
Pierce	17	3	2	5	8	0	0	0
Weir	34	2	3	5	56	0	0	0
Dunn	13	1	4	5	2	0	0	0
Brownschidle	17	1	4	5	5	1	0	0
Paterson	13	1	3	4	24	0	0	0
Jensen	13	0	4	4	6	0	0	0
Marty Howe	19	1	1	2	10	0	0	0
Liut	12	0	0	0	2	0	0	0
Hoffman	1	0	0	0	0	0	0	0
Staniowski	1	0	0	0	0	0	0	0
Weeks	24	0	0	0	0	0	0	0
Millen	44	0	0	0	4	0	0	0
Bench	80	--	--	--	18	--	--	--
Totals	80	268	450	718	1602	72	6	30
Opponents	80	318	548	866	1680	62	12	41

Goaltending	GP	Mts	W	L	T	Sho	GGA
Liut	12	731	4	7	1	1	2.95
Staniowski	1	20	0	0	0	0	3.00
Weeks	24	1457	10	12	2	2	3.79
Millen	44	2659	16	22	6	1	4.22
Totals	80	4867	30	41	9	4	3.93
Opponents	80	4867	41	30	9	6	3.30

April 29, 1986

MONTREAL - The bones of many old goaltenders can be found under the ice at the Montreal Forum.

The Whalers know that feeling Tuesday, losing 2-1 in sudden-death overtime to Montreal in Game 7.

Forward Claude Lemieux, a rookie who has a penchant for big-play goals, gained control of a loose puck behind the Hartford cage, then circled in front of goalie Mike Liut. His backhand shot caught the short side at 5:55 to end this epic battle between the Adams Division rivals.

The seven-game series was a classic. Montreal outscored Hartford 16-13 with three games decided by one goal.

A short-handed goal by Mike McPhee at 18:47

Lemieux, Habs deep-six Whalers in overtime, 2-1

of the first period gave the Habs a 1-0 lead, a cushion goalie Patrick Roy nursed until 2:48 remained when Dave Babych, taking a drop pass from Dean Evason, powered one home from the left point forcing overtime.

Kevin Dineen was among the standouts for the Whalers. He went 6-7-13 with two game-winners including the only tally in Game 6 as Hartford extended the series to the limit.

1985-86 League Standings

CLARENCE CAMPBELL CONFERENCE

Norris Division

Team	GP	W	L	T	GF	GA	PTS
Chicago	80	39	33	8	351	349	86
Minnesota	80	38	33	9	327	305	85
St. Louis	80	37	34	9	302	291	83
Toronto	80	25	48	7	311	386	57
Detroit	80	17	57	6	266	415	40

Smythe Division

Edmonton	80	56	17	7	426	310	119
Calgary	80	40	31	9	354	315	89
Winnipeg	80	26	47	7	295	372	59
Vancouver	80	23	44	13	282	333	59
Los Angeles	80	23	49	8	284	389	54

PRINCE OF WALES CONFERENCE

Adams Division

Quebec	80	43	31	6	330	289	92
Montreal	80	40	33	7	330	280	87
Boston	80	37	31	12	311	288	86
Hartford	80	40	36	4	332	302	84
Buffalo	80	37	37	6	296	291	80

Patrick Division

Philadelphia	80	53	23	4	335	241	110
Washington	80	50	23	7	315	272	107
NY Islanders	80	39	29	12	327	284	90
NY Rangers	80	36	38	6	280	276	78
Pittsburgh	80	34	38	8	313	305	76
New Jersey	80	28	49	3	300	374	59

League Champion: Montreal Canadiens
League MVP: Wayne Gretzky, Edmonton
Top Scorer: Wayne Gretzky, Edmonton

(52-163-215 in 80 GP)

It's Wales night

February 4, 1986

HARTFORD - All eyes were on Hartford Tuesday night and the players involved in the NHL's 38th annual All-Star Game seemed more star-struck than the sellout crowd of 15,126 that came to watch the greatest players battle.

Usually the midseason exhibition is a high-scoring affair yet for 28 minutes, the greatest goal-scorers gathered on the same ice surface were unable to solve either Grant Fuhr of the Campbell Conference or Mario Gosselin of the Wales Conference.

At the finish, the dynamic tandem

Whalers Season Statistics

		Regular Season							Playoffs							
	GP	G	A	Pts	PM	PP	SH	GW	GP	G	A	Pts	PM	PP	SH	GW
Turgeon	76	45	34	79	88	13	0	5	9	2	3	5	4	0	0	2
Ferraro	76	30	47	77	57	14	0	0	10	3	6	9	4	3	0	0
Francis	53	24	53	77	24	7	1	4	10	1	2	3	4	0	0	0
Dineen	57	33	35	68	124	6	0	8	10	6	7	13	18	1	0	2
Gavin	76	26	29	55	51	3	5	4	10	4	1	5	13	0	0	0
D.Babych	62	10	43	53	38	7	1	2	8	1	3	4	14	0	0	0
Evason	55	20	28	48	65	5	2	4	10	1	4	5	10	0	0	0
Lawless	64	17	21	38	20	5	0	1	1	0	0	0	0	0	0	0
Robertson	76	13	24	37	358	3	0	0	10	1	0	1	67	0	0	0
Tippett	80	14	20	34	18	0	2	1	10	2	2	4	4	0	1	0
Crawford	57	14	20	34	16	4	0	2	--	--	--	--	--	--	--	--
Siltanen	52	8	22	30	30	6	0	1	--	--	--	--	--	--	--	--
W.Babych	37	11	17	28	59	2	0	2	10	0	1	1	2	0	0	0
Anderson	14	8	17	25	2	1	0	1	10	5	8	13	0	3	0	0
Jarvis	57	8	16	24	20	0	3	0	10	0	0	0	10	0	0	0
Murzyn	78	3	23	26	125	0	0	1	4	0	0	0	12	0	0	0
Quenneville	71	5	20	25	83	1	0	1	10	0	2	2	12	0	0	0
Samuelsson	80	5	19	24	172	0	1	1	10	1	2	3	38	0	0	1
MacDermid	74	13	10	23	160	0	0	2	10	2	1	3	20	0	0	1
Neufeld	16	5	10	15	40	3	0	0	--	--	--	--	--	--	--	--
Malone	22	6	7	13	24	1	0	0	--	--	--	--	--	--	--	--
Pettersson	23	5	5	10	2	2	0	0	--	--	--	--	--	--	--	--
Bothwell	62	2	8	10	53	0	0	0	10	0	0	0	10	0	0	0
Kleinendorst	41	2	7	9	62	0	0	0	10	0	1	1	18	0	0	0
Gardner	18	1	8	9	4	0	0	0	--	--	--	--	--	--	--	--
McEwen	10	3	2	5	6	0	0	0	8	0	4	4	6	0	0	0
Hoffman	6	1	2	3	2	0	0	0	--	--	--	--	--	--	--	--
Shaw	8	0	2	2	4	0	0	0	--	--	--	--	--	--	--	--
Zuke	17	0	2	2	12	0	0	0	--	--	--	--	--	--	--	--
Liut	57	0	2	2	0	0	0	0	8	0	0	0	0	0	0	0
Weeks	27	0	1	1	9	0	0	0	3	0	0	0	0	0	0	0
Fenton	1	0	0	0	0	0	0	0	--	--	--	--	--	--	--	--
Cote	2	0	0	0	0	0	0	0	--	--	--	--	--	--	--	--
Newberry	3	0	0	0	0	0	0	0	--	--	--	--	--	--	--	--
Paterson	5	0	0	0	5	0	0	0	--	--	--	--	--	--	--	--
Brownschidle	9	0	0	0	4	0	0	0	--	--	--	--	--	--	--	--
Bench	80	--	--	--	24	--	--	10	--	--	--	--	0	--	--	--
Totals	80	332	554	886	1760	83	13	40	10	29	50	79	256	7	1	6
Opponents	80	302	482	784	1745	72	8	36	10	23	34	57	267	10	3	4

Goaltending	GP	Mts	W	L	T	Sho	GGA	GP	Mts	W	L	Sho	GAA
Liut	57	3282	27	34	4	2	3.61	8	441	5	2	1	1.90
Weeks	27	1544	13	13	0	1	3.85	3	169	1	2	0	2.84
Totals	80	4826	40	36	4	3	3.69	10	610	6	4	1	2.26
Opponents	80	4817	36	40	4	3	4.06	10	610	4	6	0	2.85

of the New York Islanders, Bryan Trottier and Mike Bossy, worked their magic with "Trotts" scoring at 3:05 of overtime to cap a 4-3 win for the Wales.

Brian Propp had two goals and Peter Stastny also tallied for the Wales. Tony Tanti, Wayne Gretzky and Dale Hawerchuk scored for the Campbells.

Whalers capture Adams flag

April 4, 1987

HARTFORD - The sellout crowd was primed and so were the players.

With 15,126 cheering their first-place heroes on, the Whalers struck for three goals within a 3:20 span of the third period to overtake the Rangers and post a 5-3 victory.

Captain Ron Francis notched three helpers in the decision, a milestone achievement for Hartford which officially clinched the Adams Division flag - the first division crown for the club since the early years of the World Hockey Association.

"There was a time when we'd get standing ovations for just coming close," Francis said.

"Our goal was to win the division outright," netminder Mike Liut said after turning away 25 shots for the victory. "To fulfill that is great."

Down 3-2 entering the final 20 minutes, Sylvain Turgeon scored at 4:53 to tie the game. John Anderson, assisted by Francis, made it 4-3 at 6:24. Paul MacDermid, with helpers from Francis and Dave Tippett, iced it at 8:17.

The Whalers, 43-29-7, have stayed atop the pack since the puck was dropped back in October. Hartford winds up the year at Buffalo on Sunday and will dock in with the league's fourth best mark.

The playoffs open April 8 with the Quebec Nordiques.

1986-87 League Standings

CLARENCE CAMPBELL CONFERENCE

Norris Division

Team	GP	W	L	T	GF	GA	PTS
St. Louis	80	32	33	15	281	293	79
Detroit	80	34	36	10	260	274	78
Chicago	80	29	37	14	290	310	72
Toronto	80	32	42	6	286	319	70
Minnesota	80	30	40	10	296	314	70

Smythe Division

Edmonton	80	50	24	6	372	284	106
Calgary	80	46	31	3	318	289	95
Winnipeg	80	40	32	8	279	271	88
Los Angeles	80	31	41	8	318	341	70
Vancouver	80	29	43	8	282	314	66

PRINCE OF WALES CONFERENCE

Adams Division

Hartford	80	43	30	7	287	270	93
Montreal	80	41	29	10	277	241	92
Boston	80	39	34	7	301	276	85
Quebec	80	31	39	10	267	276	72
Buffalo	80	28	44	8	280	308	64

Patrick Division

Philadelphia	80	46	26	8	310	245	100
Washington	80	38	32	10	285	278	86
NY Islanders	80	35	33	12	279	281	82
NY Rangers	80	34	38	8	307	323	76
Pittsburgh	80	30	38	12	297	290	72
New Jersey	80	29	45	6	293	368	64

League Champion: Edmonton Oilers
League MVP: Wayne Gretzky, Edmonton
Top Scorer: Wayne Gretzky, Edmonton
(62-121-183 in 79 GP)

Jarvis is now "Ironman"

December 26, 1986

HARTFORD - Center Doug Jarvis established a record for consecutive games played in Hartford's 1-1 stalemate with Montreal.

The veteran defensive forward played in his 915th consecutive game to break Garry Unger's standard.

"A highlight, something I will treasure," Jarvis said.

Whalers Season Statistics

				Regular Season								Playoffs					
	GP	G	A	Pts	PM	PP	SH	GW	GP	G	A	Pts	PM	PP	SH	GW
Francis	75	30	63	93	45	7	0	7	6	2	2	4	6	1	0	0
Dineen	78	40	39	79	110	11	0	7	6	2	1	3	31	1	0	0
Anderson	76	31	44	75	19	7	0	5	6	1	2	3	0	1	0	0
Ferraro	80	27	32	59	42	14	0	2	6	1	1	2	8	0	0	0
Evason	80	22	37	59	67	7	2	2	5	3	2	5	35	0	0	0
Lawless	60	22	32	54	14	4	0	2	2	0	2	2	2	0	0	0
Gavin	79	20	22	42	28	3	2	4	6	2	4	6	10	0	0	0
D.Babych	66	8	33	41	44	7	0	1	6	1	1	2	14	1	0	0
Turgeon	41	23	13	36	45	6	0	4	6	1	2	3	6	0	0	0
Samuelsson	78	2	31	33	162	0	0	0	5	0	1	1	41	0	0	0
Tippett	80	9	22	31	42	0	3	2	6	0	2	2	4	0	0	0
Murzyn	74	9	19	28	95	1	0	0	6	2	1	3	29	1	0	1
Jarvis	80	9	13	22	20	0	2	2	6	0	0	0	4	0	0	0
MacDermid	72	7	11	18	202	0	0	1	6	2	1	3	34	0	0	1
McEwen	48	8	8	16	32	5	0	2	1	1	1	2	0	1	0	0
Semenko	52	4	8	12	87	0	0	0	4	0	0	0	15	0	0	0
Kleinendorst	66	3	9	12	130	0	0	0	4	1	3	4	20	0	0	0
Quenneville	37	3	7	10	24	0	1	1	6	0	0	0	0	0	0	0
Cote	67	2	8	10	20	0	0	0	6	0	2	2	0	0	0	0
Ladouceur	36	2	3	5	51	0	0	0	6	0	2	2	12	0	0	0
Barr	30	2	4	6	19	0	1	0	--	--	--	--	--	--	--	--
Millar	10	2	2	4	0	1	0	0	--	--	--	--	--	--	--	--
Liut	59	0	2	2	4	0	0	0	6	0	0	0	2	0	0	0
Bothwell	4	1	0	1	0	0	0	0	--	--	--	--	--	--	--	--
Robertson	20	1	0	1	98	0	0	1	--	--	--	--	--	--	--	--
Gardner	8	0	1	1	0	0	0	0	--	--	--	--	--	--	--	--
Churla	20	0	1	1	78	0	0	0	2	0	0	0	42	0	0	0
Britz	1	0	0	0	0	0	0	0	--	--	--	--	--	--	--	--
Hughes	2	0	0	0	2	0	0	0	3	0	0	0	0	0	0	0
Shaw	2	0	0	0	0	0	0	0	--	--	--	--	--	--	--	--
W.Babych	4	0	0	0	4	0	0	0	--	--	--	--	--	--	--	--
Courteau	4	0	0	0	0	0	0	0	--	--	--	--	--	--	--	--
Sherven	7	0	0	0	0	0	0	0	--	--	--	--	--	--	--	--
Weeks	25	0	0	0	0	0	0	0	1	0	0	0	0	0	0	0
Bench	80	--	--	--	12	--	--	--	6	--	--	--	2	--	--	--
Totals	80	287	464	751	1496	73	11	43	6	19	30	49	315	6	0	2
Opponents	80	270	457	727	1657	66	10	31	6	27	49	76	294	12	1	4

Goaltending	GP	Mts	W	L	T	Sho	GAA	GP	Mts	W	L	Sho	GAA
Liut	59	3471	31	22	5	4	3.23	6	332	2	4	0	4.52
Weeks	25	1367	12	8	2	1	3.42	1	36	0	0	0	1.67
Totals	80	4838	43	31	7	5	3.29	6	368	2	4	0	4.40
Opponents	80	4829	31	43	7	2	3.44	6	368	4	2	0	3.10

1987-88 League Standings

CLARENCE CAMPBELL CONFERENCE
Norris Division

Team	GP	W	L	T	GF	GA	PTS
Detroit	80	41	28	11	322	269	93
St. Louis	80	34	38	8	278	294	76
Chicago	80	30	41	9	284	328	69
Toronto	80	21	49	10	273	345	52
Minnesota	80	19	48	13	242	349	51

Smythe Division

Calgary	80	48	23	9	397	305	105
Edmonton	80	44	25	11	363	288	99
Winnipeg	80	33	36	11	292	310	77
Los Angeles	80	30	42	8	318	359	68
Vancouver	80	25	46	9	272	320	59

PRINCE OF WALES CONFERENCE
Adams Division

Montreal	80	45	22	13	298	238	103
Boston	80	44	30	6	300	251	94
Buffalo	80	37	32	11	283	305	85
Hartford	80	35	38	7	249	267	77
Quebec	80	32	43	5	271	306	69

Patrick Division

NY Islanders	80	39	31	10	308	267	88
Washington	80	38	33	9	281	249	85
Philadelphia	80	38	33	9	292	292	85
New Jersey	80	38	36	6	295	296	82
NY Rangers	80	36	34	10	300	283	82
Pittsburgh	80	36	35	9	319	316	81

League Champion: Edmonton Oilers
League MVP: Mario Lemieux, Pittsburgh
Top Scorer: Mario Lemieux, Pittsburgh
(70-98-168 in 76 GP)

Real estate baron buys Whalers for $31 million

June 22, 1988

HARTFORD - When the hockey team began operations in 1972, organizers were in an attorney's office in Boston, gearing to sign documents when Bob Schmertz spotted a circular, ebony-colored object.

"What's that?" Schmertz asked.

When the main founder of the New England Whalers explained to Schmertz that the item on the desk was a puck used in playing the game, the capitalist led the men in a roar of laughter.

One of the men in the room that day was Howard L. Baldwin. He launched the hockey club for $25,000 and eventually gained the fiscal support of Schmertz. The partnership enabled Baldwin to realize a promoting dream.

That's how the Whalers began and the amazing saga of a team reached a new chapter in its 17-year history when it was sold for $31 million.

Real estate developer Richard Gordon and former insurance executive Donald Conrad are the new owners.

The new partners are friends today.

They may have a few more now among the Board of Governors.

"We may have done a favor for all the other owners," Conrad says. "I don't think, 10 years ago, anyone could have imagined this team being worth what it is today. Even three years ago, I couldn't have imagined it."

The sale to Gordon-Conrad team shifts the Whalers from a consortium of Hartford-area companies that absorbed the ups-and-downs over the years to individual ownership. It may be the natural evolution of a franchise but it also ended Baldwin's long run as the Head Whale.

The Whalers moved from Boston to Hartford in the WHA in 1974-75. Conrad, then with Aetna Insurance and a major force on the Hartford Downtown Council, pushed to get a major tenant for the new Civic Center complex.

"When we brought the team here, it was for corporate responsibilities," Conrad says. "We wanted to revitalize the downtown area. New restaurants opened. The town has never been the same. It's a mecca of activity now."

Gordon, 48, has been a season ticket holder for many years. He acquired his wealth as an entrepreneur in real estate, amassing property in the city as well as suburban areas in Greater Hartford.

"I own plenty of buildings," Gordon says. "There are only 90 or so professional teams."

Whalers Season Statistics

	Regular Season								Playoffs							
	GP	G	A	Pts	PM	PP	SH	GW	GP	G	A	Pts	PM	PP	SH	GW
Francis	80	25	50	75	87	11	1	3	6	2	5	7	2	1	0	0
Dineen	74	25	25	50	217	5	0	4	6	4	4	8	8	1	0	1
Ferraro	68	21	29	50	81	6	0	2	6	1	1	2	6	1	0	0
Babych	71	14	36	50	54	10	0	2	6	3	2	5	2	0	0	0
Turgeon	71	23	26	49	71	13	0	3	6	0	0	0	4	0	0	0
Anderson	63	17	32	49	20	9	0	3	--	--	--	--	--	--	--	--
Samuelsson	76	8	33	41	159	3	0	2	5	0	0	0	8	0	0	0
Wilson	36	18	20	38	22	7	1	5	6	2	4	6	2	1	0	1
Tippett	80	16	21	37	31	1	2	2	6	0	2	2	0	0	0	0
MacDermid	80	20	15	35	139	4	0	2	6	0	5	5	14	0	0	0
Evason	77	10	18	28	115	6	0	0	6	1	1	2	2	0	0	0
Cote	67	7	21	28	30	0	1	1	6	1	1	2	4	1	0	0
Gavin	56	11	10	21	59	2	3	1	6	2	2	4	4	0	0	0
Millar	28	7	7	14	6	4	0	2	--	--	--	--	--	--	--	--
Robertson	63	2	8	10	293	0	0	0	6	0	1	1	6	0	0	0
Carson	27	5	4	9	30	1	1	1	5	1	2	3	0	0	0	0
Lawless	28	4	5	9	16	0	0	1	--	--	--	--	--	--	--	--
Kleinendorst	44	3	6	9	86	0	0	0	3	1	1	2	0	1	0	0
Peterson	52	2	7	9	40	0	0	0	4	0	0	0	2	0	0	0
Quenneville	77	1	8	9	44	0	0	0	6	0	2	2	0	0	0	0
Ladouceur	68	1	7	8	91	0	0	1	6	1	1	2	4	0	0	0
Murzyn	33	1	6	7	45	1	0	0	--	--	--	--	--	--	--	--
Reeds	38	0	7	7	31	0	0	0	--	--	--	--	--	--	--	--
Williams	26	6	0	6	87	1	0	1	--	--	--	--	--	--	--	--
N.Sheehy	26	1	4	5	116	0	0	0	1	0	0	0	7	0	0	0
Martin	5	1	2	3	14	0	0	0	--	--	--	--	--	--	--	--
McEwen	9	0	3	3	10	0	0	0	2	0	2	2	0	0	0	0
Liut	60	0	1	1	4	0	0	0	3	0	0	0	0	0	0	0
Bourgeois	1	0	0	0	0	0	0	0	--	--	--	--	--	--	--	--
Sherven	1	0	0	0	0	0	0	0	--	--	--	--	--	--	--	--
Shaw	1	0	0	0	0	0	0	0	--	--	--	--	--	--	--	--
Sidorkiewicz	1	0	0	0	0	0	0	0	--	--	--	--	--	--	--	--
Jarvis	2	0	0	0	0	0	0	0	--	--	--	--	--	--	--	--
Churla	2	0	0	0	14	0	0	0	--	--	--	--	--	--	--	--
Vellucci	2	0	0	0	11	0	0	0	--	--	--	--	--	--	--	--
Young	7	0	0	0	2	0	0	0	4	1	0	1	0	0	0	0
Brodeur	6	0	0	0	2	0	0	0	4	0	0	0	0	0	0	0
Weeks	18	0	0	0	0	0	0	0	--	--	--	--	--	--	--	--
Bench	80	--	--	8	--	--	--	--	6	--	--	0	--	--	--	--
Totals	80	249	411	660	2046	84	9	35	6	20	34	54	81	6	0	2
Opponents	80	267	434	701	2232	67	17	31	6	23	40	63	80	6	0	4

Goaltending	GP	Mts	W	L	T	Sho	GAA	GP	Mts	W	L	Sho	GAA
Brodeur	6	340	4	2	0	0	2.65	4	200	1	3	0	3.60
Liut	60	3525	25	28	5	2	3.18	3	160	1	1	0	4.13
Weeks	18	918	6	7	2	0	3.59	--	--	--	--	--	---
Sidorkiewicz	1	60	0	1	0	0	6.00	--	--	--	--	--	---
Totals	80	4843	35	38	7	2	3.26	6	360	2	4	0	3.84
Opponents	80	4837	38	35	7	4	3.08	6	360	4	2	0	3.33

1988-89 League Standings

CLARENCE CAMPBELL CONFERENCE
Norris Division

Team	GP	W	L	T	GF	GA	PTS
Detroit	80	34	34	12	313	316	80
St. Louis	80	33	35	12	275	285	78
Minnesota	80	27	37	16	258	278	70
Chicago	80	27	41	12	297	335	66
Toronto	80	28	46	6	259	342	62

Smythe Division

	GP	W	L	T	GF	GA	PTS
Calgary	80	54	17	9	354	226	117
Los Angeles	80	42	31	7	376	335	91
Edmonton	80	38	34	8	325	306	84
Vancouver	80	33	39	8	251	253	74
Winnipeg	80	26	42	12	300	355	64

PRINCE OF WALES CONFERENCE
Adams Division

	GP	W	L	T	GF	GA	PTS
Montreal	80	53	18	9	315	218	115
Boston	80	37	29	14	289	256	88
Buffalo	80	38	35	7	291	299	83
Hartford	80	37	38	5	299	290	79
Quebec	80	27	46	7	269	342	61

Patrick Division

	GP	W	L	T	GF	GA	PTS
Washington	80	41	29	10	305	259	92
Pittsburgh	80	40	33	7	347	349	87
NY Rangers	80	37	35	8	310	307	82
Philadelphia	80	36	36	8	307	285	80
New Jersey	80	27	41	12	281	325	66
NY Islanders	80	28	47	5	265	325	61

League Champion: Calgary Flames
League MVP: Wayne Gretzky, Los Angeles
Top Scorer: Mario Lemieux, Pittsburgh
(85-114-199 in 76 GP)

Turgeon goes to Devils for Verbeek

June 17, 1989

MINNEAPOLIS - The Whalers may not have forward Bobby Holik in the lineup for the 1989-90 season opener but they will have Pat Verbeek, one of the grittiest and better two-way forwards in the NHL for the lidlifter.

Hartford tabbed Holik, a bruising center from Czechoslovakia, with the 10th overall pick in the Entry Draft.

While the Whalers went on to selection 10 other hopefuls, they also swung a trade with New Jersey to land Verbeek, an inspirational-type winger who has a knack for scoring goals.

In exchange for Verbeek, the Whalers sent one-time 45-goal scorer Sylvain Turgeon to the Devils. Turgeon, who has been troubled by stomach and leg injuries, was the club's top pick (second overall) in 1983.

Whalers Season Statistics

		Regular Season								Playoffs						
	GP	G	A	Pts	PM	PP	SH	GW	GP	G	A	Pts	PM	PP	SH	GW
Dineen	79	45	44	89	167	20	1	4	4	1	0	1	10	0	0	0
Francis	69	29	48	77	36	8	0	4	4	0	2	2	0	0	0	0
Ferraro	80	41	35	76	86	11	0	7	4	2	0	2	4	0	0	0
Young	76	19	40	59	27	6	0	2	4	2	0	2	4	0	0	0
Babych	70	6	41	47	54	4	0	2	4	1	5	6	2	0	0	0
MacDermid	74	17	24	44	144	5	0	3	4	1	1	2	16	0	0	0
Lawton	35	10	16	26	28	7	0	2	3	1	0	1	0	0	0	0
Tippett	80	17	24	41	45	1	2	1	4	0	1	1	0	0	0	0
Anderson	62	16	24	40	28	1	0	0	4	0	1	1	2	0	0	0
Samuelsson	71	9	26	35	181	3	0	2	4	0	2	2	4	0	0	0
J.Hull	60	16	18	34	10	6	0	2	1	0	0	0	2	0	0	0
Turgeon	42	16	14	30	40	7	0	1	4	0	2	2	4	0	0	0
Evason	67	11	17	28	60	0	0	0	4	1	2	3	10	0	1	0
Maciver	37	1	22	23	24	1	0	0	1	0	0	0	2	0	0	0
Cote	78	8	9	17	49	1	0	0	3	0	1	1	4	0	0	0
Peterson	66	4	13	17	61	0	0	2	2	0	1	1	4	0	0	0
Maloney	21	3	11	14	23	1	0	1	4	0	0	0	8	0	0	0
Martin	38	7	6	13	113	0	0	1	1	0	0	0	4	0	0	0
Jennings	55	3	10	13	159	0	0	0	4	1	0	1	17	1	0	0
Quenneville	69	4	7	11	32	0	0	0	4	0	3	3	4	0	0	0
Ladouceur	75	2	5	7	95	0	0	0	1	0	0	0	10	0	0	0
Sidorkiewicz	44	0	3	3	0	0	0	0	2	0	0	0	0	0	0	0
Gaume	4	1	1	2	0	0	0	0	--	--	--	--	--	--	--	--
Whitmore	3	0	2	2	0	0	0	0	2	0	0	0	0	0	0	0
Reeds	7	0	2	2	6	0	0	0	--	--	--	--	--	--	--	--
Kastelic	10	0	2	2	15	0	0	0	--	--	--	--	--	--	--	--
Shaw	3	1	0	1	0	1	0	0	3	1	0	1	0	0	0	0
Thomson	5	0	0	0	14	0	0	0	--	--	--	--	--	--	--	--
Pavese	5	0	0	0	5	0	0	0	1	0	0	0	0	0	0	0
Yake	2	0	0	0	0	0	0	0	--	--	--	--	--	--	--	--
Tuer	4	0	0	0	23	0	0	0	--	--	--	--	--	--	--	--
Burt	5	0	0	0	6	0	0	0	--	--	--	--	--	--	--	--
Liut	35	0	0	0	0	0	0	0	--	--	--	--	--	--	--	--
Bench	80	--	--	--	10	--	--	--	4	--	--	--	0	--	--	--
Totals	80	299	483	782	1672	87	3	37	4	11	21	32	111	1	1	2
Opponents	80	290	484	774	1859	70	18	38	4	18	25	43	125	4	1	4

Goaltending	GP	Mts	W	L	T	Sho	GAA	GP	Mts	W	L	Sho	GAA
Whitmore	3	180	2	1	0	0	3.33	2	135	0	2	0	4.44
Liut	35	2004	13	19	1	1	4.25	--	--	--	--	--	---
Sidorkiewicz	44	2634	22	18	4	4	3.03	2	124	0	2	0	3.87
Totals	80	4818	37	38	5	5	3.55	4	259	0	4	0	4.17
Opponents	80	4815	38	37	5	1	3.61	4	259	4	0	0	2.55

Turgeon went 178-150-328 in 370 games during his tenure with the Whalers. His biggest goal came in overtime on April 9, 1986, a strike at 2:36 to give Hartford a 3-2 win over Quebec, the franchise's first ever Stanley Cup victory.

Verbeek, who slumped to 22 goals, appears to be a key fit on Hartford's top line with center Ron Francis and winger Kevin Dineen.

Though general manager Eddie Johnston announced the deal at the Entry Draft, insiders say that Emile Francis, who was pushed upstairs to team president when owner Richard Gordon opted to hire EJ, set the groundwork for the swap weeks before it became official.

1989-90 League Standings

CLARENCE CAMPBELL CONFERENCE
Norris Division

Team	GP	W	L	T	GF	GA	PTS
Chicago	80	41	33	6	316	294	88
St. Louis	80	37	34	9	295	279	83
Toronto	80	38	38	4	337	358	80
Minnesota	80	36	40	4	284	291	76
Detroit	80	28	38	14	288	323	70

Smythe Division

Calgary	80	42	23	15	348	265	99
Edmonton	80	38	28	14	315	283	90
Winnipeg	80	37	32	11	298	290	85
Los Angeles	80	34	39	7	338	337	75
Vancouver	80	25	41	14	245	306	64

PRINCE OF WALES CONFERENCE
Adams Division

Boston	80	46	25	9	289	232	101
Buffalo	80	45	27	8	286	248	98
Montreal	80	41	28	11	288	234	93
Hartford	80	38	33	9	275	268	85
Quebec	80	12	61	7	240	407	31

Patrick Division

NY Rangers	80	36	31	13	279	267	85
New Jersey	80	37	34	9	295	288	83
Washington	80	36	38	6	284	275	78
NY Islanders	80	31	38	11	281	288	73
Pittsburgh	80	32	40	8	318	359	72
Philadelphia	80	30	39	11	290	297	71

League Champion: Edmonton Oilers
League MVP: Mark Messier, Edmonton
Top Scorer: Wayne Gretzky, Los Angeles
(40-102-142 in 73 GP)

Game 4 loss may swing series

April 11, 1990

HARTFORD - Every team takes a game to the grave.

For the Whalers, Game 4 could be that one, a loss that will be etched in the soul for a long time.

Leading 5-2 with one period remaining, Hartford blew a chance to take a commanding 3-1 lead in the best-of-seven series with the hated Bruins, allowing Boston to stage one of the more gallant comebacks in Stanley Cup history.

Dave Poulin, fighting off a check from Ulf Samuelsson, scored what proved to be the game-winner, the finale of four by the B's in the final period, to scuttle the Whalers 6-5 and knot the series at 2-2.

Poulin was able to free a hand despite Samuelsson tactics and then squeezed the puck past goalie Peter Sidorkiewicz with 1:44 remaining in regulation.

The Whalers had goals from Dean Evason, Brad Shaw, Kevin Dineen, Yvon Corriveau and Ron Francis. Hartford had seized a 5-2 lead after 40 minutes on the strength of a four-goal uprising in the second period.

Whalers Season Statistics

	Regular Season								Playoffs							
	GP	G	A	Pts	PM	PP	SH	GW	GP	G	A	Pts	PM	PP	SH	GW
Francis	80	32	69	101	73	15	1	5	7	3	3	6	8	1	0	0
Verbeek	80	44	45	89	228	14	0	5	7	2	2	4	26	1	0	1
Dineen	67	25	41	66	164	8	2	2	6	3	2	5	18	0	0	1
Young	80	24	40	64	47	10	2	5	7	2	0	2	2	0	0	0
Ferraro	79	25	29	54	109	7	0	4	7	0	3	3	2	0	0	0
Evason	78	18	25	43	138	2	2	2	7	2	2	4	22	0	0	0
Babych	72	6	37	43	62	4	0	1	7	1	2	3	0	0	0	0
Andersson	50	13	24	37	6	1	2	2	5	0	3	3	2	0	0	0
Shaw	64	3	32	35	30	3	0	0	7	2	5	7	0	1	0	0
Krygier	58	18	12	30	52	5	1	3	7	2	1	3	4	0	0	0
Tippett	66	8	19	27	32	0	1	3	7	1	3	4	2	0	0	0
Tomlak	70	7	14	21	48	1	1	2	7	0	1	1	2	0	0	0
Cunneyworth	43	9	9	18	41	2	0	1	4	0	0	0	2	0	0	0
J.Hull	38	7	10	17	21	2	0	0	5	0	1	1	2	0	0	0
Ladouceur	71	3	12	15	126	0	0	0	7	1	0	1	10	0	0	1
Samuelsson	55	2	11	13	177	0	0	0	7	1	0	1	18	0	0	0
Burt	63	4	8	12	105	1	0	0	2	0	0	0	0	0	0	0
Jennings	64	3	6	9	171	0	0	0	7	0	0	0	17	0	0	0
Kastelic	67	6	2	8	198	0	0	0	2	0	0	0	0	0	0	0
Cote	28	4	2	6	14	1	0	1	5	0	0	0	2	0	0	0
Corriveau	13	4	1	5	20	0	0	0	4	1	0	1	0	0	0	0
Quenneville	44	1	4	5	34	0	0	0	--	--	--	--	--	--	--	--
Lawton	13	2	1	3	6	1	0	0	--	--	--	--	--	--	--	--
Martin	21	1	2	3	37	0	0	0	--	--	--	--	--	--	--	--
Yake	2	0	1	1	0	0	0	0	--	--	--	--	--	--	--	--
Culhane	6	0	1	1	6	0	0	0	--	--	--	--	--	--	--	--
Whitmore	9	0	1	1	4	0	0	0	--	--	--	--	--	--	--	--
Govedaris	12	0	1	1	6	0	0	0	2	0	0	0	0	0	0	0
Sidorkiewicz	46	0	1	1	4	0	0	0	7	0	0	0	0	0	0	0
Black	1	0	0	0	0	0	0	0	--	--	--	--	--	--	--	--
Bodak	1	0	0	0	7	0	0	0	--	--	--	--	--	--	--	--
Tuer	2	0	0	0	6	0	0	0	--	--	--	--	--	--	--	--
McKenzie	5	0	0	0	4	0	0	0	--	--	--	--	--	--	--	--
Dykstra	9	0	0	0	2	0	0	0	--	--	--	--	--	--	--	--
Laforge	9	0	0	0	43	0	0	0	--	--	--	--	--	2	--	--
Bench	80	--	--	--	12	--	--	--	7	--	--	--	--	--	--	--
Totals	80	275	459	734	2102	80	12	38	7	21	28	49	143	3	0	3
Opponents	80	268	450	718	2169	74	10	33	7	23	38	61	193	8	1	4

Goaltending	GP	Mts	W	L	T	Sho	GAA	GP	Mts	W	L	Sho	GAA
Liut	29	1683	15	12	1	3	2.64	--	--	--	--	--	---
Whitmore	9	442	4	2	1	0	3.53	--	--	--	--	--	---
Sidorkiewicz	46	2703	19	19	7	1	3.57	7	429	3	4	0	3.22
Totals	80	4843	38	33	9	4	3.32	7	429	3	4	0	3.22
Opponents	80	4843	33	38	9	2	3.41	7	433	4	3	0	2.91

Whalers deal Francis, Ulfie to Pens

Cullen, Zalapski come to Hartford from Pittsburgh

March 4, 1991

HARTFORD - They took away the captaincy a short time ago.

The other shoe dropped today when the Hartford Whalers, in general manager Eddie Johnston's master plan to gut the organization of popular players, shipped Ron Francis and Ulf Samuelsson, two of the most identifiable members of the team, to Pittsburgh for center John Cullen and defenseman Zarley Zalapski.

Also involved in the blockbuster deal are defenseman Grant Jennings who leaves Hartford for rugged winger Jeff Parker. Jennings is best remembered for a tremendous check on Ray Bourque in last year's playoff series, an avalanche hit that sidelined the Boston captain for most of the seven-game set.

Francis had played in more games (714) than any player in club history and holds a litany of team records. He was Hartford's first round pick in 1981 (fourth overall).

1990-91 League Standings

CLARENCE CAMPBELL CONFERENCE
Norris Division

Team	GP	W	L	T	GF	GA	PTS
Chicago	80	49	23	8	284	211	106
St. Louis	80	47	22	11	310	250	105
Detroit	80	34	38	8	273	298	76
Minnesota	80	27	39	14	256	266	68
Toronto	80	23	46	11	241	318	57

Smythe Division

Los Angeles	80	46	24	10	340	254	102
Calgary	80	46	26	8	344	263	100
Edmonton	80	37	37	6	272	272	80
Vancouver	80	28	43	9	243	315	65
Winnipeg	80	26	43	11	260	288	63

PRINCE OF WALES CONFERENCE
Adams Division

Boston	80	44	24	12	299	264	100
Montreal	80	39	30	11	273	249	89
Buffalo	80	31	30	19	292	278	81
Hartford	80	31	38	11	238	276	73
Quebec	80	16	50	14	236	354	46

Patrick Division

Pittsburgh	80	41	33	6	342	305	88
NY Rangers	80	36	31	13	297	265	85
Washington	80	37	36	7	258	258	81
New Jersey	80	32	33	15	272	264	79
Philadelphia	80	33	37	10	252	267	76
NY Islanders	80	25	45	10	223	290	60

League Champion: Pittsburgh Penguins
League MVP: Brett Hull, St. Louis
Top Scorer: Wayne Gretzky, Los Angeles
(41-122-163 in 78 GP)

Samuelsson, an All-Star presence at the blue line, was among the team's most gusto-type players. Besides a needler, Samuelsson is an effective penalty-killer.

Cullen, currently among the league's Top 10 scorers this season, should give the Whalers a creative, goal-scorer in the pivot. Though smaller than Francis, Cullen is a better skater. How effective he is without the puck is debatable considering the Pittsburgh club, an offensive powerhouse, hardly plays low-scoring games.

Zalapski, a defenseman who has loads of potential, was taken fourth overall by the Penguins in 1986 when Johnston was running the front office in Pittsburgh. Zalapski should supply goals, particularly on the power play where Hartford has struggled all season.

Jennings and Parker are role-type players. Jennings played 163 games for Hartford with 7-20-27 with 412 penalty minutes. Parker, who has also played for Buffalo, has 16-19-35 in 137 NHL games.

Whalers Season Statistics

			Regular Season							Playoffs					
	GP	G	A	Pts	PM	PP	SH	GW	GP	G	A	Pts	PM	PP	SH	GW
Verbeek	80	43	39	82	246	15	0	5	6	3	2	5	40	2	0	0
Francis	67	21	55	76	51	10	1	6	--	--	--	--	--	--	--	--
Brown	44	18	24	42	101	10	0	2	5	1	0	1	7	1	0	1
Dineen	61	17	30	47	104	4	0	2	6	1	0	1	16	0	0	0
Holik	78	21	22	43	113	8	0	3	6	0	0	0	7	0	0	0
Shaw	72	4	28	32	29	2	0	1	6	1	2	3	2	0	0	0
Krygier	72	13	17	30	95	3	0	2	6	0	2	2	0	0	0	0
Evason	75	6	23	29	170	1	0	0	6	0	4	4	29	0	0	0
Cyr	70	12	13	25	107	0	1	2	6	1	0	1	10	0	0	0
Crossman	41	4	19	23	19	2	0	0	--	--	--	--	--	--	--	--
Wilson	45	8	15	23	16	4	0	1	--	--	--	--	--	--	--	--
Samuelsson	62	3	18	21	174	0	0	0	--	--	--	--	--	--	--	--
Cote	73	7	12	19	17	1	0	0	6	0	2	2	2	0	0	0
Cullen	13	8	8	16	18	4	0	1	6	2	7	9	10	0	0	0
Tomlak	64	8	8	16	55	0	1	0	3	0	0	0	2	0	0	0
Young	34	6	9	15	8	3	1	2	--	--	--	--	--	--	--	--
Cunneyworth	32	9	5	14	49	0	0	1	1	0	0	0	0	0	0	0
Andersson	41	4	7	11	8	0	0	0	--	--	--	--	--	--	--	--
Burt	42	2	7	9	63	1	0	1	--	--	--	--	--	--	--	--
Hunter	11	4	3	7	40	1	0	0	6	5	1	6	17	3	0	0
Ferraro	15	2	5	7	18	1	0	0	--	--	--	--	--	--	--	--
McKenzie	41	4	3	7	108	0	0	0	6	0	0	0	8	0	0	0
Zalapski	11	3	3	6	6	3	0	0	6	1	3	4	8	0	0	1
Babych	8	0	6	6	4	0	0	0	--	--	--	--	--	--	--	--
Yake	19	1	4	5	10	0	0	0	6	1	1	2	16	0	1	0
Jennings	44	1	4	5	82	0	0	0	--	--	--	--	--	--	--	--
Kastelic	45	2	2	4	211	0	0	0	--	--	--	--	--	--	--	--
Govedaris	14	1	3	4	4	0	0	1	--	--	--	--	--	--	--	--
Ladouceur	67	1	3	4	118	0	0	0	6	1	4	5	6	0	0	0
Richards	2	0	4	4	2	0	0	0	6	0	0	0	2	0	0	0
Sidorkiewicz	52	0	4	4	6	0	0	0	6	0	1	1	2	0	0	0
Houda	19	1	2	3	41	0	0	0	6	0	0	0	8	0	0	0
Tancill	9	1	1	2	4	0	1	0	--	--	--	--	--	--	--	--
Corriveau	23	1	1	2	18	0	0	0	--	--	--	--	--	--	--	--
Baca	9	0	2	2	14	0	0	0	--	--	--	--	--	--	--	--
Sanderson	2	1	0	1	0	0	0	0	3	0	0	0	0	0	0	0
Picard	5	1	0	1	2	0	0	0	--	--	--	--	--	--	--	--
Stevens	14	0	1	1	11	0	0	0	--	--	--	--	--	--	--	--
Whitmore	18	0	1	1	4	0	0	0	--	--	--	--	--	--	--	--
Black	1	0	0	0	0	0	0	0	--	--	--	--	--	--	--	--
McKay	1	0	0	0	0	0	0	0	--	--	--	--	--	--	--	--
Chapman	3	0	0	0	29	0	0	0	--	--	--	--	--	--	--	--
Bergevin	4	0	0	0	4	0	0	0	--	--	--	--	--	--	--	--
Greig	4	0	0	0	0	0	0	0	--	--	--	--	--	--	--	--
Parker	4	0	0	0	2	0	0	0	--	--	--	--	--	--	--	--
Reaugh	20	0	0	0	4	0	0	0	--	--	--	--	--	--	--	--
Bench	80	--	--	--	24	--	--	--	6	--	--	--	0	--	--	--
Totals	80	238	411	649	2219	73	5	31	6	17	29	46	192	6	1	2
Opponents	80	276	458	734	2185	72	9	38	6	24	40	64	145	10	0	4

Goaltending	GP	Mts	W	L	T	Sho	GAA	GP	Mts	W	L	Sho	GAA
Reaugh	20	1010	7	7	1	1	3.15	--	--	--	--	--	---
Whitmore	18	850	3	9	3	0	3.67	--	--	--	--	--	---
Sidorkiewicz	52	2953	21	22	7	1	3.33	6	359	2	4	0	4.01
McKay	1	35	0	0	0	0	5.14	--	--	--	--	--	---
Totals	80	4848	31	38	11	2	3.42	6	359	2	4	0	4.01
Opponents	80	4851	38	31	11	3	2.88	6	360	4	2	0	2.83

Whalers expire in double overtime

May 1, 1992

MONTREAL - It was the longest game in team annals, 85 minutes and 26 seconds.

The results, however, were familiar to other Stanley Cup series involving the Whalers when it comes to a winner-take-all Game 7.

One goal short.

Russ Courtnall, held scoreless throughout the previous six games, connected at 5:26 of the second extra session to give Montreal a 3-2 win over Hartford and clinch the series.

While the Habs advance now to play Boston, the loss marked the sixth consecutive year that the Whalers were ousted in the first round. Hartford is 1-8 in postseason play including 0-5 against Montreal.

Moments before Courtnall connected on a shot that somehow eluded acrobatic goalie Frank Pietrangelo who turned aside 53 tries from all angles, Hartford missed on two great chances against Patrick Roy.

"Ike" Corriveau, who notched the OT winner in Game 6 to push the series to the limit, had a breakaway at 3:01, only to be robbed by Roy.

Then at 4:45, Steve Konroyd had an opportunity only to have his tricky backhander stopped, Roy's 39th save.

When play shifted down ice, Gilbert Dionne, Shayne Corson and Courtnall did a great job forechecking. At the finish, Dionne's feed from the baseline to the low slot set up Courtnall.

Maybe preparing for a quick shot, Pietrangelo had slipped to his knees. Courtnall delayed pulling the trigger and when he did, his wrister caromed off Pietrangelo's stick and trickled through his pads.

The packed Forum of 17,718 went wild.

Geoff Sanderson, a rookie, and Andrew Casssels, the ex-Hab, had the goals for the Whalers. Mathieu Schneider and Dionne hit for the Canadiens.

Murray Craven, Cassels and Corriveau proved coach Jimmy Roberts' best trio in the series.

1991-92 League Standings

CLARENCE CAMPBELL CONFERENCE
Norris Division

Team	GP	W	L	T	GF	GA	PTS
Detroit	80	43	25	12	320	256	98
Chicago	80	36	29	15	257	236	87
St. Louis	80	36	33	11	279	266	83
Minnesota	80	32	42	6	246	278	70
Toronto	80	30	43	7	234	294	67

Smythe Division

	GP	W	L	T	GF	GA	PTS
Vancouver	80	42	26	12	285	250	96
Los Angeles	80	35	31	14	287	296	84
Edmonton	80	36	34	10	295	297	82
Winnipeg	80	33	32	15	251	244	81
Calgary	80	31	37	12	296	305	74
San Jose	80	17	58	5	219	359	39

PRINCE OF WALES CONFERENCE
Adams Division

	GP	W	L	T	GF	GA	PTS
Montreal	80	41	28	11	267	207	93
Boston	80	36	32	12	270	275	84
Buffalo	80	31	37	12	289	299	74
Hartford	80	26	41	13	247	283	65
Quebec	80	20	48	12	255	318	52

Patrick Division

	GP	W	L	T	GF	GA	PTS
NY Rangers	80	50	25	5	321	246	105
Washington	80	45	27	8	330	275	98
Pittsburgh	80	39	32	9	343	308	87
New Jersey	80	38	31	11	289	259	87
NY Islanders	80	34	35	11	291	299	79
Philadelphia	80	32	37	11	252	273	75

League Champion: Pittsburgh Penguins
League MVP: Mark Messier, NY Rangers
Top Scorer: Mario Lemieux, Pittsburgh
(44-87-131 in 64 GP)

EJ gets the gate

May 12, 1992

HARTFORD - Eddie Johnston, who sent plenty of players packing during his three-year hitch as general manager of the Whalers, was fired today.

"We need a change," team owner Richard Gordon said. "EJ did a lot of good things but we need someone to take us to the next level."

Whalers Season Statistics

	Regular Season								Playoffs							
	GP	G	A	Pts	PM	PP	SH	GW	GP	G	A	Pts	PM	PP	SH	GW
Cullen	77	26	51	77	141	10	0	4	7	2	1	3	12	1	0	1
Craven	61	24	30	54	38	8	4	1	7	3	3	6	6	0	1	0
Verbeek	76	22	35	57	243	10	0	3	7	0	2	2	12	0	0	0
Zalapski	79	20	37	57	116	4	0	3	7	2	3	5	6	0	0	0
Andersson	74	18	29	47	14	1	3	1	7	0	2	2	6	0	0	0
Holik	76	21	24	45	44	1	0	2	7	0	1	1	6	0	0	0
Cassels	67	11	30	41	18	2	2	3	7	2	4	6	6	1	0	0
Sanderson	64	13	18	31	18	2	0	1	7	1	0	1	2	0	0	0
R.Brown	42	16	15	31	39	13	0	2	--	--	--	--	--	--	--	--
Shaw	62	3	22	25	44	0	0	0	3	0	1	1	4	0	0	0
Burt	66	9	15	24	93	4	0	1	2	0	0	0	0	0	0	0
Bergevin	75	7	17	24	64	4	1	1	5	0	0	0	2	0	0	0
Hunter	63	10	13	23	159	5	0	0	4	0	0	0	6	0	0	0
Corriveau	38	12	8	20	36	3	0	0	7	3	2	5	18	2	0	1
Cunneyworth	39	7	10	17	71	0	0	0	7	3	0	3	9	1	1	0
Konroyd	33	2	10	12	32	1	0	0	7	0	1	1	2	0	0	0
Black	30	4	6	10	14	1	0	1	--	--	--	--	--	--	--	--
Ladouceur	74	1	9	10	127	0	0	0	7	0	1	1	11	0	0	0
Houda	56	3	6	9	125	1	0	1	6	0	2	2	13	0	0	0
Picard	25	3	5	8	6	1	0	0	--	--	--	--	--	--	--	--
McKenzie	67	5	1	6	87	0	0	0	--	--	--	--	--	--	--	--
Dineen	16	4	2	6	23	1	0	1	--	--	--	--	--	--	--	--
Greig	17	0	5	5	6	0	0	0	--	--	--	--	--	--	--	--
Pederson	5	2	2	4	0	1	0	0	--	--	--	--	--	--	--	--
Kastelic	25	1	3	4	61	0	0	0	--	--	--	--	--	--	--	--
Stevens	21	0	4	4	19	0	0	0	--	--	--	--	--	--	--	--
Cyr	17	0	3	3	19	0	0	0	--	--	--	--	--	--	--	--
Day	24	0	3	3	10	0	0	0	--	--	--	--	--	--	--	--
Shank	13	2	0	2	18	0	0	0	5	0	0	0	22	0	0	0
Yake	15	1	1	2	4	0	0	0	--	--	--	--	--	--	--	--
Gillis	12	0	2	2	48	0	0	0	5	0	1	1	0	0	0	0
Sidorkiewicz	35	0	1	1	2	0	0	0	--	--	--	--	--	--	--	--
Whitmore	45	0	1	1	16	0	0	0	1	0	0	0	0	0	0	0
Baca	1	0	0	0	0	0	0	0	--	--	--	--	--	--	--	--
Keczmer	1	0	0	0	0	0	0	0	--	--	--	--	--	--	--	--
Poulin	1	0	0	0	2	0	0	0	7	2	1	3	0	1	0	0
Pietrangelo	5	0	0	0	0	0	0	0	7	0	0	0	0	0	0	0
Tomlak	6	0	0	0	0	0	0	0	--	--	--	--	--	--	--	--
Norwood	6	0	0	0	16	0	0	0	--	--	--	--	--	--	--	--
Richards	6	0	0	0	2	0	0	0	5	0	3	3	4	0	0	0
Tancil	10	0	0	0	2	0	0	0	--	--	--	--	--	--	--	--
Bench	80	--	--	--	16	--	--	--	7	--	--	--	0	--	--	--
Totals	80	247	418	665	1793	73	10	26	7	18	28	46	147	6	2	3
Opponents	80	283	476	759	1699	86	11	41	7	21	35	56	169	8	1	4

Goaltending	GP	Mts	W	L	T	Sho	GAA	GP	Mts	W	L	Sho	GAA
Sidorkiewicz	35	1995	9	19	6	2	3.34	--	--	--	--	--	-----
Whitmore	45	2567	14	21	6	3	3.62	1	19	0	0	0	3.16
Pietrangelo	5	306	3	1	1	0	2.35	7	425	3	4	0	2.68
Totals	80	4868	26	41	13	5	3.43	7	444	3	4	0	2.70
Opponents	80	4868	41	26	13	4	3.01	7	444	4	3	0	2.43

1992-93 League Standings

CLARENCE CAMPBELL CONFERENCE
Norris Division

Team	GP	W	L	T	GF	GA	PTS
Chicago	84	47	25	12	279	230	106
Detroit	84	47	28	9	369	280	103
Toronto	84	44	29	11	288	241	99
St. Louis	84	37	36	11	282	278	85
Minnesota	84	36	38	10	272	293	82
Tampa Bay	84	23	54	7	245	332	53

Smythe Division

	GP	W	L	T	GF	GA	PTS
Vancouver	84	46	29	9	346	278	101
Calgary	84	43	30	11	322	282	97
Los Angeles	84	39	35	10	338	340	88
Winnipeg	84	40	37	7	322	320	87
Edmonton	84	26	50	8	242	337	60
San Jose	84	11	71	2	218	414	24

League Champion: Montreal Canadiens
League MVP: Mario Lemieux, Pittsburgh

PRINCE OF WALES CONFERENCE
Adams Division

	GP	W	L	T	GF	GA	PTS
Boston	84	51	26	7	332	268	109
Quebec	84	47	27	10	351	300	104
Montreal	84	48	30	6	326	280	102
Buffalo	84	38	36	10	335	297	86
Hartford	84	26	52	6	284	369	58
Ottawa	84	10	70	4	202	395	24

Patrick Division

	GP	W	L	T	GF	GA	PTS
Pittsburgh	84	56	21	7	367	268	119
Washington	84	43	34	7	325	286	93
NY Islanders	84	40	37	7	335	297	87
New Jersey	84	40	37	7	308	299	87
Philadelphia	84	36	37	11	319	319	83
NY Rangers	84	34	39	11	304	308	79

Top Scorer: Mario Lemieux, Pittsburgh (69-91-160 in 60 GP)

Whalers Season Statistics

Regular Season

	GP	G	A	Pts	PM	PP	SH	GW			GP	G	A	Pts	PM	PP	SH	GW
Sanderson	82	46	43	89	28	21	2	4		Kerr	22	0	6	6	7	0	0	0
Cassels	84	21	64	85	62	8	3	1		Leach	19	3	2	5	2	0	0	0
Verbeek	84	39	43	82	197	16	0	6		Pedersen	59	1	4	5	60	0	0	0
Zalapski	83	14	51	65	94	8	1	0		Gillis	21	1	1	2	40	0	0	0
Craven	67	25	42	67	20	6	3	2		Burke	50	0	2	2	25	0	0	0
Yake	66	22	31	53	46	4	1	2		Govedaris	7	1	0	1	0	0	0	0
Poulin	81	20	31	51	37	4	0	2		Gosselin	16	0	1	1	2	0	0	0
Weinrich	79	7	29	36	76	0	2	2		D'Alessio	1	0	0	0	0	0	0	0
Nylander	59	11	22	33	36	3	0	1		S.Daniels	1	0	0	0	19	0	0	0
Janssens	76	12	17	29	237	0	0	1		Nieckar	2	0	0	0	2	0	0	0
Kypreos	75	17	10	27	325	0	0	2		Lenarduzzi	3	0	0	0	0	0	0	0
Burt	65	6	14	20	116	0	0	0		Agnew	16	0	0	0	68	0	0	0
Konroyd	59	3	11	14	63	0	0	0		Pietrangelo	30	0	0	0	4	0	0	0
Corriveau	37	5	5	10	14	1	0	1		Bench	84	--	--	--	24	--	--	--
Cullen	19	5	4	9	58	3	0	0		**Totals**	84	284	481	765	2345	78	12	26
Cunneyworth	39	5	4	9	63	0	0	1		**Opponents**	84	369	606	975	2244	107	13	52
Petrovicky	42	3	6	9	45	0	0	0										
McKenzie	64	3	6	9	202	0	0	0										
Keczmer	23	4	4	8	28	2	0	1										
Houda	60	2	6	8	167	0	0	0										
Greig	22	1	7	8	27	0	0	0										
Day	24	1	7	8	47	0	0	0										
Kron	13	4	2	6	4	2	0	0										
Ladouceur	62	2	4	6	109	0	0	0										

Goaltending	GP	Mts	W	L	T	Sho	GAA
D'Alessio	1	11	0	0	0	0	0.00
Lenarduzzi	3	168	1	1	1	0	3.21
Gosselin	16	867	5	9	1	0	3.94
Burke	50	2656	16	27	3	0	4.16
Pietrangelo	30	1373	4	15	1	0	4.85
Totals	84	5075	26	52	6	0	4.27
Opponents	84	5078	52	26	6	2	3.31

New color for Whalers: navy blue

June 2, 1992

HARTFORD - The uniform will be new. Hopefully, the results will be as well.

The Whalers will support a new look this coming season. The predominate color in the sweater will be navy blue rather than the traditional green.

"Bigger and tougher," new GM Brian Burke says. "That's the look."

Whalers ship Holik in swap for Burke

August 29, 1992

HARTFORD - The theory in hockey suggests successful teams are strong between the pipes. Once you have a big-play netminder, the foundation is set.

"You build a championship team from the net out," Whalers general manager Brian Burke said. "With this trade, we've acquired one of the top goaltenders in his age class in the National Hockey League.

"I think Sean Burke is better than what we have here. We think he can be a workhosre. He can play at a high level."

Burke, 25, became a Whaler in a swap with New Jersey, a team where the lanky netminder shined following the 1988 Olympiad, going 10-1 down the stretch and anchoring a drive by the Devils to the Cup semifinals.

Since then, Burke has feuded with management over playing time and a contract. He opted to go the Olympic route again this past winter, then signed with San Diego of the International League. He openly pushed for a deal.

"It's a fresh start," Burke said. "My last year in New Jersey, I never got untracked. When I did play well, there was no guarantee I'd play the next night."

Along with Burke, the Whalers added mobile blue liner Eric Weinrich. The cost for both was 21-year old Bobby Holik, Hartford's top pick in 1989 and a couple of draft choices.

"It will be a challenge," Holik said. "I didn't show everything in Hartford and I had a new coach every year."

Holik, a 6-foot-3 forward who loves to forecheck, had back-to-back 21-goal

seasons but was used more in a checking role than a scoring role.

Weinrich, 25, has 13-66-79 in 173 games but was deemed expendable by the Devils to make room for promising rear guard Scott Niedermayer.

Burke's arrival certainly clouds the status of Frank Pietrangelo. The latter, 27, was exceptional down the stretch last season, not only helping to the Whalers clinch a playoff berth but taking the squad to double-overtime in Game 7 against Montreal.

"I'm surprised this trade was made," Pietrangelo said. "I think I earned the opportunity to come to camp and get a shot at playing No. 1 after last season. Management obviously feels it needs another goaltender."

1993-94 League Standings

EASTERN CONFERENCE
Northeast Division

Team	GP	W	L	T	GF	GA	PTS
Pittsburgh	84	44	27	13	299	285	101
Boston	84	42	29	13	289	252	97
Montreal	84	41	29	14	283	248	96
Buffalo	84	43	32	9	282	218	95
Quebec	84	34	42	8	277	292	76
Hartford	84	27	48	9	227	288	63
Ottawa	84	14	61	9	201	397	37

Atlantic Division

NY Rangers	84	52	24	8	299	231	112
New Jersey	84	47	25	12	306	220	106
Washington	84	39	35	10	277	263	88
NY Islanders	84	36	36	12	282	264	84
Florida	84	33	34	17	233	233	83
Philadelphia	84	35	39	10	294	314	80
Tampa Bay	84	30	43	11	224	251	71

WESTERN CONFERENCE
Central Division

Detroit	84	46	30	8	356	275	100
Toronto	84	43	29	12	280	243	98
Dallas	84	42	29	13	286	265	97
St. Louis	84	40	33	11	270	283	91
Chicago	84	39	36	9	254	240	87
Winnipeg	84	24	51	9	245	344	57

Pacific Division

Calgary	84	42	29	13	302	256	97
Vancouver	84	41	40	3	279	276	85
San Jose	84	33	35	16	252	265	82
Anaheim	84	33	46	5	229	251	71
Los Angeles	84	27	45	12	294	322	66
Edmonton	84	25	45	14	261	305	64

League Champion: New York Rangers
League MVP: Sergei Federov, Detroit
Top Scorer: Wayne Gretzky, Los Angeles

(38-92-130 in 81 GP)

Suspension for brawlers?

March 25, 1994

BUFFALO - Several Whalers were involved in a nightclub fight early Thursday in a cafe owned by football's Jim Kelly.

"Knowing them," quipped a disgruntled fan at how another season has melted into disaster, "they lost."

They have, if team owner Richard Gordon has his way.

Six players in the fracas will be suspended, Gordon has announced. The brawlers, however, will likely appeal since the NHL Players Association will take exception to the idea.

Whalers Season Statistics

Regular Season

	GP	G	A	Pts	PM	PP	SH	GW
Verbeek	84	37	38	75	177	15	1	3
Sanderson	82	41	26	67	42	15	1	6
Cassels	79	16	42	58	37	8	1	3
Kron	77	24	26	50	8	2	1	3
Nylander	58	11	33	44	24	4	0	1
Zalapski	56	7	30	37	56	0	0	0
Pronger	81	5	25	30	113	2	0	0
Propp	65	12	17	29	44	3	1	2
Patrick	47	8	20	28	32	4	1	2
Burt	63	1	17	18	75	0	0	0
Cunneyworth	63	9	8	17	87	0	1	1
Storm	68	6	10	16	27	1	0	0
Chibirev	37	4	11	15	2	0	0	1
Turcotte	19	2	11	13	4	0	0	0
Godynyuk	43	3	9	12	40	1	0	0
Janssens	84	2	10	12	137	0	0	0
Petrovicky	33	6	5	11	39	1	0	0
Marchment	42	3	7	10	124	0	1	1
Greig	31	4	5	9	31	0	0	0
Sandlak	27	6	2	8	32	2	0	1
Lemieux	16	6	1	7	82	0	0	2
Drury	16	1	5	6	10	0	0	0
McCrimmon	65	1	5	6	72	0	0	0
Smyth	21	3	2	5	10	0	0	0
Potvin	51	2	3	5	246	0	0	0
Kucera	16	1	3	4	14	1	0	0
Poulin	9	2	1	3	11	1	0	0
Crowley	21	1	2	3	10	1	0	0
McKenzie	26	1	2	3	67	0	0	0
Stevens	9	0	3	3	4	0	0	0
Ranheim	15	0	3	3	2	0	0	0
McGill	30	0	3	3	46	0	0	0
Weinrich	8	1	1	2	2	1	0	0
Harkins	28	1	0	1	49	0	0	0
Keczmer	12	0	1	1	12	0	0	0
Reese	19	0	1	1	0	0	0	0
Tomlak	1	0	0	0	0	0	0	0
Lenarduzzi	1	0	0	0	0	0	0	0
Corriveau	3	0	0	0	0	0	0	0
Houda	7	0	0	0	23	0	0	0
Gosselin	7	0	0	0	0	0	0	0
Pedersen	7	0	0	0	9	0	0	0
Kypreos	10	0	0	0	37	0	0	0
Pietrangelo	19	0	0	0	2	0	0	0
Burke	47	0	0	0	16	0	0	0
Bench	84	--	--	--	22	--	--	--
Totals	84	227	388	615	1811	61	8	27
Opponents	84	288	495	783	1920	88	9	48

Goaltending	GP	Mts	W	L	T	Sho	GAA
Lenarduzzi	1	21	0	0	0	0	2.86
Burke	47	2750	17	24	5	2	2.99
Reese	19	1057	5	9	3	1	3.18
Pietrangelo	19	984	5	11	1	0	3.60
Gosselin	7	239	0	4	0	0	5.27
Totals	84	5081	27	48	9	3	3.24
Opponents	84	5088	48	27	9	5	2.63

1994-95 League Standings

EASTERN CONFERENCE
Northeast Division

Team	GP	W	L	T	GF	GA	PTS
Quebec	48	30	13	5	185	134	65
Pittsburgh	48	29	16	3	181	158	61
Boston	48	27	18	3	150	127	57
Buffalo	48	22	19	7	130	119	51
Hartford	48	19	24	5	127	141	43
Montreal	48	18	23	7	125	148	43
Ottawa	48	9	34	5	117	174	23

Atlantic Division

Team	GP	W	L	T	GF	GA	PTS
Philadelphia	48	28	16	4	150	132	60
New Jersey	48	22	18	8	136	121	52
Washington	48	22	18	8	136	120	52
NY Rangers	48	22	23	3	139	134	47
Florida	48	20	22	6	115	127	46
Tampa Bay	48	17	28	3	120	144	37
NY Islanders	48	15	28	5	126	158	35

WESTERN CONFERENCE
Central Division

Team	GP	W	L	T	GF	GA	PTS
Detroit	48	33	11	4	180	117	70
St. Louis	48	28	15	5	178	135	61
Chicago	48	24	19	5	156	115	53
Toronto	48	21	19	8	135	146	50
Dallas	48	17	23	8	136	135	42
Winnipeg	48	16	25	7	157	177	39

Pacific Division

Team	GP	W	L	T	GF	GA	PTS
Calgary	48	24	17	7	163	135	55
Vancouver	48	18	18	12	153	148	48
San Jose	48	19	25	4	129	161	42
Los Angeles	48	16	23	9	142	174	41
Edmonton	48	17	27	4	136	183	38
Anaheim	48	16	27	5	125	164	37

League Champion: New Jersey Devils
League MVP: Eric Lindros, Philadelphia
Top Scorer: Jaromir Jagr, Pittsburgh (32-38-70 in 48 GP)

Whalers Season Statistics

Regular Season

1994-95	GP	G	A	Pts	PM	PP	SH	GW
Cassels	48	7	30	37	18	1	0	1
Turcotte	47	17	18	35	22	3	1	3
Sanderson	46	18	14	32	24	4	0	4
Rice	40	11	10	21	61	4	0	1
Ranheim	47	6	14	20	10	0	0	1
Kucera	48	3	17	20	30	0	0	1
Carson	38	9	10	19	29	4	0	3
Kron	37	10	8	18	10	3	1	1
Verbeek	29	7	11	18	53	3	0	0
Nikolishin	39	8	10	18	10	1	1	0
Burt	46	7	11	18	65	3	0	1
Wesley	48	2	14	16	50	1	0	1
Pronger	43	5	9	14	54	3	0	1
Lemieux	41	6	5	11	32	0	0	1
Drury	34	3	6	9	21	0	0	0
Janssens	46	2	5	7	93	0	0	0
Glynn	43	1	6	7	32	0	0	0
Smyth	16	1	5	6	13	0	0	0
Chibirev	8	3	1	4	0	0	0	0

	GP	G	A	Pts	PM	PP	SH	GW
Chase	28	0	4	4	141	0	0	0
Featherstone	13	1	1	2	32	0	0	0
Storm	6	0	3	3	0	0	0	0
S.Daniels	12	0	2	2	55	0	0	0
Malik	1	0	1	1	0	0	0	0
McCrimmon	33	0	1	1	42	0	0	0
Burke	42	0	1	1	8	0	0	0
Petrovicky	2	0	0	0	0	0	0	0
Reese	11	0	0	0	0	0	0	0
Sandlak	13	0	0	0	0	0	0	0
Godynyuk	14	0	0	0	8	0	0	0
Bench	48	--	--	--	2	--	--	--
Totals	48	127	221	348	915	31	3	19
Opponents	48	141	236	377	863	37	6	24

Goaltending	GP	Mts	W	L	T	Sho	GAA
Reese	11	477	2	5	1	0	3.27
Burke	42	2418	17	19	4	0	2.68
Totals	48	2893	19	24	5	0	2.78
Opponents	48	2902	24	19	5	0	2.78

Call them CompuWhale'

June 28, 1994

HARTFORD - The Hartford Whalers host the Entry Draft today at the Hartford Civic Center.

By the time young hopefuls from around the global are selected and touted as the next franchise savior in some NHL ports until contract negotiations begin, there will be a different signature on future checks for employees of the Whalers.

The NHL club was officially sold to Compuware Corporation, a computer-based company in Michigan.

Terms of the sale are complex because of a deal brokered by the Connecticut Development Authority. The purchase price is $45 million, a tad less than the going rate for expansion clubs.

The new ownership group is headed by Peter Karmanos. Other partners are Thomas Thewes and Jim Rutherford, a former NHL goalie. The latter will be the club's general manager.

Karmanos has sought a team for several years and hopes to go full bore into youth hockey in the area, modeling programs much like the Compuware hockey clubs in suburban Detroit.

Unlike local real estate developer Richard Gordon who was a novice in professional sports yet spurned offers to sell the club to out-of-state interests, the clock is ticking on a four-year lease Karmanos agreed to in purchasing the hockey club.

"We're the only game in town and we're determined to make things work here," Karmanos said. "I see no reason why this franchise, with a couple of player moves, can't be one of the better teams in the league."

Upward movement in the standings remains the toughest thing for the Whalers to do lately besides sell season tickets. Hartford may be able to do a little bit of both in the coming months should it secure center Jeff O'Neill, a promising center at Guelph.

Rutherford is high on O'Neill and would select the youngster fifth overall when Hartford gets its turn. The top pick is owned by Florida and defenseman Ed Jovanovski is the consensus No. 1 pick.

"O'Neill has a chance," Rutherford said. "He's in the mold of Doug Gilmour. We like him."

1995-96 League Standings

EASTERN CONFERENCE
Northeast Division

Team	GP	W	L	T	GF	GA	PTS
Pittsburgh	82	49	29	4	362	284	102
Boston	82	40	31	11	282	269	91
Montreal	82	40	32	10	265	248	90
Hartford	82	34	39	9	237	259	77
Buffalo	82	33	42	7	247	262	73
Ottawa	82	18	59	5	191	291	41

Atlantic Division

Team	GP	W	L	T	GF	GA	PTS
Philadelphia	82	45	24	13	282	208	103
NY Rangers	82	41	27	14	272	237	96
Florida	82	41	31	10	254	234	92
Washington	82	39	32	11	234	204	89
Tampa Bay	82	38	32	12	238	248	88
New Jersey	82	37	33	12	215	202	86
NY Islanders	82	22	50	10	229	315	54

WESTERN CONFERENCE
Central Division

Team	GP	W	L	T	GF	GA	PTS
Detroit	82	62	13	7	325	181	131
Chicago	82	40	28	14	273	220	94
Toronto	82	34	36	12	247	252	80
St. Louis	82	32	34	16	219	248	80
Winnipeg	82	36	40	6	275	291	78
Dallas	82	26	42	14	227	280	66

Pacific Division

Team	GP	W	L	T	GF	GA	PTS
Colorado	82	47	25	10	326	240	104
Calgary	82	34	37	11	241	240	79
Vancouver	82	32	35	15	278	278	79
Anaheim	82	35	39	8	234	247	78
Edmonton	82	30	44	8	240	304	68
Los Angeles	82	24	40	18	256	302	66
San Jose	82	20	55	7	252	357	47.

League Champion: Colorado Avalanche
League MVP: Mario Lemieux, Pittsburgh
Top Scorer: Mario Lemiuex, Pittsburgh (69-92-161 in 70 GP)

Whalers Season Statistics

Regular Season

	GP	G	A	Pts	PM	PP	SH	GW
Shanahan	74	44	34	78	125	17	2	6
Sanderson	81	34	31	65	40	6	0	7
Cassels	81	20	43	63	39	6	0	1
Emerson	81	29	29	58	78	12	2	5
J.Brown	48	7	31	38	38	5	0	0
Nikolishin	61	14	37	51	34	4	1	3
Kron	77	22	28	50	6	8	1	3
Ranheim	73	10	20	30	14	0	1	1
O'Neill	65	8	19	27	40	1	0	1
Wesley	68	8	14	24	88	6	0	1
Rice	59	10	12	22	47	1	0	2
Burt	78	4	9	13	121	0	0	1
Featherstone	68	2	10	12	138	0	0	1
Dineen	20	2	7	9	67	0	0	0
Diduck	79	1	9	10	88	0	0	0
Kapanen	35	5	4	9	6	0	0	0
McCrimmon	58	3	6	9	62	0	1	0
Janssens	81	2	7	9	155	0	0	0
S.Daniels	53	3	4	7	254	0	0	0
Chase	55	2	4	6	220	0	0	1
Burke	66	0	5	6	16	0	0	0

	GP	G	A	Pts	PM	PP	SH	GW
Martins	23	1	3	4	8	0	0	0
Glynn	54	0	4	4	44	0	0	0
Smyth	21	2	1	3	8	1	0	0
Carson	11	1	0	1	0	0	0	0
Godynyuk	3	0	0	0	2	0	0	0
McBain	3	0	0	0	0	0	0	0
Malik	7	0	0	0	4	0	0	0
Muzzatti	22	0	0	0	33	0	0	0
Kucera	30	2	6	8	10	0	0	1
Lemieux	29	1	2	3	31	0	0	0
Reese	7	0	0	0	0	0	0	0
Bench	82	--	--	--	18	--	--	--
Totals	82	237	382	619	1844	67	8	34
Opponents	82	259	437	696	1677	83	9	39

Goaltending	GP	Mts	W	L	T	Sho	GAA
Muzzatti	22	1013	4	8	3	1	2.90
Reese	7	275	2	3	0	1	3.05
Burke	66	3669	28	28	6	4	3.11
Totals	82	4953	34	39	9	6	3.06
Opponents	82	4964	39	34	9	8	2.80

Whalers net Shanahan from Blues for Pronger

July 25, 1995

HARTFORD - The Whalers have lacked a big-play goal scorer since Blaine Stoughton was filling the net with 50 or so a season back in the early 1980s.

The annual search for a sniper ended today when Hartford obtained Brendan Shanahan from the Blues for second-year defenseman Chris Pronger.

It seems Whalers general manager Jim Rutherford may have scored a major one here. . . unless Shanahan is the fraud that St. Louis coach Mike Keenan has portrayed him this past season.

Whalers pull up anchor in Hartford

March 26, 1997

HARTFORD - Will it be Columbus, St. Paul, Nashville or Houston?

Wherever Peter Karmanos gets a deal to satisfy his ego and stuff his wallet, you can bet his hockey team will play next season.

The only place it will not play in 1997-98 will be Hartford.

Karmanos, at a press conference, announced that negotiations with the state of Connecticut ended without a workable agreement for a new complex.

"Given the realities of this market," Karmanos said, "the offer does not enable the Whalers to be viable. Therefore, the Whalers, with great reluctance, are announcing that we will concentrate our efforts to find a new home for the team."

Connecticut governor John Rowland had been negotating with Karmanos since January in an effort to hammer out a deal. The state had agreed to build a new complex and guarantee the Whalers revenue streams of $50 million annually but Karmanos, possibly because of his ego and the NHL looking to deep-six small market cities, declined.

A year ago, a season ticket drive resulted in over 9,000 sales, the best in team history.

What dogged the Whalers throughout their history was the product on ice failed more often than it won. That was the biggest drawback to attract ticket buyers as much as coaxing players that Hartford was a desired location.

"We came in here and we tried to turn the situation around," Karmanos said. "No matter which way you cut it, we weren't able to do it. We don't like to lose and this means we lost."

The Whalers will leave one year earlier on a four-year lease agreement. They will also pay the state a $20.5 million exit fee to walk the plank.

1996-97 League Standings

EASTERN CONFERENCE
Northeast Division

	GP	W	L	T	GF	GA	PTS
Buffalo	82	40	30	12	237	208	92
Ottawa	82	38	36	8	285	280	84
Montreal	82	31	36	15	226	234	77
Hartford	82	31	36	11	226	256	77
Boston	82	26	47	9	234	300	61

Atlantic Division

	GP	W	L	T	GF	GA	PTS
New Jersey	82	45	23	14	231	182	104
Philadelphia	82	45	24	13	274	217	103
Florida	82	35	28	19	221	201	89
NY Rangers	82	38	34	10	258	231	86
Washington	82	33	40	9	214	231	75

WESTERN CONFERENCE
Central Division

	GP	W	L	T	GF	GA	PTS
Dallas	82	48	26	8	252	198	104
Detroit	82	38	26	18	253	197	94
Phoenix	82	38	37	7	240	243	83
St. Louis	82	36	35	11	236	239	83
Chicago	82	34	35	13	223	210	81
Toronto	82	30	44	8	230	273	68

Pacific Division

	GP	W	L	T	GF	GA	PTS
Colorado	82	49	24	9	277	205	107
Anaheim	82	36	33	13	245	233	85
Edmonton	82	36	37	9	252	247	81
Vancouver	82	35	40	7	257	273	77
Calgary	82	32	41	9	214	239	73
Los Angeles	82	28	43	11	214	268	67
San Jose	82	27	47	8	211	278	62

League Champion: Detroit Red Wings
League MVP: Dominik Hasek, Buffalo
Top Scorer: Mario Lemieux, Pittsburgh (50-72-122 in 76 GP)

Whalers Season Statistics

Regular Season

1996-97	GP	G	A	Pts	PM	PP	SH	GW
Sanderson	80	36	31	67	29	12	1	4
Cassels	81	22	44	66	46	8	0	2
Primeau	75	26	25	51	161	6	3	2
Dineen	78	19	29	48	141	8	0	5
Emerson	66	9	29	38	34	2	1	2
Rice	78	21	14	35	59	5	0	2
Wesley	68	6	26	32	40	3	1	0
O'Neill	72	14	16	30	40	2	1	2
Kapanen	45	13	12	25	6	3	0	2
Kron	68	10	12	22	10	2	0	4
Ranheim	67	10	11	21	18	0	3	1
Leschyshyn	64	4	13	17	30	1	1	1
Chiasson	18	3	11	14	7	3	0	0
Burt	71	2	11	13	79	0	0	0
Manderville	44	6	5	11	18	0	0	1
Diduck	56	1	10	11	40	0	0	1
Coffey	20	3	5	8	18	1	0	1
Haller	35	2	6	8	48	0	0	0
Nikolishin	12	2	5	7	2	0	0	0
Featherstone	41	2	5	7	87	0	0	0
Godynyuk	55	1	6	7	41	0	0	0
King	12	3	3	6	2	1	0	0
Janssens	54	2	4	6	90	0	0	0
Malik	47	1	5	6	50	0	0	1
Grimson	75	2	2	4	218	0	0	0
K.Brown	11	0	4	4	6	0	0	0
Domenichelli	13	2	1	3	7	1	0	0
Chase	28	1	2	3	122	0	0	0
Murray	8	1	1	2	10	0	0	0
Pratt	9	0	2	2	6	0	0	0
J.Daniels	10	0	2	2	0	0	0	0
Burke	51	0	2	2	14	0	0	0
Glynn	1	1	0	1	2	0	0	0
Shananhan	2	1	0	1	0	0	1	0
Martins	2	0	1	1	0	0	0	0
Muzzatti	31	0	1	1	18	0	0	0
J.Brown	1	0	0	0	0	0	0	0
McBain	6	0	0	0	0	0	0	0
Giguere	8	0	0	0	0	0	0	0
Bench	82	--	--	--	18	--	--	--
Totals	82	226	356	582	1513	58	12	32
Opponents	82	256	418	699	1481	51	11	39

Goaltending	GP	Mts	W	L	T	Sho	GGA
Burke	51	2985	22	22	6	4	2.69
Muzzatti	31	1591	9	13	5	0	3.43
Giguere	8	394	1	4	0	0	3.65
Totals	82	4996	32	39	11	4	3.07
Opponents	82	5002	39	32	11	7	2.80

Whalers sign off with 2-1 victory

April 13, 1997

HARTFORD - The fans embraced their heroes for the last time.

And the Whalers gave their faithful one final hour of hockey, a 2-1 win over Tampa Bay, the final entry in the club's 25-year history.

The day "Brass Bonanza" died will forever be one remembered as the misty-eyed matinee at Trumbull Street.

Moments after the foghorn sounded at 4:03 p.m. (EST), the Whalers gathered around goalie Sean Burke and hoisted their sticks in tribute to the 14,660 patrons. The players in the Hartford colors then circled the rink, saluting the fans who had watched the NHL team close out its run at the Civic Center Coliseum.

"You knew this day was coming for a while, but you can never be prepared for something like this," Burke said.

"This is it," Burke said. "We walk out this door today and get in our cars and go home. That's the last time we'll ever come to this building and see these people together. And that's sad."

The crowd roared throughout the game. Whalers captain Kevin Dineen, who scored the last goal in team annals, could only reflect.

"I hope us leaving doesn't hurt this town badly," Dineen said. "I worry about it. I'll always be part of it. Man, this is an empty, empty feeling. I can't believe this is over. These fans are the greatest fans in the world"

The Whalers may have been underachievers for most of their history, but that never mattered to the faithful. There was always next year.

There is no next year.

Transactions at a Glance (1972-1997)

1972-73

Nov. 1, 1972: Whalers obtain Brit Selby from Quebec for Bob Brown

Jan. 8, 1973: Whalers sell Dick Sarrazin to Philadelphia

Feb. 16, 1973: Whalers obtain Mike Byers from Los Angeles for Mike Hyndman

1973-74

June 6, 1973: Whalers acquire Bob Charlebois from Toronto for Brit Selby

May 15, 1974: Whalers obtain Guy Dufour from Quebec for Ric Jordan

1974-75

July 31, 1974: Whalers sell Hugh Harris to Phoenix

Sept. 19, 1974: Whalers obtain Wayne Carleton from Toronto for second-round draft choice and future considerations

Dec. 31, 1974: Whalers ship Jim Dorey to Toronto to complete Carleton trade

Feb. 15, 1975: Whalers acquire Ron Climie from Edmonton for Tim Sheehy

1975-76

July 1, 1975: Whalers obtain Rosaire Paiement from Denver for first-round draft selection

July 15, 1975: Whalers sign Danny Arndt

Sept. 24, 1975: Whalers sign underage junior Gordie Roberts

Oct. 15, 1975: Whalers sell John French to San Diego

Oct. 28, 1975: Whalers sell Terry Caffery to Calgary

Jan. 15, 1976: Whalers acquire Ralph Backstrom and Don Borgeson when Denver/Ottawa franchise suspends operations

Jan. 19, 1976: Whalers acquire Mike Rogers from Edmonton for Wayne Carleton

Jan. 29, 1976: Whalers acquire Kerry Ketter and Steve Carlyle from Edmonton for Paul Hurley

Feb. 24, 1976: Whalers sell Mike Byers to Cincinnati

March 9, 1976: Whalers sign Ron Busnick when Minnesota franchise ceases operations

*In 1980-81 when he had 40-65-105 points, Hartford's **Mike Rogers** became just the fourth player in NHL annals to net 100 points for a non-playoff club. The previous ones were Bobby Hull of Chicago and Gordie Howe of Detroit in 1968-69 and Marcel Dionne of Los Angeles in 1974-75.*

1976-77

May 26, 1976: Whalers obtain Greg Carroll and Bryan Maxwell from Cincinnati for the collegiate draft rights to Mike Liut

June 29, 1976: Whalers obtain the rights to Wayne Connelly from Florida for Fred O'Donnell and Steve Carlson

June 30, 1976: Whalers sign Dave Hynes

June 30, 1976: Whalers release Nick Fotiu who signs with New York Rangers (NHL)

Nov. 12, 1976: Whalers obtain Gary MacGregor from Indianapolis for Rosaire Paiement

Nov. 29, 1976: Whalers purchase Dale Smedsmo from Cincinnati

Jan. 19, 1977: Whalers sell Gary MacGregor to Calgary

Jan. 19, 1977: Whalers obtain Dave Keon, John McKenzie, Jack Carlson, Steve Carlson and Dave Dryden from Edmonton for Danny Arndt, the draft rights to Dave Debol and financial considerations

Feb. 1, 1977: Whalers ship Dale Smedsmo to Cincinnati in exchange for a draft pick

Feb. 5, 1977: Whalers obtain Bill Butters and Mike Antonovich from Edmonton in exchange for Ron Busnick and Brett Callighen

1977-78

May 23, 1977: Whalers sign Gordie Howe, Mark Howe and Marty Howe who were free agents

June 16, 1977: Whalers obtain Louis Levasseur from Edmonton for the rights to Dave Dryden, Brett Callighen and a draft choice

Aug. 5, 1977: Whalers sign Al Smith

Aug. 28, 1977: Whalers sign Andre Peloffy

Feb. 9, 1978: Whalers release Bryan Maxwell who signs with Minnesota (NHL)

Feb. 11, 1978: Whalers obtain Ron Plumb from Cincinnati for Greg Carroll

Feb. 11, 1978: Whalers claim Tim Sheehy on waivers from Detroit (NHL)

Feb. 16, 1978: Whalers sell Bill Butters to Minnesota (NHL)

1978-79

May 18, 1978: Whalers sign underage draft picks Jordy Douglas and Jeff Brubaker

July 10, 1978: Whalers release Danny Bolduc who signs with Detroit (NHL)

Aug. 15, 1978: Whalers sign free agent Andre Lacroix

Sept. 14, 1978: Whalers lose Steve Carlson and Jim Mayer who are claimed on waivers by Edmonton

Sept. 18, 1978: Whalers obtain John Garrett from Birmingham for future considerations

Sept. 20, 1978: Whalers obtain Warren Miller from Quebec for Louis Levasseur

Dec. 13, 1978: Whalers obtain Blaine Stoughton and Dave Inkpen from financially troubled Indianapolis franchise

Jan. 15, 1979: Whalers sell Jack Carlson to Minnesota (NHL)

March 13, 1979: Whalers release Brad Selwood who signs with Montreal (NHL)

Transactions at a Glance (1972-1997)

1979-80

June 3, 1979: In the WHA Dispersal Draft, the Whalers claimed Steve Alley, Tony Cassolato, Paul Henderson and Bob Stephenson from Birmingham; Dave Debol, Chuck Luksa and Byron Shutt from Cincinnati

June 3, 1979: Whalers name John Garrett (Chicago), Mark Howe (Boston) and Jordy Douglas (Toronto) as Priority Selections in NHL Expansion Draft; lost in Reclamation Draft were George Lyle (to Detroit), Jim Mayer and Warren Miller (both to New York Rangers)

June 3, 1979: Whalers select 16 players in NHL Expansion Draft: Alan Hangsleben (Montreal), Nick Fotiu (New York Rangers), Rick Ley (Toronto), Al Sims (Boston), Jean Savard (Chicago), Ralph Klassen (Colorado), Rick Hodgson (Atlanta), Kevin Kemp (Toronto), Bill Bennett (Boston), Bernie Johnston (Philadelphia), Brian Hill (Atlanta), Dave Given (Buffalo), Maynard Schurman (Philadelphia), Nick Beverly (Colorado), Norm LaPointe (Vancouver), Don Kozak (Vancouver)

June 14, 1979: Whalers acquire Terry Richardson from the New York Islanders for Ralph Klassen

Dec. 13, 1979: Whalers sign Larry Giroux as a free agent

Dec. 24, 1979: Whalers obtain Pat Boutette from Toronto for Bob Stephenson

Jan. 17, 1980: Whalers obtain Tom Rowe from Washington for Alan Hangsleben

Feb. 27, 1980: Whalers obtain Bobby Hull from Winnipeg for future considerations

1980-81

May 23, 1980: Whalers sign free agent Thommy Abrahamsson

June 5, 1980: Whalers obtain Rick Meagher from Montreal for flip-flop of third and fifth round picks in 1981 Entry Draft (Montreal took Dieter Hegen and Steve Rooney; Hartford took Paul MacDermid and Dan Bourbonnais)

June 19, 1980: Whalers obtain Mike Veisor from Chicago for a second-round draft pick in 1981 Entry Draft (Chicago took Kevin Griffin)

Aug. 7, 1980: Whalers reclaim Warren Miller for $100 from the New York Rangers

Nov. 21, 1980: Whalers obtain Norm Barnes and Jack McIlhargey from Philadelphia for a second-round draft choice in 1982 Entry Draft

Norm Barnes played 74 games for the Whalers over two winters in the early 1980s after coming over in a deal with the Flyers. A series of knee and leg injuries prematurely ended the defenseman's career.

Dec. 16, 1980: Whalers acquire Mike Fidler from Minnesota for Gordie Roberts

Jan. 15, 1980: Whalers ship Nick Fotiu to the New York Rangers for a fifth-round draft choice in 1981 Entry Draft (Hartford took Bill Maguire)

Feb. 20, 1980: Whalers obtain Gilles Lupien from Pittsburgh for a 1981 draft choice (Pittsburgh took Paul Edwards)

1981-82

June 15, 1981: Whalers sign free agent Greg Millen with compensation owed to Pittsburgh

June 29, 1981: Whalers send Pat Boutette and Kevin McClelland to Pittsburgh as compensation for signing Greg Millen

July 3, 1981: Whalers acquire Rick MacLeish, Blake Wesley and Don Gillen from Philadelphia in exchange for Ray Allison and Fred Arthur. The teams also flip-flopped first and third-round selections in the 1982 Entry Draft and the Whalers also reacquired their No. 2 pick in 1982 in prior deal with Flyers involving Norm Barnes and Jack McIlhargey (Philadelphia took Ron Sutter, transferred the second-round pick to Toronto which took Peter Ihnacak, and Miroslav Dvorak; Hartford took Paul Lawless, Mark Paterson and Kevin Dineen).

Oct. 1, 1981: Whalers sign Paul Shmyr as a free agent

Oct. 2, 1981: Whalers obtain Doug Sulliman, Chris Kotsopoulos and Gerry McDonald from the New York Rangers for Mike Rogers

Oct. 2, 1981: Whalers obtain Garry Howatt from the New York Islanders for fifth-round pick in 1983 (NY Islanders took Bob Caulfield)

Oct. 6, 1981: Whalers secure Rob McClanahan (from Buffalo) and Mike McDougal (from New York Rangers) in NHL Waiver Draft, lose Jeff Brubaker (to Montreal)

Nov. 13, 1981: Whalers claim George Lyle on waivers from Detroit

Dec. 21, 1981: Whalers obtain Pierre Larouche from Montreal in a flip-flop of first-round 1984 Entry Draft picks, a second-round pick in 1984 and a flip-flop of third-round picks in 1985 (Montreal took Petr Svoboda and transferred the second-round pick to St. Louis which took Brian Benning; Montreal took Rockey Dundas in 1985; Hartford selected Sylvain Cote)

Dec. 29, 1981: Whalers obtain Russ Anderson from Pittsburgh for Rick MacLeish

Jan. 31, 1982: Whalers lose Tom Rowe on waivers to Washington

Feb. 2, 1982: Whalers deal Rob McClanahan to the New York Rangers for 10th round draft pick in 1983 (Reine Karlsson)

1982-83

Aug. 19, 1982: Whalers obtain Ken Linseman, Greg Adams, first-round (David A. Jensen) and third-round (Leif Karlsson) draft choices in 1984 from Philadelphia for Mark Howe and a third-round pick (Derrick Smith) in 1984; Whalers then shuttle Ken Linseman and Don Nachbaur to Edmonton for Risto Siltanen and Brent Loney.

Aug. 24, 1982: Whalers sign free agent Bob Sullivan

Oct. 1, 1982: Whalers obtain Mark Johnson and Kent-Erik Andersson from Minnesota in exchange for Jordy Douglas and a 1984 fifth-round draft choice (Jiri Poner)

Oct. 1, 1982: Whalers acquire Ed Hospodar from the New York Rangers for Kent-Erik Andersson

Oct. 1, 1982: Whalers loan Marty Howe to Boston

Oct. 5, 1982: Whalers obtain Paul Marshall from Pittsburgh for a 10th-round draft choice (Greg Rolston) in 1983

Oct. 15, 1982: Whalers acquire Merlin Malinowski and Scott Fusco from New Jersey for Garry Howatt and Rick Meagher

Dec. 3, 1982: Whalers obtain Pierre Lacroix from Quebec for Blake Wesley

1983-84

May 29, 1983: Whalers sign Marty Howe as free agent

July 4, 1983: Whalers acquire Norm Dupont from Winnipeg for a 1984 fourth-round draft choice (Scott Schneider)

July 5, 1983: Whalers obtain Joel Quenneville and Richie Dunn from Calgary for Mickey Volcan

Sept. 30, 1983: Whalers acquire Greg Malone from Pittsburgh for a third-round draft pick (Bruce Racine) in 1985

Oct. 3, 1983: Whalers claim Mike Zuke, Mike Crombeen and Bob Crawford (all from St. Louis) in Waiver Draft, lose Mark Renaud (to Buffalo)

Oct. 3, 1983: Whalers net Torrie Robertson from Washington in exchange for Greg Adams

Nov. 11, 1983: Whalers obtain Ed Staniowski from Winnipeg for Mike Veisor

Jan. 21, 1984: Whalers sign free agent Tony Currie

Feb. 25, 1984: Whalers sign free agents Dave Tippett and Mark Fusco following the Winter Olympic Games

Feb. 27, 1984: Whalers obtain Scot Kleinendorst from the New York Rangers for Blaine Stoughton

March 2, 1984: Whalers claim Jack Brownschilde on waivers from St. Louis

May 29, 1984: Whalers deal an eighth-round draft choice (Urban Nordin) in 1984 to Detroit for rights to Brad Shaw

1984-85

Aug. 31, 1984: Whalers obtain Steve Weeks from New York Rangers for future considerations

Oct. 9, 1984: Whalers claim Dave Lumley (from Edmonton) and Wally Weir (from Quebec) in Waiver Draft

Nov. 16, 1984: Whalers obtain Pat Boutette from Pittsburgh for Ville Siren

Dec. 5, 1984: Whalers claim Tony Currie on waivers from Edmonton

Feb. 6, 1985: Whalers lose Dave Lumley on waivers to Edmonton

Feb. 22, 1985: Whalers acquire Mike Liut and future considerations (Jorgen Pettersson) from St. Louis for Greg Millen and Mark Johnson

March 1, 1985: Whalers lose Wally Weir on waivers to Pittsburgh

*Mark Johnson was part of a 2-for-2 swap with the Blues on February 22, 1985, a deal that brought goalie **Mike Liut** to Hartford from St. Louis. Also involved in the trade were **Greg Millen** and **Jorgen Pettersson**.*

March 12, 1985: Whalers obtain Dean Evason and Peter Sidorkiewicz from Washington for David A. Jensen

April 16, 1985: Whalers obtain Jorgen Pettersson from St. Louis to complete the trade made on Feb. 22.

1985-86

Oct. 4, 1985: Whalers purchase Tim Bothwell from St. Louis

Oct. 7, 1985: Whalers obtain Stewart Gavin from Toronto for Chris Kotsopoulos

Nov. 21, 1985: Whalers acquire Dave Babych from Winnipeg for Ray Neufeld

Dec. 7, 1985: Whalers secure Doug Jarvis from Washington for Jorgen Pettersson

Jan. 17, 1986: Whalers obtain Wayne Babych from Quebec for Greg Malone

Feb. 3, 1986: Whalers acquire Bill Gardner from Chicago for third-round pick (Mike Dagenais) in 1987 Entry Draft.

March 8, 1988: Whalers obtain John Anderson from Quebec for Risto Siltanen

March 11, 1986: Whalers net Mike McEwen from the New York Rangers for Bob Crawford

1986-87

Sept. 17, 1986: Whalers select Joe Tracy (Ohio State) in Supplemental Draft

Oct. 6, 1986: Whalers claim Gord Sherven (from Edmonton) in Waiver Draft

Oct. 6, 1986: Whalers obtain Yves Courteau from Calgary for Mark Paterson

Oct. 21, 1986: Whalers acquire Dave Barr from St. Louis for Tim Bothwell

Dec. 11, 1986: Whalers obtain Dave Semenko from Edmonton for a third-round pick (Trevor Sim) in 1988 Entry Draft

Jan. 14, 1986: Whalers acquire Randy Ladouceur from Detroit for Dave Barr

March 10, 1986: Whalers obtain Pat Hughes from St. Louis for a 10th-round pick (Andy Cesarski) in the 1987 Entry Draft.

1987-88

June 13, 1987: Whalers select Ken Lovesin (U of Saskatoon) in Supplemental Draft

Sept. 8, 1987: Whalers obtain Bill Root from Toronto for Dave Semenko

Oct. 5, 1987: Whalers claim Doug Wickenheiser (from St. Louis) and Brent Peterson (from Vancouver in Waiver Draft, lose Wickenheiser (to Vancouver) and Bill Root (to St. Louis)

Oct. 15, 1987: Whalers purchase Dave "Tiger" Williams from Los Angeles

Jan. 3, 1988: Whalers obtain Carey Wilson, Neil Sheehy and the rights to Lane MacDonald from Calgary for Shane Churla and Dana Murzyn

Jan. 22, 1988: Whalers obtain Lindsay Carson from Philadelphia for Paul Lawless

March 8, 1988: Whalers obtain Charles Bourgeois from St. Louis with a third-round draft choice (Blair Atcheynum) in 1989 in exchange for Hartford's 1989 second-round draft choice (Rick Corriveau) to complete an Oct. 5, 1986 deal between the teams involving Mark Reeds

June 10, 1988: Whalers select Todd Krygier (U of Connecticut) in Supplemental Draft

1988-89

July 6, 1988: Whalers obtain Ed Kastelic and Grant Jennings from Washington for Neil Sheehy and Mike Millar

Oct. 3, 1988: Whalers lose Stewart Gavin and Tom Martin (both to Minnesota) in Waiver Draft

Nov. 23, 1988: Whalers release Richard Brodeur

Dec. 26, 1988: Whalers obtain Don Maloney, Brian Lawton and Norm Maciver from the New York Rangers for Carey Wilson and a fifth-round pick (Lubos Rob) in the 1989 Entry Draft

March 6, 1988: Whalers acquire Jim Thomson from Washington for Scot Kleinendorst

March 7, 1988: Whalers obtain Jim Pavese from Detroit for Torrie Robertson

1989-90

June 17, 1989: Whalers obtain Pat Verbeek from New Jersey for Sylvain Turgeon

Oct. 2, 1989: Whalers select Mikael Andersson (from Buffalo) in Waiver Draft

Oct. 7, 1989: Whalers obtain Mike Berger from Minnesota for Kevin Sullivan

Oct. 10, 1989: Whalers obtain Jim Ennis from Edmonton for Norm Maciver

Oct. 16, 1989: Whalers sign free agent Steve Dykstra

Oct. 31, 1989: Whalers acquire Chris Cichocki from New Jersey for Jim Thomson

Dec. 1, 1989: Whalers lose Brian Lawton (to Quebec) on waivers for $20,000

Dec. 13, 1989: Whalers obtain Randy Cunneyworth from Winnipeg for Paul MacDermid

March 3, 1990: Whalers obtain Jeff Sirkka from Boston for Steve Dykstra

March 5, 1990: Whalers acquire Yvon Corriveau from Washington for Mike Liut

March 6, 1990: Whalers obtain Cam Brauer from Edmonton for Marc Laforge

1990-91

June 15, 1990: Whalers select Jim Crozier (Cornell) in Supplemental Draft

July 9, 1990: Whalers acquire Carey Wilson and a third-round draft pick (Michael Nylander) in the 1991 Entry Draft from the New York Rangers for Jody Hull

Oct. 1, 1990: Whalers deal Dave Tippett to Washington for a sixth-round draft choice (Jarrett Reid) in the 1992 Entry Draft

*Ray Neufeld, a fourth-round pick by Hartford in 1979, moved on to Winnipeg on November 21, 1985 in a swap that brought **Dave Babych** to the Whalers.*

Oct. 3, 1990: Whalers deal Joel Quenneville to Washington for future considerations

Oct. 11, 1990: Whalers purchase Todd Richards from Montreal for cash

Oct. 31, 1990: Whalers acquire Marc Bergevin from the New York Islanders for a fifth-round pick (Ryan Duthie) in the 1992 Entry Draft

Nov. 13, 1990: Whalers obtain Doug Crossman from the New York Islanders for Ray Ferraro

Dec. 21, 1990: Whalers acquire Rob Brown from Pittsburgh for Scott Young

Feb. 20, 1991: Whalers obtain Doug Houda from Detroit for Doug Crossman

March 4, 1991: Whalers acquire John Cullen, Zarley Zalapski and Jeff Parker from Pittsburgh for Ron Francis, Ulf Samuelsson and Grant Jennings

March 5, 1991: Whalers obtain Mark Hunter from Calgary for Carey Wilson

1991-92

Aug. 26, 1991: Whalers obtain Paul Fenton to Calgary for a sixth-round pick (Joel Bochard) in the 1992 Entry Draft

Sept. 8, 1991: Whalers send Sylvain Cote to Washington for a second-round pick (Andrei Nikolishin) in the 1992 Entry Draft

Sept. 17, 1991: Whalers obtain Andrew Cassels from Montreal for a second-round pick (Valeri Bure) in the 1992 Entry Draft

Oct. 2, 1991: Whalers obtain Dan Keczmer from San Jose for Dean Evason

Oct. 3, 1991: Whalers obtain Lee Norwood from New Jersey for future considerations

Oct. 4, 1991: Whalers send Todd Krygier to Washington (transferred to Calgary) for a fourth-round draft pick (Jason Smith) in the 1993 Entry Draft

Oct. 18, 1991: Whalers obtain Mike McHugh from San Jose for Paul Fenton

Nov. 13, 1991: Whalers obtain Murray Craven and a fourth-round draft pick (Kevin Smyth) in the 1992 Entry Draft from Philadelphia for Kevin Dineen

Nov. 13, 1991: Whalers send Lee Norwood to St. Louis for future considerations

Nov. 14, 1991: Whalers send Barry Pederson to Boston for future considerations

Nov. 21, 1991: Whalers obtain Jukka Suomalainen from Minnesota to complete a 1991 Entry Draft deal

Dec. 18. 1991: Whalers obtain Daniel Shank from Detroit for Chris Tancil

Jan. 24, 1992: Whalers obtain Steve Konroyd from Chicago for Rob Brown

Jan. 27, 1992: Whalers acquire Paul Gillis from Chicago for future considerations

March 10, 1992: Whalers obtain Frank Pietrangelo from Pittsburgh for a third-round pick (Sven Butenschon) and a seventh-round pick (Serge Aubin) in the 1994 Entry Draft (the deal was subject to an arbitrator's ruling which was made on Sept. 14, 1992)

1992-93

June 15, 1992: Whalers sell Brad Shaw to New Jersey

June 15, 1992: Whalers obtain Nick Kypreos from Washington for Mark Hunter and future considerations

June 16, 1992: Whalers obtain Allen Pederson from Minnesota for sixth-round draft choice (Rick Mrozik) in 1993

June 24, 1992: Whalers sign free agent Jim Agnew

July 8, 1992: Whalers obtain Tim Kerr from the New York Rangers for future considerations

Aug. 6, 1992: Whalers sign free agent Trevor Stienburg

Aug. 20, 1992: Whalers sent Yvon Corriveau to Washington to complete the June 15, 1992 deal between the two teams

Aug. 28, 1992: Whalers obtain Sean Burke and Eric Weinrich from New Jersey for Bobby Holik, a second-round pick (Jay Pandolfo) in the 1993 Entry Draft and future considerations

Sept. 3, 1992: Whalers acquire Mark Janssens from Minnesota for James Black

Oct. 1, 1992: Whalers obtain Corrie D'Alessio from Vancouver for Kay Whitmore

Oct. 9, 1992: Whalers ship Michel Picard to San Jose for future considerations

Nov. 21, 1992: Whalers claim Jamie Leach on waivers from Pittsburgh

Nov. 24, 1992: Whalers ship John Cullen to Toronto (transferred to San Jose) for a second-round choice (Vlastimil Kroupa) in the 1993 Entry Draft

Jan. 21, 1993: Whalers acquire Yvon Corriveau from San Jose as future considerations to complete the Oct. 9, 1992 deal between the two teams

Jan. 21, 1993: Tim Kerr announces his retirement

March 22, 1993: Whalers obtain Robert Kron and future considerations (Jim Sandlak, May 17, 1993) from Vancouver for Murray Craven and a fifth-round draft choice (previously acquired by Hartford — Vancouver took Scott Walker) in 1993 Entry Draft

June 1, 1993: Whalers obtain Brad McCrimmon from Detroit for a sixth-round draft choice (Tim Spitzig) in the 1993 Entry Draft

June 20, 1993: Whalers acquire Sergei Makarov from Calgary for a fourth-round draft pick (Jason Smith) in the 1993 Entry Draft which Hartford had previously obtained from Washington on Oct. 4, 1991

June 25, 1993: Whalers select Kent Fearns (Colorado) in Supplemental Draft

June 26, 1993: Whalers swing Entry Draft Day deal with San Jose to select Chris Pronger (second overall). Hartford traded its first-round choice (Victor Kozlov), second-round choice (Vlastimil Kroupa) and third-round choice (Villen Peltonen) to San Jose along with Sergei Makarov. The Whalers also sent a second-round pick (obtained in a previous deal with Toronto) to Florida (later transferred to Winnipeg which took Scott Langkow) as part of the draft-day deal where the Panthers would opt not to pick Kozlov.

June 30, 1993: Emile Francis retires team president, ending a 47-year career in professional hockey

Emile "Cat" Francis spent 47 years in professional hockey, starting as a goaltender and eventually shifting to management where he coached or served tenures as general manager for three NHL clubs: the New York Rangers, St. Louis Blues and the Hartford Whalers. In 1982, he was elected to the NHL's Hall of Fame. When the Whalers conducted a "Czar Hunt" in 1983 to find a new leader, the organization hired Francis who took the helm and simply revived the slumbering franchise. The energetic and enthusiastic bossman remained in charge for six winters. He was always quotable and a historian on virtually anyone in hockey dating back to World War II. There was always a defenseman "who has more guts than a slaughterhouse" or "a hard-nosed winger" obtained in an off-season deal.

1993-94

Sept. 1, 1993: Brian Burke resigns as general manager to take a managerial post with the National Hockey League as vice president of operations

Nov. 2, 1993: Whalers obtain Bryan Marchment and Steve Larmer from Chicago for Patrick Poulin and Eric Weinrich; Whalers obtain Darren Turcotte and James Patrick from the New York Rangers for Steve Larmer, Nick Kypreos, Barry Richter and a sixth-round pick (Yuri Litvinov) in 1994 Entry Draft

Nov. 3, 1993: Whalers claim Bob McGill on waivers from the New York Islanders

Nov. 3, 1993: Whalers acquire Marc Potvin from Los Angeles for Doug Houda

Nov. 12, 1993: Whalers sell Martin Hamrlik to St. Louis

Nov. 19, 1993: Whalers obtain Jeff Reese from Calgary for Dan Keczmer

Dec. 16, 1993: Whalers acquire Alexander Godynyuk from Florida for Jim McKenzie

Jan. 24, 1994: Whalers obtain Todd Harkins from Calgary for Scott Morrow

Jan. 25, 1994: Whalers obtain Ted Crowley from Toronto for Mark Greig and a sixth-round pick (Yuri Litvinov) in the 1994 Entry Draft which was transferred to the New York Rangers

March 10, 1994: Whalers obtain Gary Suter, Ted Drury and Paul Ranheim from Calgary for Zarley Zalapski, James Patrick and Michael Nylander

March 11, 1994: Whalers obtain Frantisek Kucera and Jocelyn Lemieux from Chicago for Gary Suter, Randy Cunneyworth and a third-round pick (Larry Courville) in 1995 which was later transferred to Vancouver

March 18, 1994: Whalers deal Ken Belanger to Toronto for a ninth-round pick (Matt Ball) in the 1994 Entry Draft

1994-95

June 28, 1994: Whalers franchise is sold to Peter Karmanos, Thomas Thewes and Jim Rutherford

July 13, 1994: Whalers sign free agent Jimmy Carson

Aug. 18, 1994: Whalers sign free agent Steven Rice with compensation owed to Edmonton

Aug. 26, 1994: Whalers obtain Glen Wesley from Boston for first-round picks in 1995 (Kyle McLaren), 1996 and 1997 Entry Drafts

Aug. 30, 1994: Whalers lose Bryan Marchment to Edmonton as compensation for signing Steven Rice

Jan. 4, 1995: Whalers obtain Brian Glynn (from Vancouver) and Kelly Chase (from St. Louis) in Waiver Draft

March 23, 1995: Whalers obtain Glen Featherstone, Michael Stewart, a first-round draft pick (Jean-Sebastein Giguere) in the 1995 Entry Draft and a fourth-round pick in the 1996 Entry Draft for Pat Verbeek

May 31, 1995: Whalers send Jan Vopat to Los Angeles for a fourth-round pick (Ian MacNeil) in the 1995 Entry Draft

1995-96

July 27, 1995: Whalers obtain Brendan Shanahan from St. Louis for Chris Pronger

July 29, 1995: Whalers claim Jeff Bes on waivers from Dallas

Brendan Shanahan came to the Whalers on July 27, 1995 from St. Louis in a swap for Chris Pronger. After pushing for a trade, Hartford sent the disgruntled winger to Detroit in a multi-player swap that brought Keith Primeau to Hartford.

Aug. 1, 1995: Whalers sign free agent Gerald Diduck

Oct. 6, 1995: Whalers obtain Nelson Emerson from Winnipeg for Darren Turcotte

Oct. 6, 1995: Whalers lose Ted Drury to Ottawa in Wavier Draft

Oct. 6, 1995: Whalers claim Jason Muzzatti on waivers from Calgary

Nov. 1, 1995: Whalers assign Robert Petrovicky to Detroit of the IHL

Nov. 5, 1995: Whalers name Paul Maurice as 13th coach of the hockey club, replacing Paul Holmgren who was fired 12 games into the season

Nov. 29, 1995: Whalers deal Jeff Reese to Tampa Bay for a ninth-round pick in 1996 Entry Draft

Dec. 16, 1995: Whalers deal Jocelyn Lemieux and a second-round pick in 1998 to New Jersey for Jim Dowd and a second-round pick in 1997

Dec. 16, 1995: Whalers obtain Jeff Brown and a third-round pick in 1998 from Vancouver for Frantisek Kucera, Jim Dowd and New Jersey's second-round pick in 1997

Dec. 28, 1995: Whalers obtain Kevin Dineen from Philadelphia for future considerations

1996-97

Oct. 1, 1996
Whalers sign free agents Kent Manderville and Chris Longo and obtain Kevin Brown from Anaheim for Espen Knutsen.

Oct. 9, 1996
Whalers obtain Keith Primeau, Paul Coffey and 1997 first-round draft pick from Detroit for Brendan Shanahan and Brian Glynn.

Oct. 12, 1996
Whalers claim Stu Grimson on waivers from Detroit.

Nov. 11, 1996
Whalers acquire Curtis Leschyshyn from Washington for Andrei Nikolishin.

Dec. 15, 1996
Whalers acquire Kevin Haller, a 1997 first-round draft pick and a 1997 seventh-round draft pick from Philadelphia for Paul Coffey and a 1997 third-round pick.

March 5, 1997
Whalers obtain Steve Chiasson from Calgary and a third-round pick in 1997 for Glen Featherstone, Hnat Domenichelli, a second and third round pick in 1997.

March 18, 1997
Whalers obtain Derek King from the New York Islanders for a fifth-round pick in 1997, obtain Bates Battaglia and a fourth-round pick in 1998 from Anaheim for Mark Janssens, obtain an eighth-round in 1997 from Toronto for Kelly Chase and acquire Chris Murray from Phoenix for Gerald Diduck.

May 6, 1997
Whalers owner Peter Karmanos announces team will relocate operations to Raleigh, North Carolina for the 1997-98 season.

Whalers NHL Entry Drafts (1979-1996)

1979 Entry Draft

August 9 at Queen Elizabeth Hotel

18.	Ray Allison	RW	Brandon
39.	Stuart Smith	D	Peterborough
60.	Don Nachbaur	LW	Billings
81.	Ray Neufeld	RW	Edmonton
102.	Mark Renaud	D	Niagara Falls
123.	Dave McDonald	LW	Brandon

1980 Entry Draft

June 11 at Montreal Forum

8.	Fred Arthur	D	Cornwall
29.	Michel Galarneau	C	Hull
50.	Mickey Volcan	D	U of North Dakota
71.	Kevin McClelland	RW	Niagara Falls
92.	Darren Jensen	G	U of North Dakota
113.	Mario Cerri	C	Ottawa
134.	Mike Martin	D	Sudbury
155.	Brent Denat	LW	Michigan Tech
176.	Paul Fricker	G	U of Michigan
197.	Lorne Bokshowan	C	Saskatoon

1981 Entry Draft

June 10 at Montreal Forum

4.	Ron Francis	C	S.S. Marie
61.	Paul MacDermid	C	Windsor
67.	Mike Hoffman	LW	Brantford
93.	Bill Maguire	D	Niagara Falls
103.	Dan Bourbonnais	LW	Calgary
130.	John Mokosak	D	Victoria
151.	Denis Dore	RW	Chicoutimi
172.	Jeff Poeschl	G	N.Michigan
193.	Larry Power	C	Kitchener

1982 Entry Draft

June 9 at Montreal Forum

14.	Paul Lawless	LW	Windsor
35.	Mark Paterson	D	Ottawa
56.	Kevin Dineen	RW	Denver U
67.	Ulf Samuelsson	D	Leksands
88.	Ray Ferraro	C	Penticton B
109.	Randy Gilhen	LW	Winnipeg
130.	Jim Johannson	C	Rochester Mayo
151.	Mickey Krampotich	C	Hibbing HS
172.	Kevin Skilliter	D	Cornwall
214.	Martin Linse	D	Djurgardens
235.	Randy Cameron	D	Winnipeg

1983 Entry Draft

June 8 at Montreal Forum

2.	Sylvain Turgeon	LW	Hull Olympiques
20.	David A. Jensen	LW	Lawrence Academy
23.	Ville Siren	D	Ilves (Finland)
61.	Leif Karlsson	D	Mora (Sweden)
64.	Dave MacLean	RW	Belleville
72.	Ron Chyzowski	C	St. Albert
104.	Brian Johnson	D	Silver Bay H.S.
124.	Joe Reekie	D	North Bay
143.	Chris DuPerron	D	Chicoutimi
144.	Jamie Falle	G	Clarkson College
164.	Bill Fordy	LW	Guelph
193.	Reine Karlsson	LW	Sodertalje
204.	Allan Acton	LW	Saskatoon
224.	Darcy Kaminski	D	Lethbridge

1984 Entry Draft

June 9 at Montreal Forum

11.	Sylvain Cote	D	Quebec
110.	Mike Millar	RW	Brantford
131.	Mike Velucci	D	Belleville
173.	John Devereaux	C	Scitate HS
194.	Brent Regan	LW	St. Albert
215.	Jim Culhane	D	Western Michigan
236.	Peter Abric	G	North Bay

1985 Entry Draft

June 15 at Toronto Convention Centre

5.	Dana Murzyn	D	Calgary
26.	Kay Whitmore	G	Peterborough
68.	Gary Callaghan	C	Belleville
110.	Shane Churla	RW	Medicine Hat
131.	Chris Brandt	LW	S.S. Marie
152.	Brian Puhalsky	LW	Notre Dame HS
173.	Greg Dornach	C	Miami (Ohio) U
194.	Paul Tory	C	U of Ill-Chicago
215.	Jerry Pawloski	LW	Harvard
236.	Bruce Hill	LW	Denver U

1986 Entry Draft

June 21 at Montreal Forum

11.	Scott Young	RW	Boston U
32.	Marc Laforge	D	Kingston
74.	Brian Chapman	D	Belleville
95.	Bill Horn	G	Western Michigan
116.	Joe Quinn	RW	Calgary
137.	Steve Torrel	C	Hibbing HS
158.	Ron Hoover	C	Western Michigan
179.	Robert Glasgow	RW	Sherwood Park
200.	Sean Evoy	G	Cornwall
221.	Cal Brown	D	Penticton
242.	Brian Verbeek	C	Kingston

1987 Entry Draft

June 13 at Joe Louis Arena (Detroit)

18.	Jody Hull	RW	Peterborough
39.	Adam Burt	D	North Bay
81.	Terry Yake	C	Brandon
102.	Marc Rousseau	D	Denver U
123.	Jeff St. Cyr	D	Michigan Tech
144.	Greg Wolf	D	Buffalo Regal
165.	John Moore	C	Yale
186.	Joe Day	LW	St.Lawrence
228.	Kevin Sullivan	C	Princeton
249.	Steve Laurin	D	Dartmouth

1988 Entry Draft

June 11 at Montreal Forum

11.	Chris Govedaris	LW	Toronto
32.	Barry Richter	D	Culver Academy
74.	Dean Dyer	C	Lake Superior
95.	Scott Morrow	LW	New Hampshire
116.	Corey Beaulieu	D	Seattle
137.	Kerry Russell	RW	Michigan State
158.	Jim Burke	D	Maine
179.	Mark Hirth	C	Michigan State
200.	Wayde Bucsis	RW	Prince Albert
221.	Rob White	D	St. Lawrence
242.	Dan Slatalla	LW	Deerfield Academy

Dana Murzyn, Hartford's top pick in 1985, went 13-49-62 in 185 games over three seasons with the Whalers.

1989 Entry Draft

June 17 at Met Center (Minnesota)

10.	Robert Holik	C	Dukla Jihlava
53.	Blair Atchenyum	RW	Moose Jaw
73.	Jim McKenzie	LW	Victoria
94.	James Black	C	Portland
115.	Jerome Bechard	LW	Moose Jaw
136.	Scott Daniels	LW	Regina
157.	Raymound Saumier	RW	Trois Rivieres
178.	Michel Picard	LW	Trios Rivieres
199.	Trevor Buchanan	LW	Kamloops
220.	John Battice	D	London
241.	Peter Kasowski	C	Swift Current

1990 Entry Draft

June 16 at B.C. Place (Vancouver)

15.	Mark Greig	RW	Lethbridge
36.	Geoff Sanderson	C	Swift Current
57.	Mike Lenarduzzi	G	S.S. Marie
78.	Chris Bright	C	Moose Jaw
120.	Cory Keenan	D	Kitchener
141.	Jergus Baca	D	Kosice
162.	Martin D'Orsonnens	D	Clarkson
183.	Corey Osmak	C	Nipiwan T-II
204.	Espen Knutsen	C	Valeriengen
225.	Tommie Eriksen	D	Prince Albert
246.	Denis Chalifoux	C	Laval

Whalers NHL Entry Drafts (1979-1996)

1991 Entry Draft

June 9 at Memorial Auditorium (Buffalo)

9.	Patrick Poulin	LW	St.Hyacinthe
31.	Martin Hamrlik	D	ZPS-Zlin
53.	Todd Hall	D	Hamden HS
59.	Michael Nylander	C	Huddinge
75.	Jim Storm	LW	Michigan Tech
119.	Mike Harding	RW	Northern Michigan
141.	Brian Mueller	D	Kent Prep
163.	Steve Yule	D	Kamloops
185.	Chris Belanger	D	Western Michigan
207.	Jason Currie	G	Clarkson College
229.	Mike Santonelli	C	Matignon HS
251.	Rob Peters	D	Ohio State

1992 Entry Draft

June 20 at Montreal Forum

9.	Robert Petrovicky	C	Dukla Trencin
47.	Andrei Nikolishin	LW	Dynamo Moscow
57.	Jan Vopat	D	Litvinov
79.	Kevin Smyth	LW	Moose Jaw
81.	Jason McBain	D	Portland W.Hawks
143.	Jarret Reid	C	S.S. Marie
153.	Ken Belanger	LW	Ottawa
177.	Konstantin Korotkov	C	Spartak
201.	Greg Zwakman	D	U of Minnesota
225.	Steve Halko	D	U of Michigan
249.	Joacim Esbjors	D	V.Frolunda

1993 Entry Draft

June 26 at Le Colisee de Quebec

2.	Chris Pronger	D	Peterborough
72.	Marek Malik	D	TJ Vitkovice
84.	Trevor Roenick	RW	Jr. Bruins
115.	Nolan Pratt	D	Portland
188.	Manny Legace	G	Niagara Falls
214.	Dmitri Gorenko	LW	CSKA
240.	Wes Swinson	D	Kitchener
266.	Igor Chibirev	C	Fort Wayne

1994 Entry Draft

June 28-29 at Hartford Civic Center

5.	Jeff O'Neill	C	Guelph
83.	Hnat Domenichelli	C	Kamloops
109.	Ryan Risidore	D	Guelph
187.	Tom Buckley	C	St.Joseph
213.	Ashlin Halfnight	D	Harvard
230.	Matt Ball	RW	Detroit Jrs.
239.	Brian Regan	G	Westminster
265.	Steve Nimigon	LW	Niagara Falls

1995 Entry Draft

July 9 at Northlands Coliseum (Edmonton)

13.	J.S. Giguere	G	Halifax Jrs.
35.	Sergei Fedotov	D	Dynamo
85.	Ian MacNeil	D	Oshawa
87.	Sami Kapanen	LW	HIF Helsinki
113.	Hugh Hamilton	D	Spokane
165.	Bryon Ritchie	C	Lethridge
191.	Milan Kostolny	RW	Detroit Jrs.
217.	Mike Rucinski	D	Detroit Jrs.

Sami Kapanen, a fourth-round pick in 1995, had his best year with the Whalers in 1996-97, going 13-12-25 in 45 games. He scored the last penalty-shot goal (against Boston's Jim Carey) in Hartford history and it proved to be the game-winner in a 6-3 verdict over the last-place Bruins on March 12, 1997.

1996 Entry Draft

June 22 at Kiel Center, St. Louis

34.	Trevor Wasyluk	LW	Medicine Hat
61.	Andrei Petrunin	LW	Krylja (Russia)
88.	Craig MacDonald	C	Harvard
104.	Steve Waysiko	C	Detroit Whalers
116.	Mark McMahon	D	Kitchener
143.	Aaron Baker	G	Tri-City
171.	Greg Kuznik	D	Seattle
197.	Kevin Marsh	LW	Calgary
223.	Craig Adams	RW	Harvard

Supplemental Draft

1986	Joe Tracy	RW	Ohio State
1987	Ken Lovsin	D	Saskatchewan U
1988	Todd Krygier	LW	U of Connecticut
1989	Chris Tancil	C	U of Wisconsin
1990	Jim Crozier	G	Cornell
1991	Shaun Gravistin	G	Alaska-Anchorage
1992	No draft	. . .	
1993	Ken Feams	D	Colorado College
1994	Steve Martins	C	Harvard

All-Time Franchise Leaders

Regular Season

National Hockey League

Games Played

Ron Francis	714
Kevin Dineen	587
Adam Burt	499
Dave Tippett	483
Ulf Samuelsson	463
Joel Quenneville	457
Randy Ladouceur	452
Ray Ferraro	442
Dean Evason	434

Points

Ron Francis	821
Kevin Dineen	503
Pat Verbeek	403
Blaine Stoughton	377
Geoff Sanderson	352
Ray Ferraro	351
Andrew Cassels	350
Sylvain Turgeon	328
Dave Babych	240
Dean Evason	235

Goals

Ron Francis	264
Kevin Dineen	235
Blaine Stoughton	219
Pat Verbeek	192
Geoff Sanderson	189
Sylvain Turgeon	178
Ray Ferraro	153
Ray Neufeld	95
Dean Evason	87
Mark Johnson	85
Mike Rogers	84
Andrew Cassels	75
Dave Tippett	75
John Anderson	72

Assists

Ron Francis	557
Kevin Dineen	268
Andrew Cassels	253
Pat Verbeek	211
Dave Babych	196
Ray Ferraro	194
Geoff Sanderson	163
Blaine Stoughton	158
Sylvain Turgeon	150
Dean Evason	148
Mark Howe	147

Most Games in a Season

84 —	Andrew Cassels	1992-93
84 —	Pat Verbeek	1992-93
84 —	Pat Verbeek	1993-94
84 —	Mark Janssens	1993-94

Penalty Minutes

Torrie Robertson	1,368
Kevin Dineen	1,239
Pat Verbeek	1,144
Ulf Samuelsson	1,108
Paul MacDermid	744
Adam Burt	723
Randy Ladouceur	717
Mark Janssens	712
Dean Evason	617
Ron Francis	540
Ed Kastelic	485

Most Penalty Minutes in a Season

358 —	Torrie Robertson	1985-86
337 —	Torrie Robertson	1984-85
325 —	Nick Kypreos	1992-93
293 —	Torrie Robertson	1987-88
254 —	Scott Daniels	1995-96
246 —	Marc Potvin	1993-94
246 —	Pat Verbeek	1990-91

Most Goals in a Season

56 —	Blaine Stoughton	1979-80
52 —	Blaine Stoughton	1981-82
46 —	Geoff Sanderson	1992-93
45 —	Kevin Dineen	1988-89
45 —	Blaine Stoughton	1982-83
45 —	Sylvain Turgeon	1985-86
44 —	Mike Rogers	1979-80
44 —	Pat Verbeek	1989-90
44 —	Brendan Shanahan	1995-96
43 —	Pat Verbeek	1990-91
43 —	Blaine Stoughton	1980-81
41 —	Ray Ferraro	1988-89
41 —	Geoff Sanderson	1993-94
40 —	Mike Rogers	1980-81
40 —	Kevin Dineen	1986-87
40 —	Sylvain Turgeon	1983-84

Most Assists in a Season

69 —	Ron Francis	1989-90
65 —	Andrew Cassels	1992-93
65 —	Mike Rogers	1980-81
63 —	Ron Francis	1983-84
61 —	Mike Rogers	1979-80
60 —	Ron Francis	1983-84
59 —	Ron Francis	1982-83
57 —	Ron Francis	1984-85
56 —	Mark Howe	1979-80
55 —	Ron Francis	1990-91
53 —	Ron Francis	1985-86
52 —	Mark Johnson	1983-84
52 —	Dave Keon	1979-80
52 —	Pat Boutette	1980-81
51 —	John Cullen	1991-92
51 —	Zarley Zalapski	1992-93
50 —	Ron Francis	1987-88

GOALTENDING LEADERS

Games Played

Sean Burke	256
Mike Liut	252
Greg Millen	219
Peter Sidorkiewicz	178
John Garrett	122
Steve Weeks	94
Kay Whitmore	75
Mike Veisor	69
Jason Muzzatti	52
Frank Pietrangelo	44

Wins

Mike Liut	115
Sean Burke	100
Peter Sidorkiewicz	71
Greg Millen	62
Steve Weeks	42
John Garrett	36
Kay Whitmore	23

Losses

Greg Millen	120
Sean Burke	120
Mike Liut	111
Peter Sidorkiewicz	79
John Garrett	57
Steve Weeks	42
Mike Veisor	37

Goals-Against Average

Sean Burke	3.12
Jeff Reese	3.13
Darryl Reaugh	3.15
Jason Muzzatti	3.23
Peter Sidorkiewicz	3.33
Mike Liut	3.36

Minutes Played

Sean Burke	
Mike Liut	14,705
Greg Millen	12,963
Peter Sidorkiewicz	10,346
John Garrett	7,088

Shutouts

Mike Liut	13
Sean Burke	10
Peter Sidorkiewicz	8
Greg Millen	4
Steve Weeks	4

Assists

Sean Burke	11
Greg Millen	10
Peter Sidorkiewicz	9
Mike Liut	5
Kay Whitmore	5

Penalty Minutes

Sean Burke	79
Jason Muzzatti	51
Greg Millen	24
Kay Whitmore	24
John Garrett	16

Ties

Greg Millen	33
John Garrett	27
Peter Sidorkiewicz	24
Sean Burke	24
Mike Liut	19
Kay Whitmore	10

Most Appearances in a Season

66 —	Sean Burke	1995-96
60 —	Greg Millen	1982-83
60 —	Greg Millen	1983-84
60 —	Mike Liut	1987-88
59 —	Mike Liut	1986-87
57 —	Mike Liut	1985-86
55 —	Greg Millen	1981-82
54 —	John Garrett	1980-81
52 —	John Garrett	1979-80
52 —	Peter Sidorkiewicz	1990-91

100 Points in a Season

105 —	Mike Rogers	1979-80
105 —	Mike Rogers	1980-81
101 —	Ron Francis	1989-90
100 —	Blaine Stoughton	1979-80

Note: John Cullen had 110 points in 1990-91, a combined sum of 94 with Pittsburgh and 16 with Hartford.

Ron Francis *captained the Whalers and played in Hartford for a franchise record 714 games. He finally achieved recognition for his leadership and play-making skills when he was traded to the Pittsburgh Penguins. Drafted fourth overall in 1981, the centerman made his NHL debut in an 8-4 loss to Washington on November 14, 1981. His last game as a Whaler was a 4-4 tie at the Civic Center against Toronto.*

All-Time Franchise Leaders

*Goaltender **Mike Liut** keeps his eyes on the puck after making a save as defenseman **Risto Siltanen** arrives to clear the zone. Liut rated among the club's all-time best puckstoppers; Siltanen had a cannon of a shot from the blue line.*

Lowest GAA in a Season

2.64 — Mike Liut	1989-90
2.68 — Sean Burke	1994-95
2.69 — Sean Burke	1996-97
2.90 — Jason Muzzatti	1995-96
3.03 — Peter Sidorkiewicz	1988-89
3.11 — Sean Burke	1995-96
3.15 — Daryl Reaugh	1990-91
3.18 — Jeff Reese	1993-94
3.18 — Mike Liut	1987-88
3.23 — Mike Liut	1986-87

Most Ties in a Season

12 — John Garrett	1980-81
12 — Greg Millen	1981-82
11 — John Garrett	1979-80

Most Minutes Played in a Season

3,669 — Sean Burke	1995-96
3,576 — Greg Millen	1983-84
3,506 — Greg Millen	1982-83
3,471 — Mike Liut	1986-87
3,326 — Mike Liut	1987-88
3,282 — Mike Liut	1985-86

Most Shutouts in a Season

4 — Mike Liut	1986-87
4 — Peter Sidorkiewicz	1988-89
4 — Sean Burke	1995-96
4 — Sean Burke	1996-97

Most Wins in a Season

31 — Mike Liut	1986-87
28 — Sean Burke	1995-96
27 — Mike Liut	1985-86
25 — Mike Liut	1987-88
22 — Peter Sidorkiewicz	1988-89
22 — Sean Burke	1996-97
21 — Greg Millen	1983-84

Most Losses in a Season

38 — Greg Millen	1982-83
30 — Greg Millen	1981-82
30 — Greg Millen	1983-84
28 — Sean Burke	1995-96
28 — Mike Liut	1987-88
27 — Sean Burke	1992-93

Whalers vs. NHL teams (1979-97)

Anaheim Mighty Ducks

	GP	W	L	T	GF	GA
1993-94	2	0	2	0	5	9
1994-95	did not play					
1995-96	2	2	0	0	7	5
1996-97	2	1	1	0	7	7

Atlanta Flames

	GP	W	L	T	GF	GA
1979-80	4	3	1	0	19	13

Boston Bruins

	GP	W	L	T	GF	GA
1979-80	4	1	2	1	16	14
1980-81	4	1	1	2	14	15
1981-82	8	2	4	2	21	35
1982-83	8	2	6	0	26	39
1983-84	8	2	5	1	21	31
1984-85	8	4	4	0	28	34
1985-86	8	3	4	1	29	26
1986-87	8	6	2	0	39	28
1987-88	8	3	4	1	21	28
1988-89	8	3	5	0	21	32
1989-90	8	3	4	1	22	27
1990-91	8	2	5	1	17	32
1991-92	8	1	6	1	27	31
1992-93	7	1	6	0	21	31
1993-94	5	1	3	1	11	15
1994-95	4	1	3	0	9	13
1995-96	6	1	5	0	11	20
1996-97	6	4	1	1	28	18
Playoffs						
1989-90	7	3	4	0	21	23
1990-91	6	2	4	0	17	24

Buffalo Sabres

	GP	W	L	T	GF	GA
1979-80	4	1	3	0	11	14
1980-81	4	1	2	1	13	16
1981-82	8	1	4	3	18	30
1982-83	8	2	5	1	28	40
1983-84	8	3	5	0	24	29
1984-85	8	0	5	3	16	30
1985-86	8	6	2	0	35	24
1986-87	8	4	4	0	23	27
1987-88	8	4	3	1	28	19
1988-89	8	5	3	0	38	23
1989-90	8	2	6	0	19	35
1990-91	8	2	4	2	26	36
1991-92	8	4	2	2	34	27
1992-93	9	1	6	2	21	39
1993-94	5	1	3	1	12	19
1994-95	4	2	2	0	10	12
1995-96	5	2	3	0	8	12
1996-97	6	4	2	0	23	21

Calgary Flames

	GP	W	L	T	GF	GA
1980-81	4	1	3	0	19	13
1981-82	3	1	2	0	8	15
1982-83	3	0	1	2	9	16
1983-84	3	0	2	1	7	13
1984-85	3	0	3	0	10	21
1985-86	3	2	1	0	18	12
1986-87	3	1	2	0	10	15
1987-88	3	0	3	0	12	19
1988-89	3	1	2	0	9	12
1989-90	3	0	2	1	11	14
1990-91	3	1	2	0	9	17
1991-92	2	1	1	0	6	6
1992-93	2	0	2	0	6	8
1993-94	2	0	2	0	4	10
1994-95	did not play					
1995-96	2	0	2	0	4	6
1996-97	2	1	1	0	4	3

Chicago Blackhawks

	GP	W	L	T	GF	GA
1979-80	4	1	1	2	15	15
1980-81	4	0	3	1	11	24
1981-82	3	1	1	1	11	15
1982-83	3	0	3	0	7	12
1983-84	3	2	1	0	11	10
1984-85	3	2	1	0	8	11
1985-86	3	1	2	0	15	17
1986-87	3	1	2	0	10	13
1987-88	3	2	1	0	8	6
1988-89	3	2	1	0	16	10
1989-90	3	1	2	0	11	15
1990-91	3	1	1	1	7	8
1991-92	3	0	3	0	3	10
1992-93	2	0	2	0	2	9
1993-94	2	2	0	0	9	4
1994-95	did not play					
1995-96	2	1	1	0	4	6
1996-97	2	0	1	1	5	6

Colorado Avalanche

	GP	W	L	T	GF	GA
1995-96	2	2	0	0	7	4
1996-97	2	0	1	1	4	8

Colorado Rockies

	GP	W	L	T	GF	GA
1979-80	4	2	1	1	20	13
1980-81	4	2	1	1	16	11
1981-82	3	0	2	1	9	12

Dallas Stars

	GP	W	L	T	GF	GA
1992-93	2	1	0	1	8	7
1993-94	2	0	1	1	3	7
1994-95	did not play					
1995-96	2	1	1	0	7	9
1996-97	2	0	2	0	1	6

*It's the NHL All-Star Game on February 4, 1986 and members of the Wales Conference, Larry Robinson of Montreal (left) and **Sylvain Turgeon** of the Whalers celebrate after Philadelphia's Brian Propp (center) scored, his first of two in the game. The Wales beat the Campbells 4-3 before a sellout crowd at the Hartford Civic Center at 3:05 of overtime on a goal by Bryan Trottier of the New York Islanders.*

Detroit Red Wings

	GP	W	L	T	GF	GA
1979-80	4	2	1	1	16	13
1980-81	4	2	0	2	15	11
1981-82	3	2	0	1	13	5
1982-83	3	0	3	0	9	17
1983-84	3	1	1	1	11	14
1984-85	3	2	1	0	15	9
1985-86	3	2	1	0	14	10
1986-87	3	1	1	1	8	9
1987-88	3	2	1	0	6	6
1988-89	3	2	1	0	10	9
1989-90	3	2	0	1	10	6
1990-91	3	2	1	0	12	8
1991-92	2	0	2	0	5	12
1992-93	2	0	2	0	2	7
1993-94	3	1	2	0	7	12
1994-95	did not play					
1995-96	2	0	2	0	4	7
1996-97	2	0	2	0	2	9

Edmonton Oilers

	GP	W	L	T	GF	GA
1979-80	4	2	1	1	13	8
1980-81	4	1	2	1	18	23
1981-82	3	0	2	1	10	16
1982-83	3	0	2	1	10	17
1983-84	3	1	2	0	17	10
1984-85	3	1	2	0	13	17
1985-86	3	0	3	0	9	16
1986-87	3	1	2	0	10	11
1987-88	3	1	2	0	7	13
1988-89	3	1	2	0	13	14
1989-90	3	1	0	2	11	9
1990-91	3	1	2	0	10	9
1991-92	3	0	3	0	3	10
1992-93	2	1	1	0	8	6
1993-94	2	1	0	1	9	6
1994-95	did not play					
1995-96	2	1	0	1	9	5
1996-97	2	0	2	0	3	6

Florida Panthers

	GP	W	L	T	GF	GA
1993-94	4	3	1	0	11	7
1994-95	4	0	3	1	9	14
1995-96	4	1	3	0	12	12
1996-97	4	2	1	1	8	10

Los Angeles Kings

	GP	W	L	T	GF	GA
1979-80	4	2	1	1	19	15
1980-81	4	0	3	1	14	20
1981-82	3	2	0	1	17	14
1982-83	3	1	2	0	10	12
1983-84	3	2	1	0	13	15
1984-85	3	0	2	1	7	10
1985-86	3	2	1	0	16	11
1986-87	3	1	2	0	11	12
1987-88	3	3	0	0	12	9
1988-89	3	1	2	0	13	20
1989-90	3	2	1	0	14	11
1990-91	3	1	2	0	9	12
1991-92	3	1	0	2	13	9
1992-93	2	0	2	0	4	14
1993-94	2	0	2	0	5	10
1994-95	did not play					
1995-96	2	2	0	0	14	6
1996-97	2	0	2	0	5	8

Minnesota North Stars

	GP	W	L	T	GF	GA
1979-80	4	0	4	0	6	22
1980-81	4	1	3	0	11	23
1981-82	3	1	2	0	13	15
1982-83	3	0	2	1	11	17
1983-84	3	3	0	0	17	10
1984-85	3	1	2	0	11	12
1985-86	3	1	2	0	8	11
1986-87	3	1	2	0	10	10
1987-88	3	3	0	0	19	4
1988-89	3	2	1	0	9	8
1989-90	3	2	1	0	13	12
1990-91	3	1	2	0	5	10
1991-92	3	1	1	1	12	12

Montreal Canadiens

	GP	W	L	T	GF	GA
1979-80	4	0	1	3	16	21
1980-81	4	1	3	0	13	20
1981-82	8	0	7	1	16	45
1982-83	8	3	5	0	24	40
1983-84	8	0	7	1	21	34
1984-85	8	2	5	1	34	47
1985-86	8	3	4	1	33	34
1986-87	8	4	3	1	24	25
1987-88	8	2	4	2	22	37
1988-89	8	1	7	0	22	36
1989-90	8	3	4	1	21	25
1990-91	8	4	4	0	26	24
1991-92	8	2	3	3	18	20
1992-93	7	2	5	0	26	38
1993-94	5	1	4	0	10	22
1994-95	4	3	0	1	14	9
1995-96	6	2	4	0	17	19
1996-97	6	3	3	0	16	14

Playoffs

	GP	W	L	T	GF	GA
1979-80	3	0	3	0	8	18
1986-87	7	3	4	0	13	16
1987-88	6	2	4	0	20	23
1988-89	4	0	4	0	11	18
1991-92	7	3	4	0	18	21

New Jersey Devils

	GP	W	L	T	GF	GA
1982-83	3	2	1	0	9	6
1983-84	3	1	1	1	13	13
1984-85	3	3	0	0	13	4
1985-86	3	2	1	0	14	15
1986-87	3	1	1	1	16	15
1987-88	3	1	1	1	6	7
1988-89	3	2	1	0	14	11
1989-90	3	2	1	0	16	9
1990-91	3	2	1	0	13	8
1991-92	3	0	3	0	4	16
1992-93	4	1	3	0	14	21
1993-94	4	1	2	1	9	12
1994-95	3	0	1	2	6	7
1995-96	4	1	2	1	8	11
1996-97	4	1	3	0	8	15

New York Islanders

	GP	W	L	T	GF	GA
1979-80	4	1	3	0	9	14
1980-81	4	0	2	2	9	21
1981-82	3	0	2	1	11	20
1982-83	3	1	2	0	7	13
1983-84	3	2	1	0	13	12
1984-85	3	1	1	1	8	8
1985-86	3	1	2	0	9	12
1986-87	3	2	1	0	11	9
1987-88	3	1	2	0	6	13
1988-89	3	2	1	0	14	8
1989-90	3	2	1	0	11	8
1990-91	3	0	2	1	4	13
1991-92	3	1	2	0	9	10
1992-93	4	1	2	1	16	18
1993-94	4	2	1	1	16	11
1994-95	3	2	1	0	12	8
1995-96	4	3	0	1	19	11
1996-97	4	1	2	1	11	14

New York Rangers

	GP	W	L	T	GF	GA
1979-80	4	1	2	1	14	18
1980-81	4	1	3	0	15	24
1981-82	3	1	1	1	10	9
1982-83	3	1	2	0	9	19
1983-84	3	2	1	0	9	8
1984-85	3	1	0	2	10	9
1985-86	3	2	1	0	14	5
1986-87	3	3	0	0	14	9
1987-88	3	1	2	0	6	9
1988-89	3	2	1	0	13	12
1989-90	3	1	2	0	11	14
1990-91	3	1	2	0	11	13
1991-92	3	1	2	0	11	17
1992-93	3	1	1	1	11	14
1993-94	4	1	3	0	7	15
1994-95	3	1	2	0	6	9
1995-96	3	1	2	0	6	9
1996-97	4	2	2	0	13	13

Ottawa Senators

	GP	W	L	T	GF	GA
1992-93	7	6	1	0	34	15
1993-94	5	4	0	1	14	9
1994-95	4	4	0	0	16	7
1995-96	5	3	1	1	13	14
1996-97	5	2	2	1	14	15

Philadelphia Flyers

	GP	W	L	T	GF	GA
1979-80	4	0	2	2	13	19
1980-81	4	0	3	1	12	21
1981-82	3	0	3	0	8	15
1982-83	3	1	2	0	15	19
1983-84	3	2	1	0	17	16
1984-85	3	0	2	1	7	13
1985-86	3	0	3	0	2	11
1986-87	3	2	1	0	10	10
1987-88	3	1	2	0	13	14
1988-89	3	1	1	1	15	14
1989-90	3	2	1	0	9	6
1990-91	3	3	0	0	8	3
1991-92	3	3	0	0	15	8
1992-93	3	0	3	0	10	15
1993-94	4	1	3	0	15	20
1994-95	3	1	2	0	9	10
1995-96	4	0	3	1	6	16
1996-97	4	0	3	1	6	14

Phoenix Coyotes

	GP	W	L	T	GF	GA
1996-97	2	2	0	0	3	1

Pittsburgh Penguins

	GP	W	L	T	GF	GA
1979-80	4	2	1	1	21	15
1980-81	4	2	2	0	20	20
1981-82	3	0	2	1	11	18
1982-83	3	0	3	0	7	14
1983-84	3	1	2	0	10	11
1984-85	3	2	1	0	13	10
1985-86	3	2	1	0	10	12
1986-87	3	3	0	0	18	12
1987-88	3	1	2	0	11	13
1988-89	3	2	1	0	20	15
1989-90	3	2	0	1	15	9
1990-91	3	2	1	0	11	10
1991-92	3	2	0	1	16	11
1992-93	3	0	3	0	8	21
1993-94	5	1	4	0	22	23
1994-95	4	2	2	0	16	15
1995-96	6	2	3	1	18	28
1996-97	5	1	2	2	17	22

Quebec Nordiques

	GP	W	L	T	GF	GA
1979-80	4	1	1	2	20	17
1980-81	4	2	2	0	16	16
1981-82	8	2	3	3	31	36
1982-83	8	2	6	0	31	51
1983-84	8	1	3	4	27	35
1984-85	8	3	5	0	20	36
1985-86	8	4	4	0	33	33
1986-87	8	3	3	2	19	23
1987-88	8	2	6	0	24	34
1988-89	8	4	3	1	28	25
1989-90	8	6	1	1	31	19
1990-91	8	1	3	4	22	35
1991-92	8	3	5	0	22	30
1992-93	7	3	3	1	23	30
1993-94	5	1	4	0	9	18
1994-95	4	1	3	0	6	11
Playoffs						
1985-86	3	3	0	0	16	7
1986-87	6	2	4	0	19	27

St. Louis Blues

	GP	W	L	T	GF	GA
1979-80	4	2	2	0	14	13
1980-81	4	0	3	1	15	25
1981-82	3	2	1	0	10	9
1982-83	3	1	2	0	10	13
1983-84	3	2	1	0	14	7
1984-85	3	2	1	0	10	5
1985-86	3	1	1	1	11	14
1986-87	3	2	1	0	13	8
1987-88	3	1	2	0	7	8
1988-89	3	1	0	2	10	6
1989-90	3	1	2	0	10	11
1990-91	3	0	3	0	5	12
1991-92	3	1	2	0	8	15
1992-93	3	1	2	0	10	13
1993-94	3	0	3	0	4	8
1994-95	did not play					
1995-96	2	0	2	0	5	10
1996-97	2	1	1	0	6	7

San Jose Sharks

	GP	W	L	T	GF	GA
1991-92	3	1	2	0	12	11
1992-93	2	2	0	0	11	7
1993-94	2	0	2	0	4	11
1994-95	did not play					
1995-96	2	1	1	0	9	11
1996-97	2	0	1	1	6	6

Tampa Bay Lightning

	GP	W	L	T	GF	GA
1992-93	2	2	0	0	7	4
1993-94	4	2	2	0	13	13
1994-95	4	1	3	0	6	11
1995-96	4	1	2	1	12	12
1996-97	4	4	0	0	19	11

Coaching Records

National League

Most Career Games
374 — Jack Evans
224 — Larry Pleau
160 — Rick Ley
160 — Paul Holmgren
152 — Paul Maurice
140 — Don Blackburn

80 — Jimmy Roberts
70 — Paul Maurice
67 — Pierre McGuire
49 — Larry Kish
13 — John Cunniff

Most Career Wins
163 — Jack Evans
81 — Larry Pleau
69 — Rick Ley
61 — Paul Maurice
54 — Paul Holmgren
42 — Don Blackburn
29 — Paul Maurice
26 — Jimmy Roberts
23 — Pierre McGuire
12 — Larry Kish
3 — John Cunniff

Most Career Losses
174 — Jack Evans
117 — Larry Pleau
93 — Paul Holmgren
72 — Paul Maurice
71 — Rick Ley
63 — Don Blackburn
41 — Jimmy Roberts
37 — Pierre McGuire
33 — Paul Maurice
32 — Larry Kish
9 — John Cunniff

Most Career Ties
37 — Jack Evans
35 — Don Blackburn
26 — Larry Pleau
20 — Rick Ley
19 — Paul Maurice
14 — Paul Holmgren
13 — Jimmy Roberts
8 — Paul Maurice
7 — Pierre McGuire
5 — Larry Kish
1 — John Cunniff

Regular Season

COACHING RECORD LEADERS

Most Wins in a Season

43 — Jack Evans	1986-87
40 — Jack Evans	1985-86
38 — Rick Ley	1989-90
37 — Larry Pleau	1988-89

Most Losses in a Season

54 — Pleau/Cunniff	1982-83
52 — Paul Holmgren	1992-93
48 — Holmgren/McGuire	1993-94
42 — Jack Evans	1983-84

Most Ties in a Season

19 — Don Blackburn	1979-80
18 — Blackburn/Pleau	1980-81
18 — Larry Pleau	1981-82
13 — Jimmy Roberts	1991-92

*The first player ever signed by the Whalers, **Larry Pleau**, took over the coaching reins of the hockey club during two different segments in the 1980s. In his last tenure with Hartford, Pleau went 37-38-5 in 1988-89. He was given the game after 17 years with the organization. "How can I complain or be angry?" Pleau reasoned after his dismissal. "I owe a lot to the Whalers because they gave me a place to call home for most of my hockey career." These days, Pleau heads up the St. Louis Blues in the position of general manager.*

Forever Whalers

Toronto Maple Leafs

	GP	W	L	T	GF	GA
1979-80	4	2	2	0	16	15
1980-81	4	1	1	2	19	17
1981-82	3	3	0	0	20	11
1982-83	3	1	2	0	11	17
1983-84	3	2	1	0	19	13
1984-85	3	2	1	0	16	12
1985-86	3	3	0	0	17	5
1986-87	3	2	1	0	13	9
1987-88	3	3	0	0	11	5
1988-89	3	2	1	0	10	7
1989-90	3	1	1	1	18	15
1990-91	3	1	0	2	9	7
1991-92	2	1	1	0	3	4
1992-93	2	0	2	0	5	9
1993-94	2	0	1	1	5	10
1994-95	did not play					
1995-96	2	2	0	0	11	4
1996-97	2	1	0	1	4	2

Vancouver Canucks

	GP	W	L	T	GF	GA
1979-80	4	1	1	2	15	15
1980-81	4	1	1	2	12	14
1981-82	3	0	2	1	5	10
1982-83	3	1	2	0	11	19
1983-84	3	0	3	0	6	16
1984-85	3	1	2	0	12	13
1985-86	3	3	0	0	14	11
1986-87	3	2	0	1	13	9
1987-88	3	1	0	2	9	7
1988-89	3	1	1	1	5	5
1989-90	3	2	1	0	5	5
1990-91	3	2	1	0	12	10
1991-92	3	0	2	1	7	12
1992-93	2	1	1	0	8	9
1993-94	2	1	1	0	5	6
1994-95	did not play					
1995-96	2	2	0	0	8	3
1996-97	2	0	2	0	5	11

Washington Capitals

	GP	W	L	T	GF	GA
1979-80	4	1	2	1	16	22
1980-81	4	1	3	0	13	19
1981-82	3	1	2	0	7	13
1982-83	3	0	2	1	10	15
1983-84	3	1	2	0	8	8
1984-85	3	2	1	0	7	4
1985-86	3	0	2	1	12	15
1986-87	3	2	1	0	9	9
1987-88	3	1	2	0	10	14
1988-89	3	0	3	0	4	12
1989-90	3	2	1	0	10	9
1990-91	3	3	0	0	10	6
1991-92	3	1	1	1	12	7
1992-93	4	1	3	0	13	18
1993-94	4	2	2	0	12	9
1994-95	4	1	2	1	8	15
1995-96	4	1	2	1	6	9
1996-97	4	1	2	1	8	9

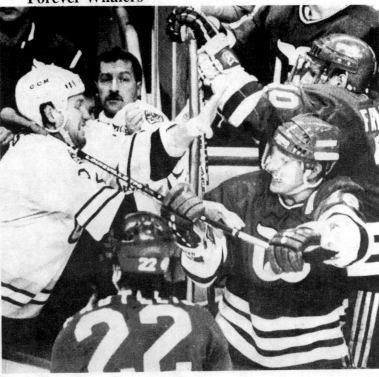

The action here occurred in Boston Garden on November 23, 1990 when **Pat Verbeek** *put the lumber to Lyndon Byers of the Bruins. Hartford won this game, 4-3, one of the rare times the Whalers escaped the cozy barn on Causeway Street with two points for 60 minutes of work.*

Winnipeg Jets

	GP	W	L	T	GF	GA
1979-80	4	2	2	0	14	16
1980-81	4	3	0	1	24	15
1981-82	3	2	1	0	15	8
1982-83	3	1	1	1	7	7
1983-84	3	0	2	1	10	15
1984-85	3	1	2	0	10	13
1985-86	3	2	1	0	19	8
1986-87	3	1	1	1	7	7
1987-88	3	2	1	0	11	12
1988-89	3	2	1	0	15	11
1989-90	3	1	2	0	7	9
1990-91	3	1	2	0	12	13
1991-92	3	2	0	1	7	5
1992-93	3	1	2	0	16	16
1993-94	2	1	0	1	6	2
1994-95	did not play					
1995-96	2	1	1	0	5	7

A couple of looks at the aftermath of the Hartford Civic Center roof collapse on January 18, 1978. The building, which seated 10,507 for hockey, was the home rink for the New England Whalers for roughly three WHA seasons until a severe winter storm destroyed the facility. Rick Sentes of the San Diego Mariners scored the first goal in the arena which opened January 11, 1975. **Don Blackburn** had the first goal for the Whalers. The last WHA tally in the building was scored by Houston's Rich Preston in a 5-4 overtime win over New England on January 14, 1978.

The loss of the Civic Center renewed the spirit and vigor of the Greater Hartford community. The I-91 Fan Club was formed as boosters and rooters of the Whalers continued to follow the team as it relocated for two seasons in nearby Springfield. The faithful cheered at the Springfield Civic Center where their heroes closed out the club's WHA history and also began their first National Hockey League campaign in 1979-80. Once the arena in Hartford was reconstructed and cleared for occupancy, the "new" Hartford Civic Coliseum seated 15,635 for hockey.

The first NHL game was February 6, 1980 and the Whalers crowned the Los Angeles Kings, 7-3. The first NHL goal scored in the building was by Mike Murphy of the Kings. **Blaine Stoughton** had the first strike for the Whalers. The last game was played April 13, 1997. Hartford went out a winner, 2-1 over Tampa Bay. The last NHL goal scored in the Coliseum was by Dino Ciccarelli of Tampa Bay. The last Hartford tally was by captain **Kevin Dineen**.

Playoffs: NHL Stanley Cup Series

1980 vs. Montreal

Date	Score	Game-Winner	Goalies & Saves	Attendance
April 8	1-6	Brian Engblom	Smith 31, Heron 20	15,992
April 9	4-8	Bob Gainey	Garrett 32, Heron 23	15,242
April 11	3-4 ot	Yvon Lambert (:29)	Smith 24, Larocque 25	14,460

1986 vs. Quebec

April 9	3-2 ot	Sylvain Turgeon (2:36)	Liut 37, Malarchuk 3	14,504
April 10	4-1	Paul MacDermid	Liut 26, Malarchuk 27	14,562
April 12	9-4	Ulf Samuelsson	Liut 29, Malarchuk 9, Gosselin 17	15,126

1986 vs. Montreal

April 17	4-1	Sylvain Turgeon	Liut 26, Roy 20	17,145
Apil 19	1-3	Guy Carbonneau	Liut 27, Roy 24	17,657
April 21	1-4	Stephane Richer	Liut 5, Weeks 13; Roy 20	15,126
April 23	2-1 ot	Kevin Dineen (1:06)	Weeks 18, Roy 22	15,126
April 25	3-5	Mike Lalor	Weeks 25, Roy 20	17,660
April 27	1-0	Kevin Dineen	Liut 32, Roy 16	15,126
April 29	1-2 ot	Claude Lemieux (5:55)	Liut 30, Roy 24	17,546

1987 vs. Quebec

April 8	3-2 ot	Paul MacDermid (2:20)	Liut 23, Gosselin 38	15,126
April 9	5-4	Dana Murzyn	Liut 20, Malarchuk 24	15,126
April 11	1-5	Peter Statsny	Liut 11, Weeks 21; Gosselin 29	14,527
April 12	1-4	Michel Goulet	Liut 33, Gosselin 28	14,389
April 14	5-5	John Ogrodnick	Liut 17, Gosselin 38	15,126
April 16	4-5 ot	Peter Stastny (6:05)	Liut 29, Gosselin 31	15,383

1988 vs. Montreal

April 6	3-4	Stephane Richer	Liut 22, Brodeur 9; Roy 21	16,523
April 7	3-7	Claude Lemieux	Brodeur 21, Roy 28	16,640
April 9	3-4	Ryan Walter	Liut 24, Roy 30	15,223
April 10	7-5	Kevin Dineen	Liut 26, Roy 36	15,061
April 12	3-1	Carey Wilson	Brodeur 22, Hayward 23	16,687
April 14	1-2	Stephane Richer	Brodeur 23, Hayward 19	15,223

1989 vs. Montreal

April 5	2-6	Ryan Walter	Whitmore 27, Roy 25	16,005
April 6	2-3	Mike McPhee	Sidorkiewicz 14, Roy 27	16,563
April 8	4-5 ot	Stephane Richer (5:01)	Sidorkiewicz 23, Hayward 23	13,363
April 9	3-4 ot	Russ Courtnall (15:12)	Whitmore 36, Roy 43	12,245

1990 vs. Boston

April 5	4-3	Randy Ladouceur	Sidorkiewicz 33, Lemelin 13	14,448
April 7	1-3	Garry Galley	Sidorkiewicz 32, Moog 25	14,448
April 9	5-3	Pat Verbeek	Sidorkiewicz 32, Moog 22	15,535
April 11	5-6	Dave Poulin	Sidorkiewicz 23; Lemelin 17, Moog 7	15,535
April 13	2-3	Cam Neely	Sidorkiewicz 15, Moog 20	14,448
April 15	3-2 ot	Kevin Dineen (12:30)	Sidorkiewicz 23, Moog 23	15,535
April 17	1-3	Craig Janney	Sidorkiewicz 22, Moog 27	14,448

1991 vs. Boston

April 3	5-2	Rob Brown	Sidorkiewicz 26, Moog 19	14,448
April 5	3-4	Cam Neely	Sidorkiewicz 36, Moog 24	14,448
April 7	3-6	Dave Christian	Sidorkiewicz 17, Moog 22	15,635
April 9	4-3	Zarley Zalapski	Sidorkiewicz 32, Moog 30	14,198
April 11	1-6	Dave Christian	Sidorkiewicz 26, Moog 25	14,448
April 13	1-3	Bob Sweeney	Sidorkiewicz 13, Mood 31	15,635

1992 vs. Montreal

April 19	0-2	Gilbert Dionne	Pietrangelo 28, Roy 32	16,624
April 21	2-5	Kirk Muller	Pietrangelo 23, Roy 15	16,627
April 23	5-2	John Cullen	Pietrangelo 32, Roy 28	6,728
April 25	3-1	Randy Cunneyworth	Pietrangelo 24, Roy 33	10,071
April 27	4-7	Kirk Muller	Pietrangelo 23, Whitmore 4; Roy 19	16,693
April 29	2-1 ot	Yvon Corriveau (:24)	Pietrangelo 42, Roy 24	8,262
May 1	2-3 ot*	Russ Courtnall (25:26)	Pietrangelo 53, Roy 39	17,718

* double overtime (longest playoff game ever played by the Whalers)

Stanley Cup Playoffs

SCORING LEADERS

Games Played

Kevin Dineen	38
Dean Evason	38
Ron Francis	33
Ray Ferraro	33
Randy Ladouceur	33
Dave Tippett	33
Dave Babych	31
Ulf Samuelsson	31
Paul MacDermid	26
Joel Quenneville	26
Sylvain Turgeon	25

Goals

Kevin Dineen	17
Dean Evason	8
Ron Francis	8
Stewart Gavin	8
Dave Babych	7
Ray Ferraro	7

Assists

Dean Evason	15
Kevin Dineen	14
Ron Francis	14
Dave Babych	13
Ray Ferraro	11
John Anderson	11

Points

Kevin Dineen	31
Dean Evason	23
Ron Francis	22
Dave Babych	20
Ray Ferraro	18
John Anderson	17
Stewart Gavin	13
John Cullen	12
Brad Shaw	12

Penalty Minutes

Ulf Samuelsson	109
Kevin Dineen	101
Paul MacDermid	84
Pat Verbeek	78
Torrie Robertson	67

Game-Winning Goals

Kevin Dineen	4
Sylvain Turgeon	2
Paul MacDermid	2

The Whalers qualified for the Stanley Cup playoffs in just eight of their 18 years in the National Hockey League. No matter the odds, there were always fans like this loyalist, hoping the Whalers could make a run at Lord Stanley's bowl of silver.

GOALTENDING LEADERS

Games Played

Mike Liut	17
Peter Sidorkiewicz	15
Frank Pietrangelo	7

Minutes Played

Mike Liut	933
Peter Sidorkiewicz	912
Frank Pietrangelo	425

Wins

Mike Liut	8
Peter Sidorkiewicz	5
Frank Pietrangelo	3

Shutouts

Mike Liut	1

Goals-Against Average

Steve Weeks	2.63
Frank Pietrangelo	2.68
Mike Liut	3.21
Richard Brodeur	3.60

Assists

Peter Sidorkiewicz	1

Penalty Minutes

Mike Liut	2
Peter Sidorkiewicz	2

Losses

Peter Sidorkiewicz	10
Mike Liut	7
Frank Pietrangelo	4

COACHING LEADERS

Games

16	Jack Evans
13	Rick Ley
10	Larry Pleau
7	Jimmy Roberts
3	Don Blackburn

Records

Jack Evans	8-8
Rick Ley	5-8
Jimmy Roberts	3-4
Larry Pleau	2-8
Don Blackburn	0-3

Whalers vs. WHA teams (1972-1979)

World Hockey Association (1972-1979)

Regular Season

SCORING LEADERS

Baltimore Blades

	GP	W	L	T	GF	GA
1974-75	2	2	0	0	8	5

Birmingham Bulls

	GP	W	L	T	GF	GA
1975-76	4	4	0	0	22	13
1976-77	9	5	4	0	38	34
1977-78	11	9	2	0	54	28
1978-79	14	5	7	2	57	56

Calgary Cowboys

	GP	W	L	T	GF	GA
1974-75	6	3	3	0	14	20
1975-76	5	3	2	0	22	19
1976-77	6	5	1	0	30	11

Chicago Cougars

	GP	W	L	T	GF	GA
1972-73	6	5	1	0	22	15
1973-74	8	4	4	0	26	27
1974-75	5	4	1	0	27	20
Playoffs						
1973-74	7	3	4	0	23	24

Cincinnati Stingers

	GP	W	L	T	GF	GA
1975-76	11	5	6	0	41	43
1976-77	9	5	4	0	30	38
1977-78	11	8	3	0	55	41
1978-79	15	6	7	2	39	45
Playoffs						
1978-79	3	2	1	0	10	10

Cleveland Crusaders

	GP	W	L	T	GF	GA
1972-73	8	5	3	0	23	20
1973-74	8	5	3	0	23	17
1974-75	6	3	2	1	16	17
1975-76	10	6	3	1	34	28
Playoffs						
1972-73	5	4	1	0	16	14
1975-76	3	3	0	0	14	6

Czechoslovakia

	GP	W	L	T	GF	GA
1977-78	1	1	0	0	5	3
1978-79	1	1	0	0	10	4

Denver Spurs-Ottawa Civics

	GP	W	L	T	GF	GA
1975-76	3	2	1	0	11	10

Games Played

Rick Ley	478
Larry Pleau	468
Brad Selwood	431
Tom Webster	352
Tommy Earl	347
Alan Hangsleben	334
Gordie Roberts	311
Mike Rogers	274
Al Smith	260
Paul Hurley	251
Tim Sheehy	232
John French	226
Ted Green	210
Thommy Abrahamsson	203
George Lyle	202

Penalty Minutes

Rick Ley	716
Brad Selwood	556
Gordie Roberts	502
Alan Hangsleben	437
Jack Carlson	334
Jim Dorey	244
Tom Webster	241
Nick Fotiu	238

Points

Tom Webster	425
Larry Pleau	372
Mike Rogers	257
Rick Ley	245
Mark Howe	198
Gordie Roberts	186
Brad Selwood	185
John French	184
Tim Sheehy	181
Dave Keon	166
George Lyle	161
Terry Caffery	152
Gordie Howe	139
Mike Antonovich	135
John McKenzie	133
Jim Dorey	131
Mike Byers	115
Don Blackburn	114
Alan Hangsleben	110
Wayne Carleton	107
Ted Green	99

Goals

Tom Webster	220
Larry Pleau	157
Mike Rogers	98
Tim Sheehy	98
George Lyle	86
Mark Howe	72
Mike Antonovich	64
Mike Byers	61
Dave Keon	60
John McKenzie	57
Terry Caffery	54
Gordie Howe	53

Assists

Larry Pleau	215
Rick Ley	210
Tom Webster	205
Mike Rogers	159
Gordie Roberts	144
Brad Selwood	143
Mark Howe	126
John French	124
Jim Dorey	113
Dave Keon	106
Terry Caffery	98

GOALTENDING LEADERS

Games Played

Al Smith	260
Bruce Landon	122
Christer Abrahamsson	102
John Garrett	41
Cap Raeder	29
Louie Levasseur	27

Minutes Played

Al Smith	15,389
Bruce Landon	6,695
C. Abrahamsson	5,739
John Garrett	2,496
Louie Levasseur	1,665
Cap Raeder	1,428

Shutouts

Al Smith	10
Christer Abrahamsson	3
Louie Levasseur	3

Wins

Al Smith	141
Bruce Landon	50
Christer Abrahamsson	41
John Garrett	20
Louie Levasseur	14
Cap Raeder	12

Losses

Al Smith	98
Bruce Landon	50
Christer Abrahamsson	46
John Garrett	17
Cap Raeder	11
Louie Levasseur	11

Ties

Al Smith	15
Bruce Landon	9
Christer Abrahamsson	7
John Garrett	4

Goals-Against Average

Cap Raeder	3.24
Al Smith	3.25
Louie Levasseur	3.46
Bruce Landon	3.46
Christer Abrahamsson	3.58
John Garrett	3.58

Assists

Al Smith	10
Bruce Landon	2
Christer Abrahamsson	2

Penalty Minutes

Al Smith	129
Bruce Landon	36
Christer Abrahamsson	18
John Garrett	6

Edmonton Oilers

	GP	W	L	T	GF	GA
1972-73	6	3	3	0	21	18
1973-74	6	3	2	1	24	21
1974-75	6	2	4	0	22	26
1975-76	4	2	1	1	11	12
1976-77	6	2	4	0	19	21
1977-78	11	7	3	1	41	35
1978-79	14	4	9	1	33	53
Playoffs						
1977-78	5	4	1	0	23	9
1978-79	7	3	4	0	29	35

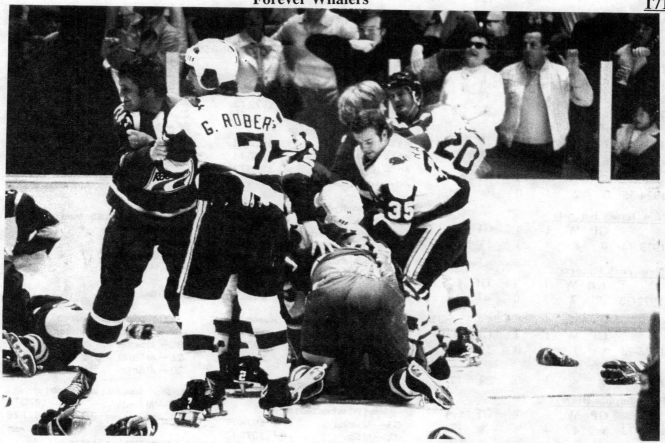

It's playoff action at the Civic Center in the spring of 1976 and members of the Whalers and Indianapolis Racers square off in this melee. New England prevailed in a seven-game series before expiring in seven games to Houston in the semifinals. The standout for the locals in the postseason was goalie Cap Raeder (center) who posted a 2.17 GAA in 14 AVCO Trophy games.

Houston Aeros

	GP	W	L	T	GF	GA
1972-73	6	2	3	1	21	26
1973-74	6	4	2	0	21	17
1974-75	6	2	4	0	18	30
1975-76	6	2	4	0	17	26
1976-77	6	3	2	1	18	17
1977-78	11	4	6	1	42	37
Playoffs						
1975-76	7	3	4	0	21	19

Indianapolis Racers

	GP	W	L	T	GF	GA
1974-75	6	5	1	0	27	10
1975-76	12	2	7	3	31	44
1976-77	10	1	7	2	25	41
1977-78	11	7	2	2	49	31
1978-79	4	3	0	1	25	16
Playoffs						
1975-76	7	4	3	0	18	15

Los Angeles Sharks

	GP	W	L	T	GF	GA
1972-73	6	3	3	0	22	20
1973-74	6	5	1	0	31	15

Coaching Records

World Hockey Association
Regular Season

Most Wins in a Season

46 — Jack Kelley	1972-73
44 — Harry Neale	1977-78
43 — Ron Ryan	1973-74
43 — Ron Ryan/Jack Kelley	1974-75

Most Losses in a Season

40 — Kelley/Blackburn/Neale 1975-76
40 — Harry Neale   1976-77

AVCO Trophy Playoffs

Games

36 — Harry Neale
21 — Jack Kelley
 9 — Don Blackburn
 7 — Ron Ryan

Records

Jack Kelley	21-14
Harry Neale	19-17
Don Blackburn	5-4
Ron Ryan	3-4

Most Ties in a Season

9 — Dineen/Blackburn   1978-79
7 — Kelley/Blackburn/Neale   1975-76

Most Career Games

173 — Harry Neale
151 — Ron Ryan
116 — Jack Kelley
 71 — Bill Dineen
 44 — Don Blackburn

Most Career Wins

84 — Harry Neale
83 — Ron Ryan
63 — Jack Kelley
33 — Bill Dineen
18 — Don Blackburn

Most Career Losses

76 — Harry Neale
59 — Ron Ryan
48 — Jack Kelley
30 — Bill Dineen
22 — Don Blackburn

Most Career Ties

12 — Harry Neale
 9 — Ron Ryan

Minnesota Fighting Saints

	GP	W	L	T	GF	GA
1972-73	6	3	3	0	23	20
1973-74	6	1	4	1	21	26
1974-75	6	2	4	0	18	26
1975-76	3	1	1	1	3	4
1976-77	6	1	3	2	16	20
Playoffs						
1974-75	6	2	4	0	18	28

Michigan Stags

	GP	W	L	T	GF	GA
1974-75	3	2	1	0	9	4

New Jersey Knights

	GP	W	L	T	GF	GA
1973-74	10	8	1	1	39	21

New York Raiders

	GP	W	L	T	GF	GA
1972-73	10	7	3	0	47	36

Ottawa Nationals

	GP	W	L	T	GF	GA
1972-73	8	5	3	0	33	29
Playoffs						
1972-73	5	4	1	0	24	17

Philadelphia Blazers

	GP	W	L	T	GF	GA
1972-73	8	4	4	0	38	33

Phoenix Roadrunners

	GP	W	L	T	GF	GA
1974-75	6	1	3	2	17	23
1975-76	6	2	3	1	24	23
1976-77	6	4	2	0	25	14

Quebec Nordiques

	GP	W	L	T	GF	GA
1972-73	8	4	3	1	36	27
1973-74	8	3	5	0	22	31
1974-75	6	4	2	0	25	28
1975-76	5	1	4	0	7	20
1976-77	11	2	8	1	33	52
1977-78	12	4	8	0	48	45
1978-79	16	11	4	1	65	54
Playoffs						
1976-77	5	1	4	0	14	23
1977-78	5	4	1	0	25	14

San Diego Mariners

	GP	W	L	T	GF	GA
1974-75	6	3	3	0	20	22
1975-76	6	3	3	0	23	27
1976-77	6	5	1	0	21	14

Soviet All-Stars

	GP	W	L	T	GF	GA
1977-78	1	1	0	0	7	2
1978-79	1	0	1	0	4	7

Toronto Toros

	GP	W	L	T	GF	GA
1973-74	8	3	4	1	33	36
1974-75	6	5	1	0	25	20

WHA Superlatives

Regular Season

100 Points in a Season

107 — Mark Howe	1972-73
103 — Tom Webster	1972-73
100 — Terry Caffery	1972-73

Most Goals in a Season

53 — Tom Webster	1972-73
43 — Tom Webster	1973-74
42 — Mark Howe	1978-79
40 — Tom Webster	1974-75

Most Penalty-Minutes in a Season

192 — Jack Carlson	1977-78
169 — Gordie Roberts	1976-77
148 — Alan Hangsleben	1978-79
148 — Rick Ley	1973-74
144 — Nick Fotiu	1974-75

SCORING LEADERS

Most Assists in a Season

65 — Mark Howe	1978-79
62 — Gordie Howe	1977-78
61 — Terry Caffery	1972-73
61 — Mark Howe	1977-78
57 — Mike Rogers	1976-77
56 — Andre Lacroix	1978-79
56 — Jim Dorey	1972-73
50 — Tom Webster	1972-73
50 — Tom Webster	1975-76

Most Games in a Season

80 — Rosaire Palement	1975-76
80 — Mike Rogers	1977-78
80 — Brad Selwood	1977-78
80 — Mike Rogers	1978-79

GOALTENDING LEADERS

Most Appearances in a Season

59 — Al Smith	1974-75
55 — Al Smith	1977-78
50 — Al Smith	1972-73
50 — Al Smith	1973-74
45 — Christer Abrahamsson	1976-77

Most Wins in a Season

33 — Al Smith	1974-75
31 — Al Smith	1972-73
30 — Al Smith	1973-74
30 — Al Smith	1977-78
20 — John Garrett	1978-79
18 — Christer Abrahamsson	1975-76

Lowest GAA in a Season

3.08 — Al Smith	1973-74
3.12 — Cap Raeder	1976-77
3.18 — Al Smith	1972-73
3.22 — Al Smith	1977-78
3.24 — C. Abrahamsson	1974-75

Most Losses in a Season

22 — Christer Abrahamsson	1976-77
21 — Al Smith	1973-74
21 — Al Smith	1974-75
20 — Al Smith	1977-78

Most Shutouts in a Season

3 — Al Smith	1972-73
3 — Louie Levasseur	1977-78

Most Ties in a Season

5 — Bruce Landon	1975-76
5 — Al Smith	1978-79

Most Assists in a Season

3 — Al Smith	1973-74
3 — Al Smith	1977-78
3 — Al Smith	1978-79

Most Penalty Minutes in a Season

37 — Al Smith	1972-73
35 — Al Smith	1978-79
33 — Al Smith	1973-74
18 — Al Smith	1974-75
12 — Christer Abrahamsson	1976-77

Vancouver Blazers

	GP	W	L	T	GF	GA
1973-74	6	4	2	0	33	27

Winnipeg Jets

	GP	W	L	T	GF	GA
1972-73	6	5	1	0	32	19
1973-74	6	3	3	0	18	22
1974-75	6	4	2	0	23	25
1975-76	4	0	4	0	7	20
1976-77	6	2	4	0	20	28
1977-78	11	3	7	1	34	47
1978-79	14	7	6	1	63	52
Playoffs						
1972-73	5	4	1	0	30	18
1977-78	4	0	4	0	8	24

Whalers 9, Jets 6

May 6, 1973
at Boston Garden

WINNIPEG	2 2 2 -	6
NEW ENGLAND	5 1 3 -	9

FIRST PERIOD

1, New England, Webster 11 (Williams, Ley), :21.

2, New England, Pleau 10 (unassisted), 4:43 (sh).

3, Winnipeg, Johnson 4 (Sutherland), 7:07.

4, New England, G.Smith (Webster, Williams), 11:47.

5, New England, Ley 3 (unassisted), 15:43.

6, Winnipeg, Beaudin 12 (Hull), 17:53.

7, New England, Sheehy 9 (Pleau, Webster), 18:41 (pp).

Penalties: Green, New England (charging), 3:04; Asmundson, Winnipeg (holding), 18:32.

SECOND PERIOD

8, New England, Webster 12 (Williams, Green), :15.

9, Winnipeg, Beaudin 13 (McDonald, Bordeleau), 3:15 (pp).

10, Winnipeg, Black 1 (Shymr), 4:02.

Penalties: Green, New England (cross-checking), 1:41; Asmundson, Winnipeg (interference), 5:49; Shmyr, Winnipeg (holding), 9:53; Green, New England (charging), 19:34.

THIRD PERIOD

11, Winnipeg, Woytowich 1 (Asmundson, Swenson), 4:59.

12, New England, Pleau 11 (Sheehy, French), 5:44.

13, New England, Pleau 12 (French, Sheehy), 7:31.

14, New England, Byers 6 (Green, Williams), 17:20 (en).

15, Winnipeg, Asmundson 1 (Cuddie, Swenson), 18:10.

Penalties: None.

SHOTS ON GOAL

Winnipeg	17	12	13 -	42
New England	11	6	10 -	27

POWER-PLAY OPPORTUNTIES

Winnipeg 1-3, New England 1-3

GOALIES

Winnipeg, Daley (17 saves)

New England, Smith (36 saves)

ATTENDANCE 11,186

REFEREE Bill Friday

*Goalie **Al Smith** reacts after the Whalers lock up the AVCO Cup. "We started from scratch and got all the way here to the victory," Smith said after New England bested Winnipeg for the WHA's first-ever championship. "We had a great season."*

Playoffs: WHA AVCO Trophy Series

1973 vs. Ottawa

Date	Score	Game-Winner	Winning Goalie	Attendance
April 7	6-3	Mike Byers	Smith	9,359
April 8	4-3 ot	Brit Selby (3:37)	Smith	6,156
April 10	2-4	Guy Trottier	Binkley	4,879
April 12	7-3	Tommy Williams	Smith	3,941
April 14	5-4 ot	Mike Byers (5:47)	Smith	12,033

1973 vs. Cleveland

April 18	3-2	Tim Sheehy	Smith	6,101
April 19	3-2	Larry Pleau	Smith	7,119
April 21	5-4	Tim Sheehy	Smith	8,391
April 22	2-5	Gary Jarrett	Cheevers	4,183
April 26	3-1	Tommy Earl	Smith	7,689

1973 vs. Winnipeg

April 29	7-2	Tommy Earl	Smith	6,526
May 2	7-4	John French	Smith	8,655
May 3	3-4	Bobby Hull	Daley	7,200
May 5	4-2	Tim Sheehy	Smith	13,697
May 6	9-6*	Larry Pleau	Smith	11,186

* Whalers win league championship

1974 vs. Chicago

April 6	6-4	Jake Danby	Smith	5,516
April 7	4-3 ot	John French (2:51)	Smith	5,516
April 9	6-8	Bob Liddington	Newton	3,125
April 10	1-2 ot	Ralph Backstrom (17:45)	Newton	3,425
April 12	2-4	Larry Mavety	Gill	5,516
April 14	2-0	John French	Smith	4,253
April 16	2-3	Jan Popiel	Gill	5,516

1975 vs. Minnesota

April 9	5-6	Wayne Connelly	Garrett	9,722
April 11	3-2 ot	Rick Ley (6:46)	Smith	10,507
April 13	3-8	Danny Walton	Garrett	10,507
April 15	5-2	Tommy Earl	Smith	11,093
April 18	0-4	Fran Huck	Garrett	10,507
April 19	1-6	Gary Gambucci	Garrett	11,870

1976 vs. Cleveland

April 9	5-3	Larry Pleau	Landon	8,417
April 10	6-1	Freddie O'Donnell	Landon	4,243
April 11	3-2	Mike Rogers	Landon	3,584

1976 vs. Indianapolis

April 16	4-1	Rosaire Paiement	Raeder	10,834
April 17	0-4	Kerry Bond	Park	10,111
April 21	3-0	Alan Hangsleben	Raeder	9,681
April 23	2-1	Mike Rogers	Raeder	10,507
April 27	0-4	Reg Thomas	Park	11,811
April 27	3-5	Dave Keon	Park	10,507
April 29	6-0	Danny Bolduc	Raeder	16,040

1976 vs. Houston

May 5	4-2	Rosaire Paiement	Raeder	9,331
May 7	2-5	Ted Taylor	Rutledge	12,739
May 9	4-1	Don Borgeson	Raeder	10,507
May 11	3-4	Terry Ruskowski	Grahame	10,507
May 13	2-4	Terry Ruskowski	Grahame	14,623
May 15	6-1	Nick Fotiu	Raeder	10,507
May 16	0-2	Poul Popiel	Grahame	14,718

1977 vs. Quebec

April 9	2-5	Francois Lacombe	Brodeur	8,090
April 12	3-7	Serge Bernier	Brodeur	8,596
April 14	3-4 ot	Paul Baxter (1:50)	Brodeur	9,223
April 16	6-4	George Lyle	Landon	10,507
April 19	0-3	Andre Boudrias	Brodeur	6,388

SCORING LEADERS

Games Played

Rick Ley	73
Larry Pleau	66
Brad Selwood	63
Alan Hangsleben	47
Mike Rogers	46
Gordie Roberts	46
Tommy Earl	46
Tom Webster	43
Al Smith	35
Tim Sheehy	35

Penalty Minutes

Rick Ley	142
Alan Hangsleben	97
Nick Fotiu	84
Brad Selwood	81
Jim Dorey	67
Garry Swain	56
George Lyle	42
Rosaire Paiement	41

Goals

Larry Pleau	29
Tom Webster	28
Mike Antonovich	17
Tim Sheehy	14
Mike Rogers	13
Mark Howe	12
John McKenzie	11
Dave Keon	11
Mike Byers	10

Assists

Rick Ley	33
Tom Webster	26
Larry Pleau	22
Jim Dorey	22
Dave Keon	21
Gordie Roberts	20
Tim Sheehy	19

Points

Tom Webster	54
Larry Pleau	51
Rick Ley	40
Mike Rogers	34
Tim Sheehy	33
Dave Keon	32
Mike Antonovich	29
John McKenzie	25
Jim Dorey	25
John French	23
Mark Howe	21
Mike Byers	21
Tommy Williams	20

Game-Winning Goals

Larry Pleau	4
Tim Sheehy	3
Tommy Earl	3
Dave Keon	3
John French	3

It's celebration time after the Whalers win the AVCO Trophy in 1973. Here, managing general partner **Howard L. Baldwin** (left) and team owner **Robert Schmertz** take part in the postgame title party.

GOALTENDING LEADERS

Games Played

Al Smith	35
Cap Raeder	15
Louie Levasseur	12
Bruce Landon	8
John Garrett	8

Minutes Played

Al Smith	1,947
Cap Raeder	879
Louie Levasseur	719
John Garrett	447
Bruce Landon	441

Goals-Against Average

Bruce Landon	1.90
Cap Raeder	2.59
Louie Levasseur	2.59
Al Smith	3.82

Shutouts

Cap Raeder	2
Bruce Landon	1

Wins

Al Smith	18
Louie Levasseur	8
Cap Raeder	7
Bruce Landon	4
John Garrett	4

Losses

Al Smith	15
Cap Raeder	8
Louie Levasseur	4
John Garrett	3
Bruce Landon	2

Assists

Al Smith	1
Cap Raeder	1
Bruce Landon	1
Louie Levasseur	1

Penalty Minutes

Al Smith	14
Cap Raeder	2
Louie Levasseur	2

1978 vs. Edmonton

April 14	6-4	Marty Howe	Levasseur	6,530
April 16	4-1	John McKenzie	Levasseur	6,784
April 19	0-2	Pierre Guite	Dryden	11,924
April 21	9-1	Gordie Howe	Levasseur	14,888
April 23	4-1	Mike Antonovich	Levasseur	6,212

1978 vs. Quebec

April 28	5-1	Ron Plumb	Levasseur	7,206
April 30	2-3	Steve Sutherland	Brodeur	7,448
May 3	5-4	Larry Pleau	Levasseur	11,690
May 5	7-3	Dave Keon	Levasseur	11,751
May 7	6-3	Gordie Howe	Levasseur	8,125

1978 vs. Winnipeg (League Finals)

May 12	1-4	Peter Sullivan	Bromley	8,125
May 14	2-5	Bob Guindon	Daley	8,125
May 19	2-10	Bobby Hull	Bromley	10,250
May 22	3-5	Anders Hedberg	Daley	10,348

1979 vs. Cincinnati

April 21	5-3	Mark Howe	Garrett	5,922
April 22	3-6	Robbie Ftorek	Liut	5,131
April 24	2-1	Blaine Stoughton	Smith	5,492

1979 vs. Edmonton

April 26	2-6	Dennis Sobchuk	Dryden	11,102
April 27	5-9	Brett Callighen	Dryden	13,349
April 29	4-1	George Lyle	Garrett	5,571
May 1	5-4	John McKenzie	Garrett	5,492
May 3	2-5	Wayne Gretzky	Dryden	11,401
May 6	8-4	Dave Keon	Garrett	7,320
May 8	3-6	"Cowboy" Bill Flett	Dryden	13,385

Composite Season Statistics (1972-1997)

Whalers, Year by Year

Season	GP	W	L	T	Pts	GF	GA	HOME W	L	T	AWAY W	L	T
1972-73	78	46	30	2	94	318	263	30	8	1	16	22	1
1973-74	78	43	31	4	90	291	260	26	11	2	17	20	2
1974-75	78	43	30	5	91	274	279	28	8	3	15	22	2
1975-76	80	33	40	7	73	255	290	22	16	2	11	24	5
1976-77	81	35	40	6	76	275	290	20	16	4	15	24	2
1977-78	80	44	31	5	93	335	269	26	14	1	18	17	4
1978-79	80	37	34	9	83	298	287	22	19	0	15	15	9
1979-80	80	27	34	19	73	303	312	22	12	6	5	22	13
1980-81	80	21	41	18	60	292	372	14	17	9	7	24	9
1981-82	80	21	41	8	60	264	351	13	17	10	8	24	8
1982-83	80	19	54	7	45	261	403	13	22	5	6	32	2
1983-84	80	28	42	10	66	288	320	19	16	5	9	26	5
1984-85	80	30	41	9	69	268	318	17	18	5	13	23	4
1985-86	80	40	36	4	84	332	302	21	17	2	19	19	2
1986-87	80	43	30	7	93	287	270	26	9	5	17	21	2
1987-88	80	35	38	7	77	249	267	21	14	5	14	24	2
1988-89	80	37	38	5	79	299	290	21	17	2	16	21	3
1989-90	80	38	33	9	85	275	268	17	18	5	21	15	4
1990-91	80	31	38	11	73	238	276	18	16	6	13	22	5
1991-92	80	26	41	13	65	247	283	13	17	10	13	24	3
1992-93	84	26	52	6	58	284	369	12	25	5	14	27	1
1993-94	84	27	48	9	63	227	288	14	22	6	13	26	3
1994-95	48	19	24	5	43	127	141	12	10	2	7	14	3
1995-96	82	34	39	9	73	237	259	22	15	4	12	24	5
1996-97	82	32	39	11	75	226	256	23	15	3	9	24	8

Regular Season and Playoff Attendance

Season	Games	Attendance	Average	Games	Attendance	Average
1972-73	39	272,255	6,981	9	79,866	8,874
1973-74	39	232,814	5,970	4	22,064	5,516
1974-75	39	305,959	7,169	3	30,521	10,174
1975-76	40	372,334	9,308	7	70,633	10,090
1976-77	40	359,263	8,982	2	19,730	9,865
1977-78	44	381,075	8,661	8	58,555	7,319
1978-79	41	284,608	6,942	6	34,928	5,821
1979-80	40	394,228	9,856	1	14,460	14,460
1980-81	40	461,008	11,525	...	....	......
1981-82	40	456,510	11,413	...	....	......
1982-83	40	427,819	10,695	...	....	......
1983-84	40	459,525	11,488	...	....	......
1984-85	40	481,059	12,026	...	....	......
1985-86	40	510,753	12,769	4	60,504	15,126
1986-87	40	569,219	14,230	3	45,378	15,126
1987-88	40	582,969	14,574	3	45,507	15,169
1988-89	40	556,823	13,921	2	25,608	12,804
1989-90	40	548,025	13,701	3	46,605	15,535
1990-91	40	497,224	12,431	3	45,468	15,156
1991-92	40	433,289	10,832	3	25,061	8,354
1992-93	41	433,918	10,583	...	....	......
1993-94	41	430,159	10,492	...	....	......
1994-95	24	284,050	11,835	...	....	......
1995-96	41	491,303	11,983	...	....	......
1996-97	41	559,939	13,657	...	....	......

	Games	Attendance	Average
WHA Totals	282	2,208,308	7,831
WHA Playoffs	39	316,297	8,110
NHL Totals	708	8,577,832	12,116
Playoff Totals	21	308,591	14,694

Whalers Playoff Records

Season	W	L	GF	GA
1972-73	12	3	70	49
1973-74	3	4	23	24
1974-75	2	4	17	28
1975-76	10	7	53	40
1976-77	1	7	14	23
1977-78	8	4	56	47
1978-79	5	5	39	45
1979-80	0	3	8	18
1980-81	...	...	...	...
1981-82	...	...	...	...
1982-83	...	...	...	...
1983-84	...	...	...	...
1984-85	...	...	...	...
1985-86	6	4	29	23
1986-87	2	4	19	27
1987-88	2	4	20	23
1988-89	0	4	11	23
1989-90	3	4	21	23
1990-91	2	4	17	24
1991-92	3	4	18	21
1992-93	...	...	...	...
1993-94	...	...	...	...
1994-95	...	...	...	...
1995-96	...	...	...	...
1996-97	...	...	...	...

The banners were many over the years at the Civic Center. Here's a look at some that were up for the final time at the Trumbull Street complex.

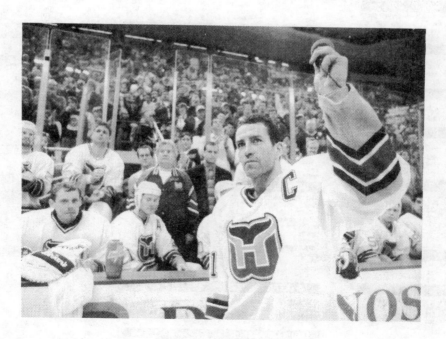

Kevin Dineen, *the last captain of the Whalers, addresses the faithful for the final time, moments after Hartford played its final game on April 13, 1997.*

'The Whale'. . . from A to Z

Coaches

DON BLACKBURN

Year	REGULAR SEASON GP	W	L	T	Pts	Pct	PLAYOFFS GP	W	L
1975-76	35	14	18	3	31	.442	...	...	...
1978-79	9	4	4	1	9	.500	10	5	5
1979-80	80	27	34	19	73	.456	3	0	3
1980-81	60	15	39	16	46	.383	...	...	...

JOHN CUNNIFF

Year	REGULAR SEASON GP	W	L	T	Pts	Pct	PLAYOFFS GP	W	L
1982-83	13	3	9	1	7	.269	...	...	...

BILL DINEEN

Year	REGULAR SEASON GP	W	L	T	Pts	Pct	PLAYOFFS GP	W	L
1978-79	71	33	30	8	74	.521	...	...	...

JACK EVANS

Year	REGULAR SEASON GP	W	L	T	Pts	Pct	PLAYOFFS GP	W	L
1983-84	80	28	42	10	66	.413	...	...	...
1984-85	80	30	41	9	69	.431	...	...	...
1985-86	80	40	36	4	84	.525	10	6	4
1986-87	80	43	30	7	93	.581	6	2	4
1987-88	54	22	25	7	51	.472	...	...	...

PAUL HOLMGREN

Year	REGULAR SEASON GP	W	L	T	Pts	Pct	PLAYOFFS GP	W	L
1992-93	84	26	52	6	58	.345	...	...	...
1993-94	17	4	11	2	10	.294	...	...	...
1994-95	48	19	24	5	43	.448	...	...	...
1995-96	12	5	6	1	11	.458	...	...	...

JACK KELLEY

Year	REGULAR SEASON GP	W	L	T	Pts	Pct	PLAYOFFS GP	W	L
1972-73	78	46	30	2	94	.602	15	12	3
1974-75	5	3	2	0	6	.600	6	2	4
1975-76	33	14	16	3	31	.469	...	...	...

LARRY KISH

Year	REGULAR SEASON GP	W	L	T	Pts	Pct	PLAYOFFS GP	W	L
1982-83	49	12	32	5	29	.296	...	...	...

RICK LEY

Year	REGULAR SEASON GP	W	L	T	Pts	Pct	PLAYOFFS GP	W	L
1989-90	80	38	33	9	85	.531	7	3	4
1990-91	80	31	38	11	73	.456	6	2	4

PAUL MAURICE

Year	REGULAR SEASON GP	W	L	T	Pts	Pct	PLAYOFFS GP	W	L
1995-96	70	29	33	8	66	.471	...	...	...
1996-97	82	32	39	11	75	.457	...	...	...

PIERRE McGUIRE

Year	REGULAR SEASON GP	W	L	T	Pts	Pct	PLAYOFFS GP	W	L
1993-94	67	23	37	7	53	.368	...	...	...

HARRY NEALE

Year	REGULAR SEASON GP	W	L	T	Pts	Pct	PLAYOFFS GP	W	L
1975-76	12	5	6	1	11	.458	17	10	7
1976-77	81	35	40	6	76	.469	5	1	4
1977-78	80	44	31	5	93	.581	14	8	6

LARRY PLEAU

Year	REGULAR SEASON GP	W	L	T	Pts	Pct	PLAYOFFS GP	W	L
1980-81	20	6	12	2	14	.350	...	...	...
1981-82	80	26	41	13	60	.406	...	...	...
1982-83	18	4	13	1	9	.250	...	...	...
1987-88	26	13	13	0	26	.500	6	2	4
1988-89	80	37	38	5	79	.494	4	0	4

JIMMY ROBERTS

Year	REGULAR SEASON GP	W	L	T	Pts	Pct	PLAYOFFS GP	W	L
1991-92	80	26	41	13	65	.406	7	3	4

RON RYAN

Year	REGULAR SEASON GP	W	L	T	Pts	Pct	PLAYOFFS GP	W	L
1973-74	78	43	31	4	90	.577	7	3	4
1974-75	73	40	28	5	85	.582	...	...	...

Goaltenders

CHRISTER ABRAHAMSSON

REGULAR SEASON

Year	GP	Min	W	L	T	So	GAA	G	A	Pts	PM
1974-75	16	870	8	6	1	1	3.24	0	0	0	0
1975-76	41	2385	18	18	2	2	3.42	0	1	1	6
1976-77	45	2484	15	22	4	0	3.84	0	1	1	2

PLAYOFFS

Year	GP	Min	W	L	So	GAA	G	A	Pts	PM
1975	1	0	0	0	0	0.00	0	0	0	0
1977	2	90	0	1	0	3.33	0	0	0	0

BILL BERGLUND

REGULAR SEASON

Year	GP	Min	W	L	T	So	GAA	G	A	Pts	PM
1973-74	3	180	2	1	0	0	3.33	0	0	0	0
1974-75	2	36	0	0	0	0	5.00	0	0	0	0

RICHARD BRODEUR

REGULAR SEASON

Year	GP	Min	W	L	T	So	GAA	G	A	Pts	PM
1987-88	6	340	4	2	0	0	2.65	0	0	0	2

PLAYOFFS

Year	GP	Min	W	L	So	GAA	G	A	Pts	PM
1988	4	200	1	3	0	3.60	0	0	0	0

SEAN BURKE

REGULAR SEASON

Year	GP	Min	W	L	T	So	GAA	G	A	Pts	PM
1992-93	50	2686	16	27	3	0	4.16	0	2	2	25
1993-94	47	2750	17	24	5	2	2.99	0	0	0	16
1994-95	42	2418	17	19	4	0	2.68	0	1	1	8
1995-96	66	3669	28	28	6	4	3.11	0	6	6	16
1996-97	51	2985	22	22	6	4	2.69	0	2	2	14

GAYE COOLEY

PLAYOFFS

Year	GP	Min	W	L	So	GAA	G	A	Pts	PM
1976	1	1	0	0	0	0.00	0	0	0	0

CORRIE D'ALESSIO

REGULAR SEASON

Year	GP	Min	W	L	T	So	GAA	G	A	Pts	PM
1992-93	1	11	0	0	0	0	0.00	0	0	0	0

JOHN GARRETT

REGULAR SEASON

Year	GP	Min	W	L	T	So	GAA	G	A	Pts	PM
1978-79	41	2496	20	17	4	2	3.58	0	0	0	6
1979-80	52	3046	16	24	11	0	3.98	0	2	2	12
1980-81	54	3145	15	27	12	0	4.60	0	1	1	2
1981-82	16	897	5	6	4	0	4.21	0	1	1	2

PLAYOFFS

Year	GP	Min	W	L	So	GAA	G	A	Pts	PM
1979	8	447	4	3	0	4.30	0	0	0	0
1980	1	60	0	1	0	8.00	0	0	0	0

JEAN-SEBASTIEN GIGUERE

REGULAR SEASON

Year	GP	Min	W	L	T	So	GAA	G	A	Pts	PM
1996-97	8	394	1	4	0	0	3.65	0	0	0	0

MARIO GOSSELIN

REGULAR SEASON

Year	GP	Min	W	L	T	So	GAA	G	A	Pts	PM
1992-93	16	867	5	9	1	0	3.94	0	1	1	0
1993-94	7	239	0	4	0	0	5.27	0	0	0	0

PAUL HOGANSON

REGULAR SEASON

Year	GP	Min	W	L	T	So	GAA	G	A	Pts	PM
1975-76	4	224	1	2	0	0	4.29	0	0	0	0

KEN HOLLAND

REGULAR SEASON

Year	GP	Min	W	L	T	So	GAA	G	A	Pts	PM
1980-81	1	60	0	1	0	0	7.00	0	0	0	0

BRUCE LANDON

REGULAR SEASON

Year	GP	Min	W	L	T	So	GAA	G	A	Pts	PM
1972-73	30	1671	15	11	1	1	3.59	0	1	1	8
1973-74	24	1386	11	9	2	0	3.55	0	0	0	24
1974-75	7	339	2	3	0	0	3.36	0	0	0	0
1975-76	38	2181	14	19	5	0	3.47	0	1	1	0
1976-77	23	1118	8	8	1	1	3.17	0	0	0	4

PLAYOFFS

Year	GP	Min	W	L	So	GAA	G	A	Pts	PM
1974	1	40	0	0	0	4.50	0	0	0	0
1976	4	197	3	0	0	2.13	0	0	0	0
1977	3	152	1	2	0	4.34	0	1	1	0

MIKE LENARDUZZI

REGULAR SEASON

Year	GP	Min	W	L	T	So	GAA	G	A	Pts	PM
1992-93	3	168	1	1	1	0	3.21	0	0	0	0
1993-94	1	21	0	0	0	0	2.86	0	0	0	0

LOUIS LEVASSEUR

REGULAR SEASON

Year	GP	Min	W	L	T	So	GAA	G	A	Pts	PM
1977-78	27	1655	14	11	2	3	3.30	0	0	0	2

PLAYOFFS

Year	GP	Min	W	L	So	GAA	G	A	Pts	PM
1978	12	719	8	4	0	2.59	0	1	1	2

MIKE LIUT

REGULAR SEASON

Year	GP	Min	W	L	T	So	GAA	G	A	Pts	PM
1984-85	12	731	4	7	1	1	3.82	0	0	0	2
1985-86	57	3282	27	23	4	2	3.61	0	2	2	0
1986-87	59	3471	31	22	5	4	3.23	0	2	2	4
1987-88	60	3326	25	28	5	2	3.18	0	1	1	4
1988-89	35	2004	13	19	1	1	4.25	0	0	0	0
1989-90	29	1683	15	12	1	3	2.21	0	0	0	0

PLAYOFFS

Year	GP	Min	W	L	So	GAA	G	A	Pts	PM
1986	8	441	5	2	1	1.90	0	0	0	0
1987	6	332	2	4	0	4.52	0	0	0	2
1988	3	160	1	1	0	4.13	0	0	0	0

ROSS McKAY

REGULAR SEASON

Year	GP	Min	W	L	T	So	GAA	G	A	Pts	PM
1990-91	1	35	0	0	0	0	5.14	0	0	0	0

GREG MILLEN

REGULAR SEASON

Year	GP	Min	W	L	T	So	GAA	G	A	Pts	PM
1981-82	55	3195	11	30	12	0	4.30	0	5	5	2
1982-83	60	3506	14	38	6	1	4.83	0	2	2	8
1984-85	60	3576	21	30	9	2	3.71	0	3	3	10
1986-87	44	2659	16	22	6	1	4.22	0	0	0	4

JASON MUZZATTI

REGULAR SEASON

Year	GP	Min	W	L	T	So	GAA	G	A	Pts	PM
1995-96	22	1013	4	8	3	1	2.90	0	0	0	33
1996-97	31	1591	9	13	5	1	3.43	0	1	1	18

TED OUIMET

REGULAR SEASON

Year	GP	Min	W	L	T	So	GAA	G	A	Pts	PM
1974-75	1	20	0	0	0	0	0.00	0	0	0	0

FRANK PIETRANGELO

REGULAR SEASON

Year	GP	Min	W	L	T	So	GAA	G	A	Pts	PM
1991-92	5	306	3	1	1	0	2.35	0	0	0	0
1992-93	30	1373	4	15	1	0	4.85	0	0	0	4
1993-94	19	984	5	11	1	0	3.60	0	0	0	2

PLAYOFFS

Year	GP	Min	W	L	So	GAA	G	A	Pts	PM
1992	7	425	3	4	0	2.68	0	0	0	0

CAP RAEDER

REGULAR SEASON

Year	GP	Min	W	L	T	So	GAA	G	A	Pts	PM
1975-76	2	100	0	1	0	0	4.80	0	0	0	0
1976-77	14	1328	12	10	1	2	3.12	0	0	0	2

PLAYOFFS

Year	GP	Min	W	L	So	GAA	G	A	Pts	PM
1976	14	819	7	7	2	2.27	0	1	1	0
1977	1	60	0	1	0	7.00	0	0	0	0

DARRYL REAUGH

REGULAR SEASON

Year	GP	Min	W	L	T	So	GAA	G	A	Pts	PM
1990-91	20	1010	7	7	1	1	3.15	0	0	0	4

JEFF REESE

REGULAR SEASON

Year	GP	Min	W	L	T	So	GAA	G	A	Pts	PM
1993-94	19	1057	5	9	3	1	3.18	0	1	1	0
1994-95	11	477	2	5	1	0	3.77	0	2	2	0
1995-96	7	275	2	3	0	1	3.05	0	0	0	0

PETER SIDORKIEWICZ

REGULAR SEASON

Year	GP	Min	W	L	T	So	GAA	G	A	Pts	PM
1987-88	1	60	0	1	0	0	6.00	0	0	0	0
1988-89	44	2634	22	18	4	4	3.03	0	3	3	0
1989-90	46	2703	19	19	7	1	3.57	0	1	1	4
1990-91	52	2953	21	22	7	1	3.33	0	4	4	6
1991-92	35	1995	9	19	6	2	3.34	0	1	1	2

PLAYOFFS

Year	GP	Min	W	L	So	GAA	G	A	Pts	PM
1989	2	124	0	2	0	3.87	0	0	0	0
1990	7	429	3	4	0	3.22	0	0	0	0
1991	6	359	2	4	0	4.01	0	1	1	2

AL SMITH

REGULAR SEASON

Year	GP	Min	W	L	T	So	GAA	G	A	Pts	PM
1972-73	51	3059	31	19	1	3	3.32	0	1	1	39
1973-74	56	3194	30	21	2	2	3.08	0	3	3	33
1974-75	59	3494	33	21	4	2	3.47	0	0	0	18
1977-78	55	3246	30	20	3	2	3.22	0	3	3	4
1978-79	40	2396	17	17	5	1	3.31	0	3	3	35
1979-80	30	1754	11	10	8	2	3.66	0	1	1	10

PLAYOFFS

Year	GP	Min	W	L	So	GAA	G	A	Pts	PM
1973	15	909	12	3	0	3.23	0	1	1	12
1974	7	399	3	4	1	3.16	0	0	0	0
1975	6	366	2	4	0	4.59	0	0	0	2
1978	3	120	0	2	0	7.00	0	0	0	0
1979	4	153	1	2	0	4.71	0	0	0	0
1980	2	120	0	2	0	5.00	0	0	0	0

ED STANIOWSKI

REGULAR SEASON

Year	GP	Min	W	L	T	So	GAA	G	A	Pts	PM
1983-84	18	1040	6	9	1	0	4.27	0	0	0	2
1984-85	1	20	0	0	0	0	3.00	0	0	0	0

MIKE VEISOR

REGULAR SEASON											
Year	GP	Min	W	L	T	So	GAA	G	A	Pts	PM
1980-81	29	1580	6	13	6	1	4.48	0	0	0	0
1981-82	13	701	5	5	2	0	4.54	0	1	1	2
1982-83	23	1277	5	16	1	1	5.54	0	0	0	2
1983-84	4	240	1	3	0	0	5.00	0	0	0	0

STEVE WEEKS

REGULAR SEASON											
Year	GP	Min	W	L	T	So	GAA	G	A	Pts	PM
1984-85	24	1457	10	12	2	2	3.79	0	0	0	0
1985-86	27	1544	13	13	0	1	3.85	0	1	1	9
1986-87	25	1368	12	8	2	1	3.42	0	0	0	0
1987-88	18	918	6	7	2	0	3.59	0	0	0	2

PLAYOFFS										
Year	GP	Min	W	L	So	GAA	G	A	Pts	PM
1986	3	169	1	2	0	2.84	0	0	0	0
1987	1	36	0	0	0	1.67	0	0	0	0

KAY WHITMORE

REGULAR SEASON											
Year	GP	Min	W	L	T	So	GAA	G	A	Pts	PM
1988-89	3	180	2	1	0	0	3.33	0	2	2	0
1989-90	9	442	4	2	1	0	3.53	0	1	1	4
1990-91	18	850	3	9	3	0	3.67	0	1	1	4
1991-92	45	2567	14	21	6	3	3.62	0	1	1	16

PLAYOFFS										
Year	GP	Min	W	L	So	GAA	G	A	Pts	PM
1989	2	135	0	2	0	4.44	0	0	0	0
1992	1	19	0	0	0	3.16	0	0	0	0

Forwards/ Defensemen

THOMMY ABRAHAMSSON

	REGULAR SEASON					PLAYOFFS				
Year	GP	G	A	Pts	PM	GP	G	A	Pts	PM
1974-75	76	8	22	30	46	6	0	0	0	0
1975-76	63	14	21	35	47	17	2	4	6	15
1976-77	64	6	24	30	33	5	0	3	3	0
1980-81	32	6	11	17	19	...	...	...	...	...

GREG ADAMS

	REGULAR SEASON					PLAYOFFS				
Year	GP	G	A	Pts	PM	GP	G	A	Pts	PM
1982-83	79	10	13	23	216	...	...	...	...	...

JIM AGNEW

	REGULAR SEASON					PLAYOFFS				
Year	GP	G	A	Pts	PM	GP	G	A	Pts	PM
1992-93	16	0	0	0	68	...	...	...	...	...

KEVIN AHEARN

	REGULAR SEASON					PLAYOFFS				
Year	GP	G	A	Pts	PM	GP	G	A	Pts	PM
1972-73	78	20	22	42	18	14	1	2	3	9

STEVE ALLEY

	REGULAR SEASON					PLAYOFFS				
Year	GP	G	A	Pts	PM	GP	G	A	Pts	PM
1979-80	7	1	1	2	0	3	0	1	1	0
1980-81	8	2	2	4	11	...	...	...	...	...

RAY ALLISON

	REGULAR SEASON					PLAYOFFS				
Year	GP	G	A	Pts	PM	GP	G	A	Pts	PM
1979-80	64	16	12	28	13	2	0	1	1	0
1980-81	6	1	0	1	0	...	...	...	...	...

JOHN ANDERSON

	REGULAR SEASON					PLAYOFFS				
Year	GP	G	A	Pts	PM	GP	G	A	Pts	PM
1985-86	14	8	17	25	2	10	5	8	13	0
1986-87	76	31	44	75	19	6	1	2	3	0
1987-88	63	17	32	49	20	...	...	...	...	...
1988-89	62	16	24	40	28	4	0	1	1	2

RUSS ANDERSON

	REGULAR SEASON					PLAYOFFS				
Year	GP	G	A	Pts	PM	GP	G	A	Pts	PM
1981-82	25	1	3	4	85	...	...	...	...	...
1982-83	57	0	6	6	171	...	...	...	...	...

MIKAEL ANDERSSON

	REGULAR SEASON					PLAYOFFS				
Year	GP	G	A	Pts	PM	GP	G	A	Pts	PM
1989-90	50	13	24	37	6	5	0	3	3	2
1990-91	41	4	7	11	8	...	...	...	...	...
1991-92	74	18	29	47	14	7	0	2	2	6

MIKE ANTONOVICH

	REGULAR SEASON					PLAYOFFS				
Year	GP	G	A	Pts	PM	GP	G	A	Pts	PM
1976-77	26	12	9	21	10	5	2	2	4	4
1977-78	75	32	35	67	32	14	10	7	17	4
1978-79	69	6	27	47	35	10	5	3	8	14
1979-80	5	0	1	1	2	...	...	...	...	...

DANNY ARNDT

	REGULAR SEASON					PLAYOFFS				
Year	GP	G	A	Pts	PM	GP	G	A	Pts	PM
1975-76	69	8	8	16	10	8	0	0	0	0
1976-77	46	8	14	22	11	...	...	...	...	...

FRED ARTHUR

	REGULAR SEASON					PLAYOFFS				
Year	GP	G	A	Pts	PM	GP	G	A	Pts	PM
1980-81	3	0	0	0	0	...	...	...	...	...

JAMES BLACK

	REGULAR SEASON					PLAYOFFS				
Year	GP	G	A	Pts	PM	GP	G	A	Pts	PM
1989-90	1	0	0	0	0	...	...	...	...	...
1990-91	1	0	0	0	0	...	...	...	...	...
1991-92	30	4	6	10	14	...	...	...	...	...

DON BLACKBURN

	REGULAR SEASON					PLAYOFFS				
Year	GP	G	A	Pts	PM	GP	G	A	Pts	PM
1973-74	75	20	39	59	18	7	2	4	6	4
1974-75	50	18	32	50	10	5	1	2	3	2
1975-76	21	2	3	5	6	...	...	...	...	...

BOB BODAK

	REGULAR SEASON					PLAYOFFS				
Year	GP	G	A	Pts	PM	GP	G	A	Pts	PM
1989-90	1	0	0	0	0	...	...	...	...	...

DANNY BOLDUC

	REGULAR SEASON					PLAYOFFS				
Year	GP	G	A	Pts	PM	GP	G	A	Pts	PM
1975-76	14	2	5	7	14	16	1	6	7	4
1976-77	33	8	3	11	15	...	...	...	...	...
1977-78	41	5	5	10	22	14	2	4	6	4

DON BORGESON

	REGULAR SEASON					PLAYOFFS				
Year	GP	G	A	Pts	PM	GP	G	A	Pts	PM
1975-76	31	9	8	17	4	3	1	1	2	0

TIM BOTHWELL

	REGULAR SEASON					PLAYOFFS				
Year	GP	G	A	Pts	PM	GP	G	A	Pts	PM
1985-86	62	2	8	10	53	10	0	0	0	8
1986-87	4	1	0	1	0	...	...	...	...	...

CHARLIE BOURGEOIS

	REGULAR SEASON					PLAYOFFS				
Year	GP	G	A	Pts	PM	GP	G	A	Pts	PM
1987-88	1	0	0	0	0	...	...	...	...	...

PAT BOUTETTE

	REGULAR SEASON					PLAYOFFS				
Year	GP	G	A	Pts	PM	GP	G	A	Pts	PM
1979-80	47	13	31	44	75	3	1	0	1	6
1980-81	80	28	52	80	160	...	...	...	...	...
1984-85	33	6	8	14	51	...	...	...	...	...

GREG BRITZ

	REGULAR SEASON					PLAYOFFS				
Year	GP	G	A	Pts	PM	GP	G	A	Pts	PM
1986-87	1	0	0	0	0	...	...	...	...	...

JEFF BROWN

	REGULAR SEASON					PLAYOFFS				
Year	GP	G	A	Pts	PM	GP	G	A	Pts	PM
1995-96	48	7	31	38	38	...	...	...	...	...
1996-97	1	0	0	0	0	...	...	...	...	...

KEVIN BROWN

	REGULAR SEASON					PLAYOFFS				
Year	GP	G	A	Pts	PM	GP	G	A	Pts	PM
1996-97	11	0	4	4	6	...	...	...	...	...

ROB BROWN

	REGULAR SEASON					PLAYOFFS				
Year	GP	G	A	Pts	PM	GP	G	A	Pts	PM
1990-91	44	18	24	42	101	5	1	0	1	7
1991-92	42	16	15	31	39	...	...	...	...	...

JACK BROWNSCHIDLE

	REGULAR SEASON					PLAYOFFS				
Year	GP	G	A	Pts	PM	GP	G	A	Pts	PM
1983-84	13	2	2	4	10	...	...	...	...	...
1984-85	17	1	4	5	5	...	...	...	...	...
1985-86	9	0	0	0	4	...	...	...	...	...

JEFF BROWNSCHIDLE

	REGULAR SEASON					PLAYOFFS				
Year	GP	G	A	Pts	PM	GP	G	A	Pts	PM
1981-82	3	0	1	1	2	...	...	...	...	...
1982-83	4	0	0	0	0	...	...	...	...	...

JEFF BRUBAKER

	REGULAR SEASON					PLAYOFFS				
Year	GP	G	A	Pts	PM	GP	G	A	Pts	PM
1978-79	12	0	0	0	19	3	0	0	0	12
1979-80	3	0	1	1	2	...	...	...	...	...
1980-81	43	5	3	8	93	...	...	...	...	...

ADAM BURT

	REGULAR SEASON					PLAYOFFS				
Year	GP	G	A	Pts	PM	GP	G	A	Pts	PM
1988-89	5	0	0	0	6	...	...	...	...	...
1989-90	63	4	8	12	105	2	0	0	0	0
1990-91	42	2	7	9	63	...	...	...	...	...
1991-92	66	9	15	24	93	2	0	0	0	0
1992-93	65	6	14	20	116	...	...	...	...	...
1993-94	63	1	17	18	75	...	...	...	...	...
1994-95	46	7	11	18	65	...	...	...	...	...
1995-96	78	4	9	13	121	...	...	...	...	...
1996-97	71	2	11	13	79	...	...	...	...	...

RON BUSNIUK

	REGULAR SEASON					PLAYOFFS				
Year	GP	G	A	Pts	PM	GP	G	A	Pts	PM
1975-76	11	0	3	3	55	17	0	2	2	14
1976-77	55	1	9	10	141	...	...	...	...	...

BILL BUTTERS

	REGULAR SEASON					PLAYOFFS				
Year	GP	G	A	Pts	PM	GP	G	A	Pts	PM
1976-77	26	1	8	9	65	5	1	0	1	15
1977-78	45	1	13	14	69	...	...	...	...	...

MIKE BYERS

	REGULAR SEASON					PLAYOFFS				
Year	GP	G	A	Pts	PM	GP	G	A	Pts	PM
1972-73	19	6	4	10	4	12	6	5	11	6
1973-74	78	29	21	50	6	7	2	4	6	12
1974-75	72	22	26	48	10	6	2	2	4	2
1975-76	21	4	3	7	0	...	...	...	...	...

TERRY CAFFERY

	REGULAR SEASON					PLAYOFFS				
Year	GP	G	A	Pts	PM	GP	G	A	Pts	PM
1972-73	74	39	61	100	14	8	3	7	10	0
1973-74	...	...	...	...	...	...	...	...	...	...
1974-75	67	15	37	52	12	...	...	...	...	...
1975-76	2	0	0	0	2	...	...	...	...	...

BRETT CALLIGHEN

	REGULAR SEASON					PLAYOFFS				
Year	GP	G	A	Pts	PM	GP	G	A	Pts	PM
1976-77	33	6	10	16	41	...	...	...	...	...

WAYNE CARLETON

	REGULAR SEASON					PLAYOFFS				
Year	GP	G	A	Pts	PM	GP	G	A	Pts	PM
1974-75	73	35	39	74	50	6	2	5	7	14
1975-76	35	12	21	33	6	...	...	...	...	...

JACK CARLSON

	REGULAR SEASON					PLAYOFFS				
Year	GP	G	A	Pts	PM	GP	G	A	Pts	PM
1976-77	35	7	5	12	81	5	1	1	2	9
1977-78	67	9	20	29	192	9	1	1	2	14
1978-79	34	2	7	9	61	...	...	...	...	...

STEVE CARLSON

	REGULAR SEASON					PLAYOFFS				
Year	GP	G	A	Pts	PM	GP	G	A	Pts	PM
1976-77	31	4	9	13	40	5	0	0	0	9
1977-78	38	6	7	13	11	13	2	7	9	2

GREG CARROLL

	REGULAR SEASON					PLAYOFFS				
Year	GP	G	A	Pts	PM	GP	G	A	Pts	PM
1977-78	48	9	14	23	27	...	...	...	...	...
1979-80	71	13	19	32	24	...	...	...	...	...

JIMMY CARSON

	REGULAR SEASON					PLAYOFFS				
Year	GP	G	A	Pts	PM	GP	G	A	Pts	PM
1994-95	38	9	10	19	29	...	...	...	...	...
1995-96	11	1	0	1	0	...	...	...	...	...

LINDSAY CARSON

	REGULAR SEASON					PLAYOFFS				
Year	GP	G	A	Pts	PM	GP	G	A	Pts	PM
1987-88	27	5	4	9	30	5	1	2	3	2

ANDREW CASSELS

	REGULAR SEASON					PLAYOFFS				
Year	GP	G	A	Pts	PM	GP	G	A	Pts	PM
1991-92	67	11	30	41	18	7	2	4	6	6
1992-93	84	21	65	86	62	...	...	...	...	...
1993-94	79	16	42	58	37	...	...	...	...	...
1994-95	46	7	30	37	18	...	...	...	...	...
1995-96	81	20	43	63	39	...	...	...	...	...
1996-97	81	22	44	66	46	...	...	...	...	...

BRIAN CHAPMAN

	REGULAR SEASON					PLAYOFFS				
Year	GP	G	A	Pts	PM	GP	G	A	Pts	PM
1990-91	3	0	0	0	29	...	...	...	...	...

BOB CHARLEBOIS

	REGULAR SEASON					PLAYOFFS				
Year	GP	G	A	Pts	PM	GP	G	A	Pts	PM
1973-74	74	4	7	11	6	7	0	0	0	4
1974-75	8	1	0	1	0	4	1	0	1	0

KELLY CHASE

	REGULAR SEASON					PLAYOFFS				
Year	GP	G	A	Pts	PM	GP	G	A	Pts	PM
1994-95	28	0	4	4	141	...	...	...	...	...
1995-96	55	2	4	6	220	...	...	...	...	...
1996-97	28	1	2	3	122	...	...	...	...	...

STEVE CHIASSON

	REGULAR SEASON					PLAYOFFS				
Year	GP	G	A	Pts	PM	GP	G	A	Pts	PM
1996-97	18	3	11	14	7	...	...	...	...	...

IGOR CHIBIREV

Year	GP	G	A	Pts	PM	GP	G	A	Pts	PM
1993-94	37	4	11	15	2	...	...	...	...	...

SHANE CHURLA

	REGULAR SEASON					PLAYOFFS				
Year	GP	G	A	Pts	PM	GP	G	A	Pts	PM
1986-87	20	0	1	1	78	2	0	0	0	42
1987-88	2	0	0	0	14	...	...	...	...	...

RON CLIMIE

	REGULAR SEASON					PLAYOFFS				
Year	GP	G	A	Pts	PM	GP	G	A	Pts	PM
1974-75	25	8	4	12	12	6	3	0	3	0
1975-76	65	25	20	45	17	...	...	...	...	...
1976-77	3	0	0	0	0	...	...	...	...	...

PAUL COFFEY

	REGULAR SEASON					PLAYOFFS				
Year	GP	G	A	Pts	PM	GP	G	A	Pts	PM
1996-97	20	3	5	8	18	...	...	...	...	...

YVON CORRIVEAU

	REGULAR SEASON					PLAYOFFS				
Year	GP	G	A	Pts	PM	GP	G	A	Pts	PM
1989-90	13	4	1	5	20	4	1	0	1	0
1990-91	23	1	1	2	18	...	...	...	...	...
1991-92	38	12	8	20	36	7	3	2	5	18
1992-93	37	5	5	10	14	...	...	...	...	...
1993-94	3	0	0	0	0	...	...	...	...	...

SYLVAIN COTE

	REGULAR SEASON					PLAYOFFS				
Year	GP	G	A	Pts	PM	GP	G	A	Pts	PM
1984-85	63	3	9	12	17	...	...	...	...	...
1985-86	2	0	0	0	0	...	...	...	...	...
1986-87	67	2	8	10	20	2	0	2	2	0
1987-88	67	7	21	28	30	6	1	1	2	4
1988-89	78	8	9	17	49	3	0	1	1	4
1989-90	28	4	2	6	14	5	0	0	0	2
1990-91	73	7	12	19	17	6	0	2	2	2

YVES COURTEAU

	REGULAR SEASON					PLAYOFFS				
Year	GP	G	A	Pts	PM	GP	G	A	Pts	PM
1986-87	4	0	0	0	0	...	...	...	...	...

MURRAY CRAVEN

	REGULAR SEASON					PLAYOFFS				
Year	GP	G	A	Pts	PM	GP	G	A	Pts	PM
1991-92	61	24	30	54	38	7	3	3	6	6
1992-93	67	25	42	67	20	...	...	...	...	...

BOB CRAWFORD

	REGULAR SEASON					PLAYOFFS				
Year	GP	G	A	Pts	PM	GP	G	A	Pts	PM
1983-84	80	36	25	61	32	...	...	...	...	...
1984-85	45	14	14	28	8	...	...	...	...	...
1985-86	57	14	20	34	16	...	...	...	...	...

MIKE CROMBEEN

Year	Regular Season GP	G	A	Pts	PM	Playoffs GP	G	A	Pts	PM
1983-84	51	1	4	5	25	...	...	...	...	...
1984-85	46	4	7	11	16	...	...	...	...	...

DOUG CROSSMAN

Year	Regular Season GP	G	A	Pts	PM	Playoffs GP	G	A	Pts	PM
1990-91	41	4	19	23	19	...	...	...	...	...

TED CROWLEY

Year	Regular Season GP	G	A	Pts	PM	Playoffs GP	G	A	Pts	PM
1993-94	21	1	2	3	10	...	...	...	...	...

JIM CULHANE

Year	Regular Season GP	G	A	Pts	PM	Playoffs GP	G	A	Pts	PM
1989-90	6	0	1	1	4	...	...	...	...	...

JOHN CULLEN

Year	Regular Season GP	G	A	Pts	PM	Playoffs GP	G	A	Pts	PM
1990-91	13	8	8	16	18	6	2	7	9	10
1991-92	77	26	51	77	14	7	2	1	3	12
1992-93	19	5	4	9	58	...	...	...	...	...

JOHN CUNNIFF

Year	Regular Season GP	G	A	Pts	PM	Playoffs GP	G	A	Pts	PM
1972-73	31	3	5	8	16	13	1	1	2	2
1973-74	30	7	5	12	14	5	1	1	2	0

RANDY CUNNEYWORTH

Year	Regular Season GP	G	A	Pts	PM	Playoffs GP	G	A	Pts	PM
1989-90	43	9	9	18	41	4	0	0	0	2
1990-91	32	9	5	14	49	1	0	0	0	0
1991-92	39	7	10	17	71	7	3	0	3	9
1992-93	39	5	4	9	63	...	...	...	...	...
1993-94	63	9	8	17	87	...	...	...	...	...

TONY CURRIE

Year	Regular Season GP	G	A	Pts	PM	Playoffs GP	G	A	Pts	PM
1983-84	32	12	16	28	4	...	...	...	...	...
1984-85	13	3	8	11	2	...	...	...	...	...

PAUL CYR

Year	GP	G	A	Pts	PM	GP	G	A	Pts	PM
1990-91	70	12	13	25	107	6	1	0	1	10

JAKE DANBY

Year	Regular Season GP	G	A	Pts	PM	Playoffs GP	G	A	Pts	PM
1972-73	72	2	2	4	6	7	1	0	1	0
1974-75	...	...	...	...	...	4	0	1	1	0
1975-76	1	0	0	0	0	...	...	...	...	...

JEFF DANIELS

1996-97	10	0	2	2	0	...	...	...	...	...

SCOTT DANIELS

Year	Regular Season GP	G	A	Pts	PM	Playoffs GP	G	A	Pts	PM
1992-93	1	0	0	0	19	...	...	...	...	...
1993-94	12	0	2	2	55	...	...	...	...	...
1995-96	53	3	4	7	254	...	...	...	...	...

JOE DAY

Year	Regular Season GP	G	A	Pts	PM	Playoffs GP	G	A	Pts	PM
1991-92	24	0	3	3	10	...	...	...	...	...
1992-93	24	1	7	8	47	...	...	...	...	...

DAVE DEBOL

Year	Regular Season GP	G	A	Pts	PM	Playoffs GP	G	A	Pts	PM
1979-80	48	12	14	26	4	3	0	0	0	0
1980-81	44	14	12	26	0					

GERALD DIDUCK

Year	Regular Season GP	G	A	Pts	PM	Playoffs GP	G	A	Pts	PM
1995-96	79	1	9	10	88	...	...	...	...	...
1996-97	56	1	10	11	40	...	...	...	...	...

KEVIN DINEEN

Year	Regular Season GP	G	A	Pts	PM	Playoffs GP	G	A	Pts	PM
1984-85	57	25	16	41	120	...	...	...	...	...
1985-86	57	33	35	68	124	10	6	7	13	18
1986-87	78	40	39	79	110	6	2	1	3	31
1987-88	74	25	25	50	217	6	4	4	8	8
1988-89	79	45	44	89	167	4	1	0	1	10
1979-80	67	25	41	66	164	6	3	2	5	18
1990-91	61	17	30	47	104	6	1	0	1	16
1991-92	16	4	2	6	23	...	...	...	...	...
1995-96	20	2	7	9	67	...	...	...	...	...
1996-97	78	19	29	48	141	...	...	...	...	...

HNAT DOMENICHELLI

Year	Regular Season GP	G	A	Pts	PM	Playoffs GP	G	A	Pts	PM
1996-97	13	2	1	3	7					

JIM DOREY

Year	Regular Season GP	G	A	Pts	PM	Playoffs GP	G	A	Pts	PM
1972-73	74	7	56	63	95	15	3	16	19	41
1973-74	77	6	40	46	134	6	0	6	6	26
1974-75	31	5	17	22	43	...	...	...	...	...

JORDY DOUGLAS

Year	Regular Season GP	G	A	Pts	PM	Playoffs GP	G	A	Pts	PM
1978-79	51	6	10	16	15	10	4	0	4	23
1979-80	77	33	24	57	39	...	...	...	...	...
1980-81	55	13	9	22	29	...	...	...	...	...
1981-82	30	10	7	17	44	...	...	...	...	...

TED DRURY

Year	Regular Season GP	G	A	Pts	PM	Playoffs GP	G	A	Pts	PM
1993-94	16	1	5	6	10	...	...	...	...	...
1994-95	34	3	6	9	21	...	...	...	...	...

RICHIE DUNN

Year	Regular Season GP	G	A	Pts	PM	Playoffs GP	G	A	Pts	PM
1983-84	63	5	20	25	30	...	...	...	...	...
1984-85	13	1	4	5	2	...	...	...	...	...

NORM DUPONT

Year	Regular Season GP	G	A	Pts	PM	Playoffs GP	G	A	Pts	PM
1983-84	40	7	15	22	12	...	...	...	...	...

STEVE DYKSTRA

Year	Regular Season GP	G	A	Pts	PM	Playoffs GP	G	A	Pts	PM
1989-90	9	0	0	0	2	...	...	...	...	...

TOMMY EARL

Year	Regular Season GP	G	A	Pts	PM	Playoffs GP	G	A	Pts	PM
1972-73	77	10	13	23	4	15	2	3	5	10
1973-74	78	10	10	20	29	7	0	2	2	2
1974-75	72	3	8	11	20	6	1	1	2	12
1975-76	66	8	11	19	26	17	0	5	5	4
1976-77	54	9	14	23	37	1	0	0	0	0

NELSON EMERSON

Year	Regular Season GP	G	A	Pts	PM	Playoffs GP	G	A	Pts	PM
1995-96	81	29	29	58	78	...	...	...	...	...
1996-97	66	9	29	38	34	...	...	...	...	...

DEAN EVASON

Year	Regular Season GP	G	A	Pts	PM	Playoffs GP	G	A	Pts	PM
1984-85	2	0	0	0	0	...	...	...	...	...
1985-86	55	20	28	48	65	10	1	4	5	10
1986-87	80	22	37	59	67	5	3	2	5	35
1987-88	77	10	18	28	115	6	1	1	2	2
1988-89	67	11	17	28	60	4	1	2	3	10
1989-90	78	18	25	43	138	7	2	2	4	22
1990-91	75	6	23	29	170	6	0	4	4	29

GLEN FEATHERSTONE

Year	Regular Season GP	G	A	Pts	PM	Playoffs GP	G	A	Pts	PM
1994-95	19	2	1	3	50	...	...	...	...	...
1995-96	68	2	10	12	138	...	...	...	...	...
1996-97	41	2	5	7	87	...	...	...	...	...

PAUL FENTON

Year	Regular Season GP	G	A	Pts	PM	Playoffs GP	G	A	Pts	PM
1984-85	33	7	5	12	10	...	...	...	...	...
1985-86	1	0	0	0	0	...	...	...	...	...

RAY FERRARO

Year	Regular Season GP	G	A	Pts	PM	Playoffs GP	G	A	Pts	PM
1984-85	44	11	17	28	40	...	...	...	...	...
1985-86	76	30	47	77	57	10	3	6	9	4
1986-87	80	27	32	59	67	6	1	1	2	8
1987-88	68	21	29	50	81	6	1	1	2	6
1988-89	80	41	35	76	86	4	2	0	2	4
1989-90	79	24	29	54	109	7	0	3	3	2
1990-91	15	2	5	7	18	...	...	...	...	...

MIKE FIDLER

Year	Regular Season GP	G	A	Pts	PM	Playoffs GP	G	A	Pts	PM
1980-81	38	9	9	18	4	...	...	...	...	...
1981-82	2	0	1	1	0	...	...	...	...	...

NICK FOTIU

Year	Regular Season GP	G	A	Pts	PM	Playoffs GP	G	A	Pts	PM
1974-75	61	2	2	4	144	4	2	0	2	27
1975-76	49	3	2	5	94	16	3	2	5	57
1979-80	74	10	8	18	107	3	0	0	0	6
1980-81	42	4	3	7	79	...	...	...	...	...

RON FRANCIS

Year	Regular Season GP	G	A	Pts	PM	Playoffs GP	G	A	Pts	PM
1981-82	59	25	43	68	51	...	...	...	...	...
1982-83	79	31	59	90	60	...	...	...	...	...
1983-84	72	23	60	83	45	...	...	...	...	...
1984-85	80	24	57	81	66	...	...	...	...	...
1985-86	53	24	53	77	24	10	1	2	3	4
1986-87	75	30	63	93	45	6	2	2	4	6
1987-88	80	25	50	75	87	6	2	5	7	2
1988-89	69	29	48	77	36	4	0	2	2	0
1989-90	80	32	69	101	73	7	3	3	6	8
1990-91	67	21	55	76	51	...	...	...	...	...

JOHN FRENCH

Year	Regular Season GP	G	A	Pts	PM	Playoffs GP	G	A	Pts	PM
1972-73	76	24	35	59	43	15	3	11	14	2
1973-74	77	24	48	72	31	7	4	2	6	2
1974-75	75	12	41	53	28	4	1	2	3	0

DAN FRIDGEN

Year	Regular Season GP	G	A	Pts	PM	Playoffs GP	G	A	Pts	PM
1981-82	2	0	1	1	0	...	...	...	...	...
1982-83	11	2	2	4	2	...	...	...	...	...

MARK FUSCO

Year	Regular Season GP	G	A	Pts	PM	Playoffs GP	G	A	Pts	PM
1983-84	17	0	4	4	2	...	...	...	...	...
1984-85	63	3	8	11	40	...	...	...	...	...

MICHEL GALARNEAU

Year	Regular Season GP	G	A	Pts	PM	Playoffs GP	G	A	Pts	PM
1980-81	30	2	6	8	9	...	...	...	...	...
1981-82	10	0	0	0	4	...	...	...	...	...
1982-83	38	5	4	9	21	...	...	...	...	...

BILL GARDNER

Year	Regular Season GP	G	A	Pts	PM	Playoffs GP	G	A	Pts	PM
1985-86	18	1	8	9	4	...	...	...	...	...
1986-87	8	0	1	1	0	...	...	...	...	...

MARTY GATEMAN

Year	Regular Season GP	G	A	Pts	PM	Playoffs GP	G	A	Pts	PM
1975-76	12	0	1	1	6	...	...	...	...	...

DALLAS GAUME

Year	Regular Season GP	G	A	Pts	PM	Playoffs GP	G	A	Pts	PM
1988-89	4	1	1	2	0	...	...	...	...	...

STEWART GAVIN

Year	Regular Season GP	G	A	Pts	PM	Playoffs GP	G	A	Pts	PM
1985-86	76	26	29	55	51	10	4	1	5	13
1986-87	79	20	22	42	28	6	2	4	6	10
1987-88	56	11	10	21	59	6	2	2	4	4

RANDY GILHEN

Year	Regular Season GP	G	A	Pts	PM	Playoffs GP	G	A	Pts	PM
1982-83	2	0	1	1	0	...	...	...	...	...

DON GILLEN

Year	Regular Season GP	G	A	Pts	PM	Playoffs GP	G	A	Pts	PM
1981-82	34	1	4	5	22	...	...	...	...	...

PAUL GILLIS

Year	Regular Season GP	G	A	Pts	PM	Playoffs GP	G	A	Pts	PM
1991-92	12	0	2	2	48	5	0	1	1	0
1992-93	21	1	1	2	40	...	...	...	...	...

LARRY GIROUX

Year	Regular Season GP	G	A	Pts	PM	Playoffs GP	G	A	Pts	PM
1979-80	47	2	5	7	44	3	0	0	0	2

BRIAN GLYNN

Year	Regular Season GP	G	A	Pts	PM	Playoffs GP	G	A	Pts	PM
1994-95	43	1	6	7	32	...	...	...	...	...
1995-96	54	0	4	4	44	...	...	...	...	...
1996-97	1	1	0	1	2	...	...	...	...	...

ALEXANDER GODYNYUK

Year	Regular Season GP	G	A	Pts	PM	Playoffs GP	G	A	Pts	PM
1993-94	43	3	9	12	40	...	...	...	...	...
1994-95	14	0	0	0	8	...	...	...	...	...
1995-96	3	0	0	0	2	...	...	...	...	...
1996-97	55	1	6	7	41	...	...	...	...	...

STU GRIMSON

Year	Regular Season GP	G	A	Pts	PM	Playoffs GP	G	A	Pts	PM
1996-97	75	2	2	4	218	...	...	...	...	...

CHRIS GOVEDARIS

Year	Regular Season GP	G	A	Pts	PM	Playoffs GP	G	A	Pts	PM
1989-90	12	0	1	1	4	2	0	1	1	2
1990-91	14	1	3	4	4	...	...	...	...	...
1992-93	7	1	0	1	0	...	...	...	...	...

TED GREEN

Year	Regular Season GP	G	A	Pts	PM	Playoffs GP	G	A	Pts	PM
1972-73	78	16	30	46	47	12	1	5	6	25
1973-74	75	7	26	33	42	7	0	4	4	2
1974-75	57	6	14	20	29	3	0	0	0	2

MARK GREIG

Year	Regular Season GP	G	A	Pts	PM	Playoffs GP	G	A	Pts	PM
1990-91	4	0	0	0	0	...	...	...	...	...
1991-92	17	0	5	5	6	...	...	...	...	...
1992-93	22	1	7	8	27	...	...	...	...	...
1993-94	31	4	5	9	31	...	...	...	...	...

KEVIN HALLER

Year	Regular Season GP	G	A	Pts	PM	Playoffs GP	G	A	Pts	PM
1996-97	28	2	6	8	48	...	...	...	...	...

ALAN HANGSLEBEN

Year	Regular Season GP	G	A	Pts	PM	Playoffs GP	G	A	Pts	PM
1974-75	26	0	4	4	8	6	0	3	3	19
1975-76	78	2	23	25	62	13	2	3	5	20
1976-77	74	13	9	22	79	4	0	0	0	9
1977-78	79	11	18	29	140	14	1	4	5	37
1978-79	77	10	19	29	148	10	1	2	3	12
1979-80	37	3	15	18	69	...	...	...	...	...

DAVE HANSON

Year	Regular Season GP	G	A	Pts	PM	Playoffs GP	G	A	Pts	PM
1976-77	1	0	0	0	9	1	0	0	0	0

TODD HARKINS

Year	Regular Season GP	G	A	Pts	PM	Playoffs GP	G	A	Pts	PM
1993-94	28	1	0	1	49	...	...	...	...	...

HUGH HARRIS

Year	Regular Season GP	G	A	Pts	PM	Playoffs GP	G	A	Pts	PM
1973-74	75	24	28	52	78	7	0	4	4	11

ARCHIE HENDERSON

Year	Regular Season GP	G	A	Pts	PM	Playoffs GP	G	A	Pts	PM
1982-83	15	2	1	3	64	...	...	...	...	...

BOB HESS

	REGULAR SEASON					PLAYOFFS				
Year	GP	G	A	Pts	PM	GP	G	A	Pts	PM
1983-84	3	0	0	0	0	...	...	...	...	...

BRIAN HILL

	REGULAR SEASON					PLAYOFFS				
Year	GP	G	A	Pts	PM	GP	G	A	Pts	PM
1979-80	19	1	1	2	4	...	...	...	...	...

RICK HODGSON

	REGULAR SEASON					PLAYOFFS				
Year	GP	G	A	Pts	PM	GP	G	A	Pts	PM
1979-80	6	0	0	0	6	1	0	0	0	0

MIKE HOFFMAN

	REGULAR SEASON					PLAYOFFS				
Year	GP	G	A	Pts	PM	GP	G	A	Pts	PM
1982-83	2	0	1	1	0	...	...	...	...	...
1984-85	1	0	0	0	0	...	...	...	...	...
1985-86	6	1	2	3	2	...	...	...	...	...

BOBBY HOLIK

	REGULAR SEASON					PLAYOFFS				
Year	GP	G	A	Pts	PM	GP	G	A	Pts	PM
1990-91	78	21	22	43	113	6	0	0	0	7
1991-92	76	21	24	45	45	7	0	1	1	6

ED HOSPODAR

	REGULAR SEASON					PLAYOFFS				
Year	GP	G	A	Pts	PM	GP	G	A	Pts	PM
1982-83	72	1	9	10	199	...	...	...	...	...
1983-84	59	0	9	9	163	...	...	...	...	...

DOUG HOUDA

	REGULAR SEASON					PLAYOFFS				
Year	GP	G	A	Pts	PM	GP	G	A	Pts	PM
1990-91	19	1	2	3	41	6	0	0	0	8
1991-92	56	3	6	9	125	6	0	2	2	13
1992-93	60	2	6	8	167	...	...	...	...	...
1993-94	7	0	0	0	23	...	...	...	...	...

GARRY HOWATT

	REGULAR SEASON					PLAYOFFS				
Year	GP	G	A	Pts	PM	GP	G	A	Pts	PM
1981-82	80	18	32	50	242	...	...	...	...	...

GORDIE HOWE

	REGULAR SEASON					PLAYOFFS				
Year	GP	G	A	Pts	PM	GP	G	A	Pts	PM
1977-78	76	34	62	96	85	14	5	5	10	15
1978-79	68	19	24	43	51	10	3	1	4	4
1979-80	80	25	26	41	42	3	1	1	2	2

MARK HOWE

	REGULAR SEASON					PLAYOFFS				
Year	GP	G	A	Pts	PM	GP	G	A	Pts	PM
1977-78	70	39	61	91	32	14	8	7	15	18
1978-79	77	42	65	107	32	6	4	2	6	6
1979-80	74	24	56	80	20	3	1	2	3	0
1980-81	63	19	46	65	54	...	...	...	...	...
1981-82	76	8	45	53	18	...	...	...	...	...

MARTY HOWE

	REGULAR SEASON					PLAYOFFS				
Year	GP	G	A	Pts	PM	GP	G	A	Pts	PM
1977-78	75	10	10	20	66	14	1	1	2	13
1978-79	66	9	15	24	31	9	0	1	1	8
1979-80	6	0	1	1	4	3	1	1	2	0
1980-81	12	0	1	1	25	...	...	...	...	...
1981-82	13	0	4	4	2	...	...	...	...	...
1983-84	69	0	11	11	34	...	...	...	...	...
1984-85	19	1	1	2	10	...	...	...	...	...

PAT HUGHES

	REGULAR SEASON					PLAYOFFS				
Year	GP	G	A	Pts	PM	GP	G	A	Pts	PM
1986-87	2	0	0	0	2	3	0	0	0	0

BOBBY HULL

	REGULAR SEASON					PLAYOFFS				
Year	GP	G	A	Pts	PM	GP	G	A	Pts	PM
1979-80	9	5	2	7	0	3	0	0	0	0

JODY HULL

	REGULAR SEASON					PLAYOFFS				
Year	GP	G	A	Pts	PM	GP	G	A	Pts	PM
1988-89	60	16	18	34	10	1	0	0	0	2
1990-91	38	7	10	17	21	5	0	1	1	2

MARK HUNTER

	REGULAR SEASON					PLAYOFFS				
Year	GP	G	A	Pts	PM	GP	G	A	Pts	PM
1990-91	11	4	3	7	40	6	5	1	6	17
1991-92	63	10	13	23	159	4	0	0	0	6

PAUL HURLEY

	REGULAR SEASON					PLAYOFFS				
Year	GP	G	A	Pts	PM	GP	G	A	Pts	PM
1986-87	77	3	15	18	60	15	0	7	7	14
1973-74	52	3	11	14	21	...	...	...	...	...
1974-75	75	3	26	29	36	6	0	1	1	4
1975-76	16	0	14	14	20	...	...	...	...	...

MIKE HYNDMAN

	REGULAR SEASON					PLAYOFFS				
Year	GP	G	A	Pts	PM	GP	G	A	Pts	PM
1972-73	59	4	14	18	21	...	...	...	...	...

DAVE HYNES

	REGULAR SEASON					PLAYOFFS				
Year	GP	G	A	Pts	PM	GP	G	A	Pts	PM
1976-77	22	5	4	9	4	...	...	...	...	...

DAVE INKPEN

	REGULAR SEASON					PLAYOFFS				
Year	GP	G	A	Pts	PM	GP	G	A	Pts	PM
1978-79	41	0	7	7	15	5	0	1	1	4

MARK JANSSENS

	REGULAR SEASON					PLAYOFFS				
Year	GP	G	A	Pts	PM	GP	G	A	Pts	PM
1992-93	76	12	17	29	237	...	...	...	...	...
1993-94	84	2	10	12	137	...	...	...	...	...
1994-95	46	2	5	7	93	...	...	...	...	...
1995-96	81	2	7	9	155	...	...	...	...	...
1996-97	54	2	6	8	137	...	...	...	...	...

DOUG JARVIS

	REGULAR SEASON					PLAYOFFS				
Year	GP	G	A	Pts	PM	GP	G	A	Pts	PM
1985-86	57	8	16	24	20	10	0	3	3	4
1986-87	80	9	13	22	20	6	0	0	0	4
1987-88	2	0	0	0	2	...	...	...	...	...

GRANT JENNINGS

	REGULAR SEASON					PLAYOFFS				
Year	GP	G	A	Pts	PM	GP	G	A	Pts	PM
1988-89	55	3	10	13	159	4	1	0	1	17
1989-90	64	3	6	9	171	7	0	0	0	17
1990-91	44	1	4	5	82	...	...	...	...	...

DAVID A. JENSEN

	REGULAR SEASON					PLAYOFFS				
Year	GP	G	A	Pts	PM	GP	G	A	Pts	PM
1984-85	13	0	4	4	6	...	...	...	...	...

MARK JOHNSON

	REGULAR SEASON					PLAYOFFS				
Year	GP	G	A	Pts	PM	GP	G	A	Pts	PM
1982-83	73	31	38	69	28	...	...	...	...	...
1983-84	79	35	52	87	27	...	...	...	...	...
1984-85	49	19	28	47	21	...	...	...	...	...

BERNIE JOHNSTON

	REGULAR SEASON					PLAYOFFS				
Year	GP	G	A	Pts	PM	GP	G	A	Pts	PM
1979-80	32	8	13	21	21	3	0	1	1	0
1980-81	25	4	11	15	8	...	...	...	...	...

RIC JORDAN

	REGULAR SEASON					PLAYOFFS				
Year	GP	G	A	Pts	PM	GP	G	A	Pts	PM
1972-73	36	1	5	6	21	4	0	0	0	0
1973-74	34	0	3	3	14	7	0	0	0	6

SAMI KAPANEN

	REGULAR SEASON					PLAYOFFS				
Year	GP	G	A	Pts	PM	GP	G	A	Pts	PM
1995-96	35	5	4	9	6	...	...	...	...	...
1996-97	45	13	12	25	2	...	...	...	...	...

AL KARLANDER

	REGULAR SEASON					PLAYOFFS				
Year	GP	G	A	Pts	PM	GP	G	A	Pts	PM
1973-74	75	20	41	61	46	7	1	3	4	2
1974-75	48	7	14	21	2	5	0	3	3	0

ED KASTELIC

	REGULAR SEASON					PLAYOFFS				
Year	GP	G	A	Pts	PM	GP	G	A	Pts	PM
1988-89	10	0	2	2	15	...	...	...	...	...
1989-90	67	6	2	8	198	2	0	0	0	0
1990-91	45	2	2	4	211	...	...	...	...	...
1991-92	25	1	3	4	61	...	...	...	...	...

DAN KECZMER

	REGULAR SEASON					PLAYOFFS				
Year	GP	G	A	Pts	PM	GP	G	A	Pts	PM
1991-92	1	0	0	0	0	...	...	...	...	...
1992-93	23	4	4	8	28	...	...	...	...	...
1993-94	12	0	1	1	12	...	...	...	...	...

MIKE KEELER

	REGULAR SEASON					PLAYOFFS				
Year	GP	G	A	Pts	PM	GP	G	A	Pts	PM
1973-74	1	0	0	0	0	1	0	0	0	0

KEVIN KEMP

	REGULAR SEASON					PLAYOFFS				
Year	GP	G	A	Pts	PM	GP	G	A	Pts	PM
1980-81	3	0	0	0	4	...	...	...	...	...

DAVE KEON

	REGULAR SEASON					PLAYOFFS				
Year	GP	G	A	Pts	PM	GP	G	A	Pts	PM
1976-77	34	14	25	39	8	5	3	1	4	0
1977-78	78	24	38	62	2	14	5	11	16	4
1978-79	79	22	43	65	2	10	3	9	12	2
1979-80	76	10	52	62	10	3	0	1	1	0
1980-81	80	13	34	47	26	...	...	...	...	...
1981-82	78	8	11	19	6	...	...	...	...	...

TIM KERR

	REGULAR SEASON					PLAYOFFS				
Year	GP	G	A	Pts	PM	GP	G	A	Pts	PM
1992-93	22	0	6	6	7	...	...	...	...	...

DEREK KING

	REGULAR SEASON					PLAYOFFS				
Year	GP	G	A	Pts	PM	GP	G	A	Pts	PM
1996-97	12	3	3	6	2	...	...	...	...	...

SCOT KLEINENDORST

	REGULAR SEASON					PLAYOFFS				
Year	GP	G	A	Pts	PM	GP	G	A	Pts	PM
1984-85	35	1	8	9	69	...	...	...	...	...
1985-86	41	2	7	9	62	10	0	1	1	18
1986-87	66	3	9	12	130	4	1	3	4	20
1987-88	44	3	6	9	86	3	1	1	2	0

STEVE KONROYD

	REGULAR SEASON					PLAYOFFS				
Year	GP	G	A	Pts	PM	GP	G	A	Pts	PM
1991-92	33	2	10	12	32	7	0	1	1	2
1992-93	59	3	11	14	63	...	...	...	...	...

CHRIS KOTSOPOULOS

	REGULAR SEASON					PLAYOFFS				
Year	GP	G	A	Pts	PM	GP	G	A	Pts	PM
1981-82	68	13	20	33	147	...	...	...	...	...
1982-83	68	6	24	30	125	...	...	...	...	...
1983-84	72	5	13	18	118	...	...	...	...	...
1984-85	33	5	3	8	53	...	...	...	...	...

ROBERT KRON

	REGULAR SEASON					PLAYOFFS				
Year	GP	G	A	Pts	PM	GP	G	A	Pts	PM
1992-83	13	4	2	6	4	...	...	...	...	...
1993-94	77	24	26	50	18	...	...	...	...	...
1994-95	37	10	8	18	10	...	...	...	...	...
1995-96	77	22	28	50	6	...	...	...	...	...
1996-97	68	10	12	22	10	...	...	...	...	...

TODD KRYGIER

	REGULAR SEASON					PLAYOFFS				
Year	GP	G	A	Pts	PM	GP	G	A	Pts	PM
1989-90	58	18	12	30	52	7	2	1	3	4
1990-91	72	13	17	30	95	6	0	2	2	0

FRANTISEK KUCERA

	REGULAR SEASON					PLAYOFFS				
Year	GP	G	A	Pts	PM	GP	G	A	Pts	PM
1993-94	16	1	3	4	14	...	...	...	...	...
1994-95	48	3	17	20	30	...	...	...	...	...
1995-96	30	2	6	8	10	...	...	...	...	...

NICK KYPREOS

	REGULAR SEASON					PLAYOFFS				
Year	GP	G	A	Pts	PM	GP	G	A	Pts	PM
1992-93	75	17	10	17	325	...	...	...	...	...
1993-94	10	0	0	0	37	...	...	...	...	...

ANDRE LACROIX

	REGULAR SEASON					PLAYOFFS				
Year	GP	G	A	Pts	PM	GP	G	A	Pts	PM
1978-79	78	32	56	88	34	10	4	4	8	0
1979-80	29	3	14	17	2	...	...	...	...	...

PIERRE LACROIX

	REGULAR SEASON					PLAYOFFS				
Year	GP	G	A	Pts	PM	GP	G	A	Pts	PM
1982-83	56	6	25	31	18	...	...	...	...	...

RANDY LADOUCEUR

	REGULAR SEASON					PLAYOFFS				
Year	GP	G	A	Pts	PM	GP	G	A	Pts	PM
1986-87	36	2	3	5	51	6	0	2	2	12
1987-88	68	1	7	8	91	6	1	1	2	4
1988-89	75	2	5	7	95	1	0	0	0	10
1988-90	71	3	12	15	126	7	1	0	1	10
1990-91	67	1	3	4	118	6	1	4	5	6
1981-92	74	1	9	10	127	7	0	1	1	11
1992-93	62	2	4	6	109	...	...	...	...	...

MARK LAFORGE

	REGULAR SEASON					PLAYOFFS				
Year	GP	G	A	Pts	PM	GP	G	A	Pts	PM
1989-90	9	0	0	0	41	...	...	...	...	...

PIERRE LAROUCHE

	REGULAR SEASON					PLAYOFFS				
Year	GP	G	A	Pts	PM	GP	G	A	Pts	PM
1981-82	45	25	25	50	12	...	...	...	...	...
1982-83	38	18	22	40	8	...	...	...	...	...

PAUL LAWLESS

	REGULAR SEASON					PLAYOFFS				
Year	GP	G	A	Pts	PM	GP	G	A	Pts	PM
1982-83	47	6	9	15	4	...	...	...	...	...
1983-84	6	0	3	3	0	...	...	...	...	...
1985-86	64	17	21	38	20	1	0	0	0	0
1986-87	60	22	32	54	14	2	0	2	2	2
1987-88	28	4	5	9	16	...	...	...	...	...

BRIAN LAWTON

	REGULAR SEASON					PLAYOFFS				
Year	GP	G	A	Pts	PM	GP	G	A	Pts	PM
1988-89	48	12	18	29	32	3	1	0	1	0

JAMIE LEACH

	REGULAR SEASON					PLAYOFFS				
Year	GP	G	A	Pts	PM	GP	G	A	Pts	PM
1992-93	19	3	2	5	4	...	...	...	...	...

JOCELYN LEMIEUX

	REGULAR SEASON					PLAYOFFS				
Year	GP	G	A	Pts	PM	GP	G	A	Pts	PM
1993-94	16	6	1	7	19	...	...	...	...	...
1994-95	41	6	5	11	32	...	...	...	...	...
1995-96	29	1	2	3	31	...	...	...	...	...

RICK LEY

	REGULAR SEASON					PLAYOFFS				
Year	GP	G	A	Pts	PM	GP	G	A	Pts	PM
1972-73	77	3	27	30	108	15	3	7	10	24
1973-74	43	6	35	41	148	7	1	5	6	18
1974-75	62	6	36	42	50	6	1	1	2	32
1975-76	67	8	30	38	78	17	1	4	5	49
1976-77	55	2	21	23	102	5	0	4	4	4
1977-78	73	3	41	44	95	14	1	8	9	4
1978-79	73	7	20	27	135	9	0	4	4	11
1979-80	65	4	16	20	92	...	...	...	...	...
1980-81	16	0	2	2	20	...	...	...	...	...

CURTIS LESCHYSHYN

	REGULAR SEASON					PLAYOFFS				
Year	GP	G	A	Pts	PM	GP	G	A	Pts	PM
1996-97	64	4	13	17	30	...	...	...	...	...

CHUCK LUKSA

	REGULAR SEASON					PLAYOFFS				
Year	GP	G	A	Pts	PM	GP	G	A	Pts	PM
1979-80	8	0	1	1	4	...	...	...	...	...

DAVE LUMLEY

	REGULAR SEASON					PLAYOFFS				
Year	GP	G	A	Pts	PM	GP	G	A	Pts	PM
1984-85	48	8	20	28	98	...	...	...	...	...

GILLES LUPIEN

	REGULAR SEASON					PLAYOFFS				
Year	GP	G	A	Pts	PM	GP	G	A	Pts	PM
1980-81	20	2	4	6	39	...	...	...	...	...
1981-82	1	0	1	1	2	...	...	...	...	...

GEORGE LYLE

	REGULAR SEASON					PLAYOFFS				
Year	GP	G	A	Pts	PM	GP	G	A	Pts	PM
1976-77	75	39	33	72	62	57	1	0	1	4
1977-78	68	30	24	54	74	12	2	1	3	13
1978-79	59	17	18	35	54	9	3	5	8	25
1981-82	14	2	12	14	9	...	...	...	...	...
1982-83	16	4	6	10	8	...	...	...	...	...

PAUL MacDERMID

	REGULAR SEASON					PLAYOFFS				
Year	GP	G	A	Pts	PM	GP	G	A	Pts	PM
1981-82	3	1	0	1	2	...	...	...	...	...
1982-83	7	0	0	0	2	...	...	...	...	...
1983-84	3	0	1	1	0	...	...	...	...	...
1984-85	31	4	7	11	29	...	...	...	...	...
1985-86	74	13	10	23	160	10	2	1	3	20
1986-87	72	7	11	18	202	6	2	1	3	34
1987-88	80	20	15	35	139	6	0	5	5	14
1988-89	74	17	27	44	141	4	1	1	2	16
1989-90	29	6	12	18	69	...	...	...	...	...

GARY MacGREGOR

	REGULAR SEASON					PLAYOFFS				
Year	GP	G	A	Pts	PM	GP	G	A	Pts	PM
1976-77	30	8	8	16	4	...	...	...	...	...

RANDY MacGREGOR

	REGULAR SEASON					PLAYOFFS				
Year	GP	G	A	Pts	PM	GP	G	A	Pts	PM
1981-82	2	1	1	2	2	...	...	...	...	...

NORM MACIVER

	REGULAR SEASON					PLAYOFFS				
Year	GP	G	A	Pts	PM	GP	G	A	Pts	PM
1988-89	37	1	22	23	24	1	0	0	0	2

RICK MacLEISH

	REGULAR SEASON					PLAYOFFS				
Year	GP	G	A	Pts	PM	GP	G	A	Pts	PM
1981-82	34	6	16	22	16	...	...	...	...	...

MAREK MALIK

	REGULAR SEASON					PLAYOFFS				
Year	GP	G	A	Pts	PM	GP	G	A	Pts	PM
1994-95	1	0	1	1	0	...	...	...	...	...
1995-96	7	0	0	0	4	...	...	...	...	...
1996-97	47	1	5	6	50	...	...	...	...	...

MERLIN MALINOWSKI

	REGULAR SEASON					PLAYOFFS				
Year	GP	G	A	Pts	PM	GP	G	A	Pts	PM
1982-83	75	5	23	28	16	...	...	...	...	...

GREG MALONE

	REGULAR SEASON					PLAYOFFS				
Year	GP	G	A	Pts	PM	GP	G	A	Pts	PM
1983-84	78	17	37	54	56	...	...	...	...	...
1984-85	76	22	39	61	67	...	...	...	...	...
1985-86	22	6	7	13	24	...	...	...	...	...

DON MALONEY

	REGULAR SEASON					PLAYOFFS				
Year	GP	G	A	Pts	PM	GP	G	A	Pts	PM
1988-89	21	3	11	14	23	4	0	0	0	8

KENT MANDERVILLE

	REGULAR SEASON					PLAYOFFS				
Year	GP	G	A	Pts	PM	GP	G	A	Pts	PM
1996-97	44	6	5	11	18	...	...	...	...	...

BRYAN MARCHMENT

	REGULAR SEASON					PLAYOFFS				
Year	GP	G	A	Pts	PM	GP	G	A	Pts	PM
1993-94	42	3	7	10	124	...	...	...	...	...

PAUL MARSHALL

	REGULAR SEASON					PLAYOFFS				
Year	GP	G	A	Pts	PM	GP	G	A	Pts	PM
1982-83	13	1	2	3	0	...	...	...	...	...

TOM MARTIN

	REGULAR SEASON					PLAYOFFS				
Year	GP	G	A	Pts	PM	GP	G	A	Pts	PM
1987-88	5	1	2	3	14	...	...	...	...	...
1988-89	38	7	6	13	113	1	0	0	0	4
1989-90	21	1	1	2	37	...	...	...	...	...

STEVE MARTINS

	REGULAR SEASON					PLAYOFFS				
Year	GP	G	A	Pts	PM	GP	G	A	Pts	PM
1995-96	23	1	3	4	8	...	...	...	...	...
1996-97	2	0	1	1	0	...	...	...	...	...

BRYAN MAXWELL

	REGULAR SEASON					PLAYOFFS				
Year	GP	G	A	Pts	PM	GP	G	A	Pts	PM
1977-78	17	2	1	3	11	...	...	...	...	...

JIM MAYER

	REGULAR SEASON					PLAYOFFS				
Year	GP	G	A	Pts	PM	GP	G	A	Pts	PM
1977-78	51	11	9	20	21	...	...	...	...	...

JASON McBAIN

	REGULAR SEASON					PLAYOFFS				
Year	GP	G	A	Pts	PM	GP	G	A	Pts	PM
1995-96	3	0	0	0	0	...	...	...	...	...
1996-97	6	0	0	0	0	...	...	...	...	...

ROB McCLANAHAN

	REGULAR SEASON					PLAYOFFS				
Year	GP	G	A	Pts	PM	GP	G	A	Pts	PM
1981-82	17	0	3	3	11	...	...	...	...	...

BRAD McCRIMMON

	REGULAR SEASON					PLAYOFFS				
Year	GP	G	A	Pts	PM	GP	G	A	Pts	PM
1993-94	65	1	5	6	72	...	...	...	...	...
1994-95	33	0	1	1	42	...	...	...	...	...
1995-96	58	3	6	9	62	...	...	...	...	...

GERRY McDONALD

	REGULAR SEASON					PLAYOFFS				
Year	GP	G	A	Pts	PM	GP	G	A	Pts	PM
1981-82	3	0	0	0	0	...	...	...	...	...
1983-84	5	0	0	0	4	...	...	...	...	...

MIKE McDOUGAL

	REGULAR SEASON					PLAYOFFS				
Year	GP	G	A	Pts	PM	GP	G	A	Pts	PM
1981-82	3	0	0	0	0	...	...	...	...	...
1982-83	55	8	10	18	43	...	...	...	...	...

MIKE McEWEN

	REGULAR SEASON					PLAYOFFS				
Year	GP	G	A	Pts	PM	GP	G	A	Pts	PM
1985-86	10	3	2	5	6	8	0	4	4	6
1986-87	48	8	8	16	32	1	1	1	2	0

BOB McGILL

	REGULAR SEASON					PLAYOFFS				
Year	GP	G	A	Pts	PM	GP	G	A	Pts	PM
1993-94	30	0	3	3	41	...	...	...	...	...

JACK McILHARGEY

	REGULAR SEASON					PLAYOFFS				
Year	GP	G	A	Pts	PM	GP	G	A	Pts	PM
1980-81	48	1	6	7	142	...	...	...	...	...
1981-82	50	1	5	6	60	...	...	...	...	...

JIM McKENZIE

	REGULAR SEASON					PLAYOFFS				
Year	GP	G	A	Pts	PM	GP	G	A	Pts	PM
1989-90	5	0	0	0	4	...	...	...	...	...
1990-91	41	4	3	7	108	6	0	0	0	8
1991-92	67	5	1	6	87	...	...	...	...	...
1992-93	64	3	6	9	202	...	...	...	...	...
1993-94	26	1	2	3	67	...	...	...	...	...

JOHN McKENZIE

	REGULAR SEASON					PLAYOFFS				
Year	GP	G	A	Pts	PM	GP	G	A	Pts	PM
1976-77	34	11	19	30	25	5	2	1	3	8
1977-78	79	27	29	56	61	14	6	6	12	16
1978-79	76	19	28	47	112	10	3	7	10	10

BOB McMANAMA

	REGULAR SEASON					PLAYOFFS				
Year	GP	G	A	Pts	PM	GP	G	A	Pts	PM
1975-76	37	3	10	13	28	12	4	3	7	4

RICK MEAGHER

	REGULAR SEASON					PLAYOFFS				
Year	GP	G	A	Pts	PM	GP	G	A	Pts	PM
1980-81	27	7	10	17	19	...	...	...	...	...
1981-82	65	24	19	43	51	...	...	...	...	...
1982-83	4	0	0	0	0	...	...	...	...	...

GLENN MERKOSKY

	REGULAR SEASON					PLAYOFFS				
Year	GP	G	A	Pts	PM	GP	G	A	Pts	PM
1981-82	7	0	0	0	2	...	...	...	...	...

GERRY METHE

	REGULAR SEASON					PLAYOFFS				
Year	GP	G	A	Pts	PM	GP	G	A	Pts	PM
1974-75	5	0	1	1	4	2	0	0	0	0

MIKE MILLAR

	REGULAR SEASON					PLAYOFFS				
Year	GP	G	A	Pts	PM	GP	G	A	Pts	PM
1992-93	10	2	2	4	0	...	...	...	...	...
1993-94	28	7	7	14	6	...	...	...	...	...

WARREN MILLER

	REGULAR SEASON					PLAYOFFS				
Year	GP	G	A	Pts	PM	GP	G	A	Pts	PM
1978-79	76	26	23	49	44	10	0	8	8	28
1980-81	77	22	22	44	37	...	...	...	...	...
1981-82	74	10	12	22	68	...	...	...	...	...
1982-83	56	1	10	11	15	...	...	...	...	...

CHRIS MURRAY

	REGULAR SEASON					PLAYOFFS				
Year	GP	G	A	Pts	PM	GP	G	A	Pts	PM
1996-97	8	1	1	2	10	...	...	...	...	...

DANA MURZYN

	REGULAR SEASON					PLAYOFFS				
Year	GP	G	A	Pts	PM	GP	G	A	Pts	PM
1985-86	78	3	23	26	125	4	0	0	0	10
1986-97	74	9	19	28	95	6	2	1	3	29
1987-88	33	1	6	7	45	...	...	...	...	...

DON NACHBAUR

	REGULAR SEASON					PLAYOFFS				
Year	GP	G	A	Pts	PM	GP	G	A	Pts	PM
1980-81	77	16	17	33	139	...	...	...	...	...
1981-82	77	5	21	26	117	...	...	...	...	...

RAY NEUFELD

	REGULAR SEASON					PLAYOFFS				
Year	GP	G	A	Pts	PM	GP	G	A	Pts	PM
1979-80	8	1	0	1	0	2	1	0	1	0
1980-81	52	5	10	15	44	...	...	...	...	...
1981-82	19	4	3	7	4	...	...	...	...	...
1982-83	80	26	31	57	86	...	...	...	...	...
1983-84	80	27	42	69	97	...	...	...	...	...
1984-85	76	22	39	61	67	...	...	...	...	...
1985-86	16	5	10	15	40	...	...	...	...	...

JOHN NEWBERRY

	REGULAR SEASON					PLAYOFFS				
Year	GP	G	A	Pts	PM	GP	G	A	Pts	PM
1985-86	3	0	0	0	0	...	...	...	...	...

BARRY NIECKAR

	REGULAR SEASON					PLAYOFFS				
Year	GP	G	A	Pts	PM	GP	G	A	Pts	PM
1992-93	2	0	0	0	0	...	...	...	...	...

ANDREI NIKOLISHIN

	REGULAR SEASON					PLAYOFFS				
Year	GP	G	A	Pts	PM	GP	G	A	Pts	PM
1994-95	39	8	10	18	10	...	...	...	...	...
1995-96	61	14	37	51	34	...	...	...	...	...
1996-97	12	2	5	7	2	...	...	...	...	...

LEE NORWOOD

	REGULAR SEASON					PLAYOFFS				
Year	GP	G	A	Pts	PM	GP	G	A	Pts	PM
1991-92	6	0	0	0	16	...	...	...	...	...

MICHAEL NYLANDER

	REGULAR SEASON					PLAYOFFS				
Year	GP	G	A	Pts	PM	GP	G	A	Pts	PM
1992-93	59	11	22	33	36	...	...	...	...	...
1993-94	58	11	33	44	24	...	...	...	...	...

FRED O'DONNELL

	REGULAR SEASON					PLAYOFFS				
Year	GP	G	A	Pts	PM	GP	G	A	Pts	PM
1974-75	76	21	15	36	84	3	0	0	0	15
1975-76	79	11	11	22	81	17	2	5	7	20

JEFF O'NEILL

	REGULAR SEASON					PLAYOFFS				
Year	GP	G	A	Pts	PM	GP	G	A	Pts	PM
1995-96	65	8	19	27	40	...	...	...	...	...
1996-97	72	14	16	30	40	...	...	...	...	...

ROSAIRE PAIEMENT

	REGULAR SEASON					PLAYOFFS				
Year	GP	G	A	Pts	PM	GP	G	A	Pts	PM
1975-76	80	28	43	71	89	17	4	11	15	41
1976-77	13	5	2	7	12	...	...	...	...	...

JEFF PARKER

	REGULAR SEASON					PLAYOFFS				
Year	GP	G	A	Pts	PM	GP	G	A	Pts	PM
1990-91	4	0	0	0	4	...	...	...	...	...

MARK PATERSON

	REGULAR SEASON					PLAYOFFS				
Year	GP	G	A	Pts	PM	GP	G	A	Pts	PM
1982-83	2	0	0	0	0	...	...	...	...	...
1983-84	9	2	0	2	4	...	...	...	...	...
1984-85	13	1	3	4	24	...	...	...	...	...
1985-86	5	0	0	0	5	...	...	...	...	...

JAMES PATRICK

	REGULAR SEASON					PLAYOFFS				
Year	GP	G	A	Pts	PM	GP	G	A	Pts	PM
1993-94	47	8	20	28	32	...	...	...	...	...

JIM PAVESE

	REGULAR SEASON					PLAYOFFS				
Year	GP	G	A	Pts	PM	GP	G	A	Pts	PM
1988-89	5	0	0	0	5	1	0	0	0	0

ALLEN PEDERSON

	REGULAR SEASON					PLAYOFFS				
Year	GP	G	A	Pts	PM	GP	G	A	Pts	PM
1992-93	59	1	4	5	60	...	...	...	...	...
1993-94	7	0	0	0	9	...	...	...	...	...

BARRY PEDERSON

	REGULAR SEASON					PLAYOFFS				
Year	GP	G	A	Pts	PM	GP	G	A	Pts	PM
1991-92	5	2	2	4	0	...	...	...	...	...

ANDRE PELOFFY

	REGULAR SEASON					PLAYOFFS				
Year	GP	G	A	Pts	PM	GP	G	A	Pts	PM
1978-79	10	2	0	2	2	...	...	...	...	...

BRENT PETERSON

	REGULAR SEASON					PLAYOFFS				
Year	GP	G	A	Pts	PM	GP	G	A	Pts	PM
1987-88	52	2	7	9	40	4	0	0	0	2
1988-89	66	4	13	17	61	2	0	1	1	2

ROBERT PETROVICKY

	REGULAR SEASON					PLAYOFFS				
Year	GP	G	A	Pts	PM	GP	G	A	Pts	PM
1992-93	42	3	6	9	45	...	...	...	...	...
1993-94	33	6	5	11	39	...	...	...	...	...
1994-95	2	0	0	0	0	...	...	...	...	...
1995-96	2	0	0	0	0	...	...	...	...	...

JORGEN PETTERSSON

	REGULAR SEASON					PLAYOFFS				
Year	GP	G	A	Pts	PM	GP	G	A	Pts	PM
1985-86	23	5	5	10	2	...	...	...	...	...

MICHEL PICARD

	REGULAR SEASON					PLAYOFFS				
Year	GP	G	A	Pts	PM	GP	G	A	Pts	PM
1990-91	5	1	0	1	2	...	...	...	...	...
1991-92	25	3	5	8	6	...	...	...	...	...

RANDY PIERRE

	REGULAR SEASON					PLAYOFFS				
Year	GP	G	A	Pts	PM	GP	G	A	Pts	PM
1983-84	17	6	3	9	9	...	...	...	...	...
1984-85	17	3	2	5	56	...	...	...	...	...

MARC POTVIN

	REGULAR SEASON					PLAYOFFS				
Year	GP	G	A	Pts	PM	GP	G	A	Pts	PM
1993-94	51	2	3	5	246	...	...	...	...	...

LARRY PLEAU

Year	REGULAR SEASON					PLAYOFFS				
	GP	G	A	Pts	PM	GP	G	A	Pts	PM
1972-73	76	39	48	87	44	15	12	7	19	15
1973-74	75	26	43	69	35	2	2	0	2	0
1974-75	78	30	34	64	50	6	2	3	5	14
1975-76	75	29	45	74	21	14	5	7	12	0
1976-77	78	11	21	32	22	5	1	0	1	0
1977-78	54	16	18	34	14	14	5	4	9	8
1978-79	28	6	6	12	10	10	2	1	3	0

RON PLUMB

Year	REGULAR SEASON					PLAYOFFS				
	GP	G	A	Pts	PM	GP	G	A	Pts	PM
1977-78	27	1	9	10	18	14	1	5	6	16
1978-79	77	4	16	20	33	9	1	3	4	0
1979-80	26	3	4	7	14	...	...	...	...	...

PATRICK POULIN

Year	REGULAR SEASON					PLAYOFFS				
	GP	G	A	Pts	PM	GP	G	A	Pts	PM
1991-92	1	0	0	0	2	7	1	2	3	0
1992-93	81	20	31	51	37	...	...	...	...	...
1993-94	9	2	1	3	11	...	...	...	...	...

CHRIS PRONGER

Year	REGULAR SEASON					PLAYOFFS				
	GP	G	A	Pts	PM	GP	G	A	Pts	PM
1993-94	81	5	25	30	113	...	...	...	...	...
1994-95	43	5	9	14	54	...	...	...	...	...

NOLAN PRATT

Year	REGULAR SEASON					PLAYOFFS				
	GP	G	A	Pts	PM	GP	G	A	Pts	PM
1996-97	9	0	2	2	6	...	...	...	...	...

KEITH PRIMEAU

Year	REGULAR SEASON					PLAYOFFS				
	GP	G	A	Pts	PM	GP	G	A	Pts	PM
1996-97	75	26	25	51	161	...	...	...	...	...

BRIAN PROPP

Year	REGULAR SEASON					PLAYOFFS				
	GP	G	A	Pts	PM	GP	G	A	Pts	PM
1993-94	65	12	17	29	44	...	...	...	...	...

JOEL QUENNEVILLE

Year	REGULAR SEASON					PLAYOFFS				
	GP	G	A	Pts	PM	GP	G	A	Pts	PM
1983-84	80	5	8	13	95	...	...	...	...	...
1984-85	79	6	16	22	96	...	...	...	...	...
1985-86	71	5	20	25	23	10	0	2	2	12
1986-87	37	3	7	10	24	6	0	0	0	0
1987-88	77	1	8	9	44	6	0	2	2	2
1988-89	69	4	7	11	32	4	0	3	3	4
1989-90	44	1	4	5	34	...	...	...	...	...

PAUL RANHEIM

Year	REGULAR SEASON					PLAYOFFS				
	GP	G	A	Pts	PM	GP	G	A	Pts	PM
1993-94	15	0	3	3	12	...	...	...	...	...
1994-95	47	6	14	20	10	...	...	...	...	...
1995-96	73	10	20	30	14	...	...	...	...	...
1996-97	67	10	11	21	18	...	...	...	...	...

MARK REEDS

Year	REGULAR SEASON					PLAYOFFS				
	GP	G	A	Pts	PM	GP	G	A	Pts	PM
1987-88	38	0	7	7	31	...	...	...	...	...
1988-89	7	0	2	2	6	...	...	...	...	...

MARK RENAUD

Year	REGULAR SEASON					PLAYOFFS				
	GP	G	A	Pts	PM	GP	G	A	Pts	PM
1979-80	13	0	2	2	4	...	...	...	...	...
1980-81	4	1	0	1	2	...	...	...	...	...
1981-82	48	1	17	18	39	...	...	...	...	...
1982-83	77	3	28	31	37	...	...	...	...	...

STEVEN RICE

Year	REGULAR SEASON					PLAYOFFS				
	GP	G	A	Pts	PM	GP	G	A	Pts	PM
1994-95	40	11	10	21	61	...	...	...	...	...
1995-96	59	10	12	22	47	...	...	...	...	...
1996-97	78	21	14	35	59	...	...	...	...	...

TODD RICHARDS

Year	REGULAR SEASON					PLAYOFFS				
	GP	G	A	Pts	PM	GP	G	A	Pts	PM
1990-91	2	0	4	4	4	5	0	3	3	4
1991-92	6	0	0	0	2	6	0	0	0	2

STEVE RICHARDSON

Year	REGULAR SEASON					PLAYOFFS				
	GP	G	A	Pts	PM	GP	G	A	Pts	PM
1975-76	6	0	0	0	0	...	...	...	...	...

DOUG ROBERTS

Year	REGULAR SEASON					PLAYOFFS				
	GP	G	A	Pts	PM	GP	G	A	Pts	PM
1975-76	76	4	13	17	51	17	1	1	2	8
1976-77	64	3	18	21	33	2	0	0	0	0

GORDIE ROBERTS

Year	REGULAR SEASON					PLAYOFFS				
	GP	G	A	Pts	PM	GP	G	A	Pts	PM
1975-76	77	3	19	22	169	5	2	2	4	6
1976-77	77	13	33	46	118	14	0	5	5	29
1977-78	78	15	46	61	113	10	0	4	4	16
1978-79	79	11	46	57	141	4	1	1	2	16
1979-80	80	8	28	36	89	3	1	1	2	2
1980-81	27	2	11	13	81	...	...	...	...	...

TORRIE ROBERTSON

Year	REGULAR SEASON					PLAYOFFS				
	GP	G	A	Pts	PM	GP	G	A	Pts	PM
1983-84	66	7	14	21	198	...	...	...	...	...
1984-85	74	11	30	41	337	...	...	...	...	...
1985-86	76	13	24	37	358	10	1	0	1	67
1986-87	20	1	0	1	98	...	...	...	...	...
1987-88	63	2	8	10	293	6	0	1	1	6

MIKE ROGERS

Year	REGULAR SEASON					PLAYOFFS				
	GP	G	A	Pts	PM	GP	G	A	Pts	PM
1975-76	36	18	14	32	10	17	5	8	13	2
1976-77	78	25	57	82	10	5	1	1	2	2
1977-78	80	28	43	71	46	14	4	6	11	8
1978-79	80	27	45	72	31	10	2	6	8	2
1979-80	80	44	61	105	32	3	0	3	3	0
1980-81	80	40	65	105	32	...	...	...	...	...

TOM ROWE

Year	REGULAR SEASON					PLAYOFFS				
	GP	G	A	Pts	PM	GP	G	A	Pts	PM
1979-80	20	6	4	10	30	3	2	0	2	0
1980-81	74	13	28	41	190	...	...	...	...	...
1981-82	21	4	0	4	36	...	...	...	...	...

PIERRE ROY

Year	REGULAR SEASON					PLAYOFFS				
	GP	G	A	Pts	PM	GP	G	A	Pts	PM
1978-79	1	0	0	0	2	...	...	...	...	...

ULF SAMUELSSON

Year	REGULAR SEASON					PLAYOFFS				
	GP	G	A	Pts	PM	GP	G	A	Pts	PM
1984-85	41	2	6	0	83	...	...	...	...	...
1985-86	80	5	19	24	172	10	1	2	3	38
1986-87	78	2	31	33	162	5	0	1	1	41
1987-88	76	8	33	41	159	5	0	0	0	8
1988-89	71	9	26	35	181	4	0	2	2	4
1989-90	55	2	11	13	177	7	1	0	1	18
1990-91	62	3	18	21	174	...	...	...	...	...

GEOFF SANDERSON

Year	REGULAR SEASON					PLAYOFFS				
	GP	G	A	Pts	PM	GP	G	A	Pts	PM
1990-91	2	1	0	1	0	3	0	0	0	0
1991-92	64	13	18	31	18	7	1	0	1	2
1992-93	82	46	43	89	28	...	...	...	...	...
1993-94	82	41	26	67	42	...	...	...	...	...
1994-95	46	18	14	32	24	...	...	...	...	...
1995-96	81	34	31	65	40	...	...	...	...	...
1996-97	82	36	31	67	29	...	...	...	...	...

JIM SANDLAK

Year	REGULAR SEASON					PLAYOFFS				
	GP	G	A	Pts	PM	GP	G	A	Pts	PM
1993-94	27	6	2	8	32	...	...	...	...	...
1994-95	13	0	0	0	0	...	...	...	...	...

DICK SARRAZIN

Year	REGULAR SEASON					PLAYOFFS				
	GP	G	A	Pts	PM	GP	G	A	Pts	PM
1972-73	35	4	7	11	0	...	...	...	...	...

JEAN SAVARD

Year	REGULAR SEASON					PLAYOFFS				
	GP	G	A	Pts	PM	GP	G	A	Pts	PM
1979-80	1	0	0	0	0	...	...	...	...	...

MAYNARD SCHURMAN

Year	REGULAR SEASON					PLAYOFFS				
	GP	G	A	Pts	PM	GP	G	A	Pts	PM
1979-80	7	0	0	0	0	...	...	...	...	...

BRIT SELBY

Year	REGULAR SEASON					PLAYOFFS				
	GP	G	A	Pts	PM	GP	G	A	Pts	PM
1972-73	65	13	29	42	48	13	3	4	7	13

BRAD SELWOOD

Year	REGULAR SEASON					PLAYOFFS				
	GP	G	A	Pts	PM	GP	G	A	Pts	PM
1972-73	75	13	21	34	110	15	3	5	8	22
1973-74	76	9	28	37	91	7	0	2	2	11
1974-75	77	4	35	39	117	5	1	0	1	11
1975-76	40	2	10	12	28	17	2	2	4	27
1976-77	41	4	12	16	71	5	0	0	0	2
1977-78	80	6	25	31	88	14	0	3	3	8
1978-79	42	4	12	16	47	...	...	...	...	...

DAVE SEMENKO

Year	REGULAR SEASON					PLAYOFFS				
	GP	G	A	Pts	PM	GP	G	A	Pts	PM
1986-87	57	4	8	12	87	...	...	...	...	...

BRENDAN SHANAHAN

Year	REGULAR SEASON					PLAYOFFS				
	GP	G	A	Pts	PM	GP	G	A	Pts	PM
1995-96	74	44	34	78	125	...	...	...	...	...
1996-97	2	1	0	1	0	...	...	...	...	...

DANIEL SHANK

Year	REGULAR SEASON					PLAYOFFS				
	GP	G	A	Pts	PM	GP	G	A	Pts	PM
1991-92	13	2	0	2	18	5	0	0	0	22

BRAD SHAW

Year	REGULAR SEASON					PLAYOFFS				
	GP	G	A	Pts	PM	GP	G	A	Pts	PM
1985-86	8	0	0	0	4	...	...	...	...	...
1986-87	2	0	0	0	0	...	...	...	...	...
1987-88	1	0	0	0	0	...	...	...	...	...
1988-89	3	1	0	1	0	...	...	...	...	...
1989-90	64	3	32	35	30	7	2	5	7	0
1990-91	72	4	28	32	29	6	1	2	3	2
1991-92	62	3	22	25	44	3	0	1	1	4

NEIL SHEEHY

Year	REGULAR SEASON					PLAYOFFS				
	GP	G	A	Pts	PM	GP	G	A	Pts	PM
1987-88	26	1	4	5	116	1	0	0	0	7

TIM SHEEHY

Year	REGULAR SEASON					PLAYOFFS				
	GP	G	A	Pts	PM	GP	G	A	Pts	PM
1972-73	78	33	38	71	30	15	9	14	23	13
1973-74	77	29	29	58	22	7	4	2	6	4
1974-75	52	20	13	33	18	...	...	...	...	...
1977-78	11	1	1	2	0	13	1	3	4	9
1979-80	12	2	1	3	0	...	...	...	...	...

GORD SHERVEN

Year	REGULAR SEASON					PLAYOFFS				
	GP	G	A	Pts	PM	GP	G	A	Pts	PM
1986-87	8	0	0	0	0	...	...	...	...	...

PAUL SHMYR

Year	REGULAR SEASON					PLAYOFFS				
	GP	G	A	Pts	PM	GP	G	A	Pts	PM
1981-82	66	1	11	12	134	...	...	...	...	...

RISTO SILTANEN

Year	REGULAR SEASON					PLAYOFFS				
	GP	G	A	Pts	PM	GP	G	A	Pts	PM
1982-83	74	5	25	30	28	...	...	...	...	...
1983-84	75	15	38	53	34	...	...	...	...	...
1984-85	76	12	33	45	30	...	...	...	...	...
1985-86	52	8	22	30	30	...	...	...	...	...

AL SIMS

Year	REGULAR SEASON					PLAYOFFS				
	GP	G	A	Pts	PM	GP	G	A	Pts	PM
1979-80	76	10	31	41	30	3	0	0	0	2
1980-81	80	16	36	52	68	...	...	...	...	...

DALE SMEDSMO

Year	REGULAR SEASON					PLAYOFFS				
	GP	G	A	Pts	PM	GP	G	A	Pts	PM
1976-77	15	2	0	2	54	...	...	...	...	...

GUY SMITH

Year	REGULAR SEASON					PLAYOFFS				
	GP	G	A	Pts	PM	GP	G	A	Pts	PM
1972-73	24	3	3	6	6	...	...	...	...	...
1973-74	16	1	5	6	25	...	...	...	...	...

STUART SMITH

Year	REGULAR SEASON					PLAYOFFS				
	GP	G	A	Pts	PM	GP	G	A	Pts	PM
1979-80	4	0	0	0	0	...	...	...	...	...
1980-81	38	1	7	8	55	...	...	...	...	...
1981-82	17	0	3	3	15	...	...	...	...	...
1982-83	18	1	0	1	25	...	...	...	...	...

KEVIN SMYTH

Year	REGULAR SEASON					PLAYOFFS				
	GP	G	A	Pts	PM	GP	G	A	Pts	PM
1993-94	21	3	2	5	10	...	...	...	...	...
1994-95	16	1	5	6	13	...	...	...	...	...
1995-96	21	2	1	3	8	...	...	...	...	...

BOB STEPHENSON

Year	REGULAR SEASON					PLAYOFFS				
	GP	G	A	Pts	PM	GP	G	A	Pts	PM
1979-80	4	0	1	1	0	...	...	...	...	...

JOHN STEVENS

Year	REGULAR SEASON					PLAYOFFS				
	GP	G	A	Pts	PM	GP	G	A	Pts	PM
1990-91	14	0	1	1	11	...	...	...	...	...
1991-92	21	0	4	4	19	...	...	...	...	...
1993-94	9	0	3	3	4	...	...	...	...	...

JIM STORM

Year	REGULAR SEASON					PLAYOFFS				
	GP	G	A	Pts	PM	GP	G	A	Pts	PM
1993-94	68	6	10	16	27	...	...	...	...	...
1994-95	6	0	3	3	0	...	...	...	...	...

STEVE STOYANOVICH

Year	REGULAR SEASON					PLAYOFFS				
	GP	G	A	Pts	PM	GP	G	A	Pts	PM
1983-84	23	3	5	8	11	...	...	...	...	...

BOB SULLIVAN

Year	REGULAR SEASON					PLAYOFFS				
	GP	G	A	Pts	PM	GP	G	A	Pts	PM
1982-83	62	18	19	37	18	...	...	...	...	...

BLAINE STOUGHTON

Year	REGULAR SEASON					PLAYOFFS				
	GP	G	A	Pts	PM	GP	G	A	Pts	PM
1978-79	36	9	3	12	6	7	4	3	7	4
1979-80	80	56	44	100	16	1	0	0	0	0
1980-81	71	43	30	73	56	...	...	...	...	...
1981-82	80	52	39	91	57	...	...	...	...	...
1982-83	72	45	31	76	27	...	...	...	...	...
1983-84	54	23	14	37	4	...	...	...	...	...

DOUG SULLIMAN

Year	REGULAR SEASON					PLAYOFFS				
	GP	G	A	Pts	PM	GP	G	A	Pts	PM
1981-82	77	29	40	69	39	...	...	...	...	...
1982-83	77	22	19	41	14	...	...	...	...	...
1983-84	67	6	12	18	20	...	...	...	...	...

GARRY SWAIN

Year	REGULAR SEASON					PLAYOFFS				
	GP	G	A	Pts	PM	GP	G	A	Pts	PM
1974-75	66	7	15	22	18	6	0	3	3	41
1975-76	79	10	16	26	46	17	3	2	5	15
1976-77	26	5	2	7	6	2	0	0	0	0

CHRIS TANCIL

Year	REGULAR SEASON					PLAYOFFS				
	GP	G	A	Pts	PM	GP	G	A	Pts	PM
1990-91	9	1	1	2	4	...	...	...	...	...
1991-92	10	0	0	0	2	...	...	...	...	...

JIM THOMSON

Year	REGULAR SEASON					PLAYOFFS				
	GP	G	A	Pts	PM	GP	G	A	Pts	PM
1988-89	5	0	0	0	14	...	...	...	...	...

DAVE TIPPETT

Year	REGULAR SEASON					PLAYOFFS				
	GP	G	A	Pts	PM	GP	G	A	Pts	PM
1983-84	17	4	2	6	2	...	...	...	...	...
1984-85	80	7	12	19	12	...	...	...	...	...
1985-86	80	14	20	34	18	10	2	2	4	4
1986-87	80	9	22	31	42	6	0	2	2	4
1987-88	80	16	21	37	32	6	0	0	0	2
1988-89	80	17	24	41	45	4	0	1	1	0
1989-90	66	8	19	27	32	7	1	3	4	2

MIKE TOMLAK

Year	REGULAR SEASON					PLAYOFFS				
	GP	G	A	Pts	PM	GP	G	A	Pts	PM
1989-90	70	7	14	21	48	7	0	1	1	2
1990-91	64	8	8	16	55	3	0	0	0	2
1991-92	6	0	0	0	0	...	...	...	...	...
1993-94	1	0	0	0	0	...	...	...	...	...

JIM TROY

Year	REGULAR SEASON					PLAYOFFS				
	GP	G	A	Pts	PM	GP	G	A	Pts	PM
1975-76	14	0	0	0	48	2	0	0	0	29
1976-77	7	0	0	0	7	...	...	...	...	...

AL TUER

Year	REGULAR SEASON					PLAYOFFS				
	GP	G	A	Pts	PM	GP	G	A	Pts	PM
1988-89	4	0	0	0	23	...	...	...	...	...
1989-90	2	0	0	0	6	...	...	...	...	...

DARREN TURCOTTE

Year	REGULAR SEASON					PLAYOFFS				
	GP	G	A	Pts	PM	GP	G	A	Pts	PM
1993-94	19	2	11	13	4	...	...	...	...	...
1994-95	47	17	18	35	22	...	...	...	...	...

SYLVAIN TURGEON

Year	REGULAR SEASON					PLAYOFFS				
	GP	G	A	Pts	PM	GP	G	A	Pts	PM
1983-84	76	40	32	72	55	...	...	...	...	...
1984-85	64	31	31	62	67	...	...	...	...	...
1985-86	76	45	34	79	88	9	2	3	5	4
1986-87	41	23	13	36	45	6	1	2	3	6
1987-88	71	23	26	49	71	6	0	0	0	4
1988-89	42	16	14	30	40	4	0	2	2	4

MIKE VELLUCCI

Year	REGULAR SEASON					PLAYOFFS				
	GP	G	A	Pts	PM	GP	G	A	Pts	PM
1987-88	2	0	0	0	11	...	...	...	...	...

PAT VERBEEK

Year	REGULAR SEASON					PLAYOFFS				
	GP	G	A	Pts	PM	GP	G	A	Pts	PM
1989-90	80	44	45	89	228	72	2	2	4	26
1990-91	80	43	39	82	246	6	3	2	5	40
1991-92	76	22	35	57	243	72	0	2	2	12
1992-93	84	39	43	82	197	...	...	...	...	...
1993-94	84	37	38	75	177	...	...	...	...	...
1994-95	29	7	11	18	53	...	...	...	...	...

MICKEY VOLCAN

Year	REGULAR SEASON					PLAYOFFS				
	GP	G	A	Pts	PM	GP	G	A	Pts	PM
1980-81	49	2	11	13	26	...	...	...	...	...
1981-82	26	1	5	6	29	...	...	...	...	...
1982-83	68	4	13	17	73	...	...	...	...	...

JIM WARNER

Year	REGULAR SEASON					PLAYOFFS				
	GP	G	A	Pts	PM	GP	G	A	Pts	PM
1978-79	41	6	9	15	20	1	0	0	0	0
1979-80	32	0	3	3	32	...	...	...	...	...

TOM WEBSTER

Year	REGULAR SEASON					PLAYOFFS				
	GP	G	A	Pts	PM	GP	G	A	Pts	PM
1972-83	77	53	50	103	83	15	12	14	26	6
1973-74	64	43	27	70	28	3	5	0	5	7
1974-85	66	40	24	64	52	3	0	2	2	0
1975-76	55	33	50	83	24	17	10	9	19	6
1976-77	70	36	49	85	43	5	1	1	2	0
1977-78	20	15	5	20	5	...	...	...	...	...

ERIC WEINRICH

Year	REGULAR SEASON					PLAYOFFS				
	GP	G	A	Pts	PM	GP	G	A	Pts	PM
1992-93	79	7	29	36	76	...	...	...	...	...
1993-94	8	1	1	2	2	...	...	...	...	...

WALLY WEIR

Year	REGULAR SEASON					PLAYOFFS				
	GP	G	A	Pts	PM	GP	G	A	Pts	PM
1984-85	34	2	3	5	56	...	...	...	...	...

BLAKE WESLEY

Year	REGULAR SEASON					PLAYOFFS				
	GP	G	A	Pts	PM	GP	G	A	Pts	PM
1981-82	78	9	18	27	123	...	...	...	...	...
1982-83	22	0	1	1	46	...	...	...	...	...

GLEN WESLEY

Year	REGULAR SEASON					PLAYOFFS				
	GP	G	A	Pts	PM	GP	G	A	Pts	PM
1994-95	48	2	14	16	50	...	...	...	...	...
1995-96	68	8	16	24	88	...	...	...	...	...
1996-97	68	6	26	32	40	...	...	...	...	...

DAVE WILLIAMS

Year	REGULAR SEASON					PLAYOFFS				
	GP	G	A	Pts	PM	GP	G	A	Pts	PM
1987-88	28	6	0	6	87	...	...	...	...	...

TOMMY WILLIAMS

Year	REGULAR SEASON					PLAYOFFS				
	GP	G	A	Pts	PM	GP	G	A	Pts	PM
1972-73	69	10	21	31	16	15	6	11	17	2
1973-74	70	21	37	58	6	4	0	3	3	10

CAREY WILSON

Year	REGULAR SEASON					PLAYOFFS				
	GP	G	A	Pts	PM	GP	G	A	Pts	PM
1987-88	36	18	20	38	22	6	2	4	6	2
1988-89	34	11	11	22	14	...	...	...	...	...
1990-92	45	8	15	23	16	...	...	...	...	...

TERRY YAKE

Year	REGULAR SEASON					PLAYOFFS				
	GP	G	A	Pts	PM	GP	G	A	Pts	PM
1988-89	2	0	0	0	0	...	...	...	...	...
1989-90	2	0	1	1	0	...	...	...	...	...
1990-91	19	1	4	5	10	6	1	1	2	16
1991-92	15	1	1	2	4	...	...	...	...	...
1992-93	66	22	31	53	46	...	...	...	...	...

ROSS YATES

Year	REGULAR SEASON					PLAYOFFS				
	GP	G	A	Pts	PM	GP	G	A	Pts	PM
1983-84	7	1	1	2	4	...	...	...	...	...

SCOTT YOUNG

Year	REGULAR SEASON					PLAYOFFS				
	GP	G	A	Pts	PM	GP	G	A	Pts	PM
1987-88	7	0	0	0	2	4	1	0	1	0
1989-90	76	19	40	59	27	4	2	0	2	4
1988-89	34	6	9	15	8	...	...	...	...	...

ZARLEY ZALAPSKI

Year	REGULAR SEASON					PLAYOFFS				
	GP	G	A	Pts	PM	GP	G	A	Pts	PM
1990-91	11	3	3	6	6	6	1	3	4	8
1991-92	79	20	37	57	120	7	2	3	5	6
1992-93	83	14	51	65	94	...	...	...	...	...
1993-94	56	7	30	37	56	...	...	...	...	...

MIKE ZUKE

Year	REGULAR SEASON					PLAYOFFS				
	GP	G	A	Pts	PM	GP	G	A	Pts	PM
1983-84	75	6	23	29	36	...	...	...	...	...
1984-85	67	4	12	16	12	...	...	...	...	...
1985-86	17	0	2	2	12	...	...	...	...	...

*Goalie **Sean Burke** makes one of his final saves as a Whaler when the hockey club concluded its 25-year history in professional sports on April 13, 1997. Burke won his 100th game in Hartford and here robs Brantt Myhres of Tampa Bay.*

Whalers Miscellaneous

Hall of Famers to Play or Coach (*denoted by asterisk) against the Whalers (from 1972 to 1997)

Bill Barber	Tony Esposito	Guy Lapointe	Steve Shutt
Andy Bathgate (WHA)	Phil Esposito	Bernie Parent (WHA)	Darryl Sittler
Mike Bossy	Bob Gainey	Brad Park	Billy Smith
Scotty Bowman *	Gordie Howe (WHA)	Gilbert Perrault	J.C. Tremblay (WHA)
Johnny Bucyk	Bobby Hull (WHA, NHL)	Jacques Plante (WHA)	Vladislav Tretiak (Exhibition)
Gerry Cheevers (WHA, NHL)	Dave Keon (WHA)	Denis Potvin	Bryan Trottier
Bobby Clarke	Guy Lafleur	Jean Ratelle	Norm Ullman
Marcel Dionne	Stan Mikita	Larry Robinson	
	Frank Mahovlich (WHA)	Glen Sather*	
	Jacque Lemaire	Serge Savard	

Members of the U.S. Hockey Hall of Fame to play against the Whalers:

Henry Boucha (WHA)   Jack McCartan (WHA)
Robbie Ftorek (WHA, NHL)   Ken Morrow (NHL)
Dave Langevin (WHA, NHL)   Tim Sheehy (WHA)

Whalers in the Hockey of Fame:

Gordie Howe
Bobby Hull
Dave Keon
Emile Francis

Whalers to coach in the NHL:

John Cunniff	Joel Quenneville
Ted Green	Al Sims
Rick Ley	Tom Webster
Larry Pleau	

All-Time Leading Scorers versus the Whalers:

	GP	G	A	Pts	PM
Ray Bourque	111	35	105	140	95
Wayne Gretzky*	61	46	78	124	60
Peter Stastny	76	41	78	119	60
Michel Goulet	83	62	40	102	122
Dale Hawerchuk	63	30	66	96	18
Dave Andreychuk	87	46	49	95	62
Bobby Smith	73	28	51	79	44

*- includes WHA totals in 1978-79 (13 GP, 11-12-23, 4)
Note: Gretzky also went 7-8-15 in 7 playoff games

Stanley Cup Leaders against the Whalers:

Scoring	GP	G	A	Pts
Cam Neely	13	9	9	18
Craig Janney	9	2	14	16
Stephane Richer	17	10	5	15
Michel Goulet	8	6	9	15
Mats Naslund	14	4	11	15
Goaltending	**GP**	**Mts.**	**GAA**	**Record**
Brian Hayward	3	184	2.28	2-1
Patrick Roy	21	1307	2.39	14-7
Andy Moog	12	691	2.43	8-4

All-Time Leading Goaltenders versus the Whalers:

	GP	Mts.	GA	Sho	GAA	Record
Dominik Hasek	23	1360	53	1	2.34	11-9-3
Patrick Roy	59	3439	155	4	2.70	32-18-6
Andy Moog	38	2169	99	1	2.74	24-7-4
Mike Richter	21	1207	58	1	2.88	13-7-1
Darren Puppa	26	1449	70	2	2.90	16-9-0

All-Time WHA Scoring Leaders versus the Whalers:

	G	A	Pts	PM
Real Cloutier	34	32	66	13
Marc Tardif	26	38	64	40
Serge Bernier	30	31	61	34
J.C. Tremblay	11	41	52	2
Bobby Hull	25	26	51	8

All-Time Leading WHA Goaltenders versus the Whalers:

	GP	Mts.	GA	GAA
Dave Dryden	26	1511	55	2.18
Richard Brodeur	27	1503	81	3.23
Joe Daley	29	1772	105	3.56

Books from Glacier Publishing

Same Game, Different Name ($19.95 U.S. Funds)

The oral, pictorial and statistical history of the World Hockey Association. Interviews with Hull, Howe, Gretzky, Messier, Neale, Gartner, Liut, Garrett, Keon, Demers and many more. Over 200 photographs to completely document pro hockey's Maverick League of the 1970s. 288 pages.   ISBN 0-9650313-1-9

Whalers Trivia Compendium ($7.95 U.S. Funds)

A look at the Hartford Whalers through 300 questions that provide an informal history of the pro hockey franchise that began in the World Hockey Association in 1972 and wrapped up its 25th season in the National Hockey League in 1997. Over 60 photographs 72 pages.   ISBN 0-9650315-0-0

Forever Whalers, From Abrahamsson to Zuke ($24.95 U.S. Funds)

An encyclopedia of hockey's Whalers, a 25-year historial tribute that salutes every player and coach in team annals from the WHA days to its final NHL season. Biographical sketches of every player, beginning with the Abrahamsson twins, Christer and Thommy, and ending with Mike Zuke. Plenty of quotes and statistics. A collectible for anyone who ever cheered on the Whale! Over 300 photographs. 190 pages.   ISBN 0-96503150-3-5

All-Star Dads ($9.95 U.S. Funds)

A parenting book with input from major league baseball players to assist fathers, coaches and mentors in building self-confidence in children and youngsters. If you know someone who wants to improve patience and anger management skills, this fast-paced book will work wonders. 96 pages.   ISBN 0-9650315-4-3

For information or book orders

Glacier Publishing
40 Oak Street
Southington, CT 06489 USA
(860) 621-7644

Books from Glacier Publishing